A Course on Finite Groups

Universitext

Universitext

For other titles in this series, go to
www.springer.com/series/223

H.E. Rose

A Course on Finite Groups

Springer

H.E. Rose
Department of Mathematics
University of Bristol
University Walk
Bristol BS8 1TW, UK
H.E.Rose@bristol.ac.uk

ISBN 978-1-84882-888-9   e-ISBN 978-1-84882-889-6
DOI 10.1007/978-1-84882-889-6
Springer London Dordrecht Heidelberg New York

British Library Cataloguing in Publication Data
A catalogue record for this book is available from the British Library

Library of Congress Control Number: 2009942898

Mathematics Subject Classification (2000): 20-01

Preface

This book is an introduction to the remarkable range and variety of finite group theory for undergraduate and beginning graduate mathematicians, and all others with an interest in the subject. My original plan was to develop the theory to the point where I could present the proofs and supporting material for some of the main results in the subject. These were to include the theorems of Lagrange, Sylow, Burnside (Normal Complement), Jordan–Hölder, Hall and Schur–Zassenhaus amongst others, and to provide an introduction to character theory developed to the point where Burnside's $p^r q^s$-theorem could be derived and Frobenius kernels and complements could be introduced. I have come to realise that this would have resulted in a rather long book and so some material would have to go. It was at this point that modern technology came to my aid. Solutions to the problems were also to be included, but these would have taken at least 90 rather dense pages and an appendix to this book was perhaps not the best place for this material. A number of textbooks now put solutions on a web site attached to the book which is maintained jointly by the author and the publisher. Extending this idea has allowed me to fulfil my original intentions and keep the printed text to manageable proportions. So the web site now attached to this book, which can be found by going to

www.springer.com

and following the product links, includes not only the Solution Appendix but also extra sections to many of the chapters and two extra web chapters. These items are listed on the contents pages, and present work that is not basic to a chapter's topic being either slightly more specialised or slightly more challenging. Also, perhaps unfortunately, all work on character theory and applications (Chapters 13 and 14) is now on the web. As this book goes to press, about half of this web material is written and 'latexed', it is hoped that the remaining half will be available when the book is published or soon after. Of course, more web items could be added later. I attended Muchio Suzuki's graduate group theory lectures given at the University of Illinois in 1974 and 1975, and so in tribute to him and the insight he gave into modern finite group theory I have ended the extended text with a discussion of his simple groups $\mathrm{Sz}(2^n)$ as an application of the Frobenius theory.

Prerequisites

This book begins with the definition of a group, and Appendices A and B give a brief résumé of the background material from Set Theory and Number Theory that is required. So in one sense, the book needs no prerequisites, only the ability to 'think-straight' and a desire to learn the subject. On the other hand, it would help if the reader had undertaken the following.

(a) We are assuming that the reader is familiar with the material of a basic abstract algebra course, and so he or she has seen at least a few examples of groups and fields, associative and commutative operations, *et cetera*, and also has had some experience working in an abstract setting.
(b) We are also assuming that the reader is familiar with the basics of linear algebra including facts about vector spaces, matrices and determinants, and the definitions of inner and Hermitian forms. We also use the elementary operations, similarity and rational canonical forms, and related topics. Most standard one-semester linear algebra textbooks provide more than is required.
(c) It would also help if the reader had undertaken a first course on analysis which included the basic set operations, elementary properties of the standard number systems: integers $\mathbb{Z}$, rational numbers $\mathbb{Q}$, real numbers $\mathbb{R}$, and the complex numbers $\mathbb{C}$, and the standard set-theoretic methods summarised in Appendix A.
(d) Lastly, some familiarity with elementary number theory would be an asset, Appendix B summarises most that is required. The Euclidean Algorithm is used widely in this book, as are the basic congruence properties.

Plan of the Book

The author of an introductory group theory text has a problem: the theory is self-contained and coherent, many topics are interconnected, and several are needed more or less from the start. On the other hand, the material in a book has perforce to be presented linearly starting at Page 1. During the planning and writing of this book, I have assumed that most readers will not read it sequentially from cover to cover, but will occasionally 'dot-about'. Hence I have allowed some 'forward reference', mostly for examples.

The essential topics that the reader should 'get to grips with' first include the basic facts about groups and subgroups, homomorphisms and isomorphisms, direct products and solubility. Also some aspects of the theory of *actions*—conjugacy, the centraliser and the normaliser—are not far behind. Of course, as noted above, although the material has to be presented linearly, it need not be read linearly, and there are considerable advantages in presenting the basic facts of a topic—homomorphisms, for example—in one place. One consequence of this fact is that the order of the chapters has some flexibility. So Chapter 7 could be read before Chapters 5 and 6 with only a small amount of back-reference in the examples. Some group-theorists may consider it essential for students to have a good grounding in the Abelian theory before the non-Abelian theory is tackled. Similarly, Chapters 10 and 11 can be read in either order with little back-reference required. So a possible non-linear reading of the text is

Sections 2.1, 2.3, 2.4 and 4.1—the basic core of the subject, then the rest of Chapter 2, Sections 4.2, 4.3, 7.1 and 11.1 in this order,

then the following sections where the reading order might be varied

Part or all of Chapters 3 and 5, Sections 7.2, 7.3, and 9.1, and Chapters 6 and 10.

Following this the remaining printed sections or possibly some of the web sections could be tackled. In the text, I have sometimes introduced topics 'early' and out of their logical order, for example, isomorphisms in Chapter 2, to deal with this point. Also, as a general rule, the 'easier' and/or more elementary parts of a topic come near the beginning of the chapter, and so the final sections often contain more specialised and/or challenging material.

Further Reading

The reader would do not harm studying any of the books listed in the bibliography, we suggest a few concentrating on the more recent titles. For a general further development of the finite theory try:

Robinson (1982), Suzuki (1982, 1986), Aschbacher (1986), Kurzweil and Stellmacher (2004), and Isaacs (2008).

Also the three volume Huppert and Blackburn (1967, 1982a, 1982b) is very comprehensive and deals with many topics not found elsewhere. For more specialised topics, the following should be read:

Doerk and Hawkes (1992) for soluble groups,
Carter (1972), the ATLAS (1985), and Conway and Sloane (1993) for finite simple groups,
James and Liebeck (1993), Huppert (1998) and Isaacs (2006) for character theory, and
Kaplansky (1969), Fuchs (1970, 1973), and Rotman (1994) for infinite Abelian groups.

Of course, some of the older books still have much to offer, these include

Burnside (1911, reprinted 2004), Kurosh (1955), Scott (1964) and Rose (1978)—no relation!

Although 45 years old, in my opinion, Scott's book remains one of the best introductions to the subject.

All errors and omissions that are still present in the text and/or web pages are entirely my fault, please contact me with details at

h.e.rose@bris.ac.uk

General comments, including comments on the correctness and/or clarity of the text, or shorter, clearer or better solutions to the problems (which could be added to the web site), are also welcome.

Acknowledgements   I have received a considerable amount of assistance during the writing of this book for which I am extremely grateful. First, from my family (especially from my wife Rita), from colleagues, both academic and computational (especially Richard Lewis and Peter Burton), at the University of Bristol, and from the staff (both editorial and 'Latex' specialist) at Springer Verlag. I have given courses based on preliminary versions of this book many times in Bristol, my students have helped me to clarify many points and I thank them for this. But my main debt of gratitude goes to those who read a final draft of the text and cleared up many inconsistencies and errors on my part. These include the referees appointed by Springer Verlag, John Bowers (formally of the University of Leeds), Robin Chapman (Universities of Bristol and Exeter), Robert Curtis for Chapter 12 (University of Birmingham), and Ben Fairbairn (University of Birmingham). These last four spent many hours going through the manuscript, and improved it greatly—they are forever in my debt.

Bristol   Harvey Rose

Contents

1   **Introduction—The Group Concept** 1

2   **Elementary Group Properties** 11
     2.1   Basic Definitions 11
     2.2   Examples 17
     2.3   Subgroups, Cosets and Lagrange's Theorem 24
     2.4   Normal Subgroups 30
     2.5   Problems 34

3   **Group Construction and Representation** 41
     3.1   Permutations 42
     3.2   Permutation Groups 48
     3.3   Matrix Groups 52
     3.4   Group Presentation 55
     3.5   Problems 59

4   **Homomorphisms** 67
     4.1   Homomorphisms and Isomorphisms 69
     4.2   Isomorphism Theorems 74
     4.3   Cyclic Groups 79
     4.4   Automorphism Groups 81
     4.5   Problems 84

5   **Action and the Orbit–Stabiliser Theorem** 91
     5.1   Actions 92
     5.2   Three Important Examples 99
     5.3   Problems 107

6   ***p*-Groups and Sylow Theory** 113
     6.1   Finite *p*-Groups 114
     6.2   Sylow Theory 119
     6.3   Applications 126
     6.4   Problems 131

7   Products and Abelian Groups 139
   7.1   Direct Products 140
   7.2   Finite Abelian Groups 146
   7.3   Semi-direct Products 151
   7.4   Problems 159

8   Groups of Order 24, Three Examples 165
   8.1   Symmetric Group S_4 165
   8.2   Special Linear Group $SL_2(3)$ 172
   8.3   Exceptional Group E 177
   8.4   Problems 183

9   Series, Jordan–Hölder Theorem and the Extension Problem 187
   9.1   Composition Series and the Jordan–Hölder Theorem 188
   9.2   Extension Problem 196
   9.3   Problems 205

10   Nilpotency 209
   10.1 Nilpotent Groups 210
   10.2 Frattini and Fitting Subgroups 217
   10.3 Problems 223

11   Solubility 229
   11.1 Soluble Groups 230
   11.2 Hall's Theorems and Solubility Conditions 236
   11.3 Problems 243

12   Simple Groups of Order Less than 10000 249
   12.1 Steiner Systems 250
   12.2 Linear Groups 254
   12.3 Unitary Groups 265
   12.4 Mathieu Groups 267
   12.5 Problems 270

Appendices A to E 277
   A   Set Theory 277
   B   Number Theory 284
   C   Data on Groups of Order 24 289
   D   Numbers of Groups with Order up to 520 293
   E   Representations of $L_2(q)$ with Order < 10000 295

Bibliography 297

Notation Index 301
        1—Symbol Index 301
        2—Notation for Classes of Groups 303
        3—Notation for Individual Groups 304

Index 305

Web Contents

3 Group Construction and Representation 321
   3.6 Representations of A_5
   3.7 Further Problems

4 Homomorphisms 331
   4.6 The Transfer
   4.7 Group Presentation, Part 2
   4.8 Further Problems

5 Action and the Orbit–Stabiliser Theorem 351
   5.4 Transitive and Primitive Permutation Groups
   5.5 Further Problems

6 p-Groups and Sylow Theory 361
   6.5 Applications 2—Burnside's Normal Complement Theorem
       and Groups with Cyclic Sylow Subgroups
   6.6 Further Problems

7 Products and Abelian Groups 371
   7.5 Infinite Abelian Groups—A Brief Introduction

9 Series, Jordan–Hölder Theorem and the Extension Problem 379
   9.4 Schur–Zassenhaus Theorem
   9.5 Further Problems

12 Simple Groups of Order Less than 10000 387
   12.6 Simple Groups of Order Less than 1000000, Iwasawa's Lemma
       and a Method for generating Steiner Systems for some Mathieu
       Groups
   12.7 Further Problems

13 Representation and Character Theory 401
   13.1 Representations and Modules

13.2 Theorems of Schur and Maschke
13.3 Characters and Orthogonality Relations
13.4 Lifts and Normal Subgroups
13.5 Problems

14 Character Tables and Theorems of Burnside and Frobenius 433
14.1 Character Tables
14.2 Burnside's $p^r q^s$-Theorem
14.3 Frobenius Groups
14.4 Problems

Solution Appendix—Answers and Solutions,
Problems 2 to 12, A and B 461
      Problem 2   461
      Problem 3   472
      Problem 4   481
      Problem 5   489
      Problem 6   497
      Problem 7   508
      Problem 8   515
      Problem 9   521
      Problem 10 524
      Problem 11 532
      Problem 12 538
      Problem A   544
      Problem B   545

Chapter 1
Introduction—The Group Concept

Groups are all-pervasive in mathematics, there is hardly a branch of the subject that does not use them in one way or another. They are also widely used in many branches of the physical sciences. In one sense, this is to be expected because groups are quite often formed when an operation like multiplication or composition is applied to a set or system. Groups occur as number systems or collections of matrices, in permutation theory, as the symmetries of geometrical objects or as sets of maps, and in many other guises. Also the theory contains many elegant, dramatic and illuminating theorems.

Group theory has developed sometimes slowly but at other times by great leaps and bounds over the past two centuries. Often ideas and results appeared first implicitly before they were explicitly written down. For example, it is thought that Galois, in the 1820s, was the first to write down the axioms of a group, but some forty years earlier Lagrange was working with permutations of the roots of equations and proved a result which led to the famous theorem that now bears his name, the comments on actions given in the Introduction to Chapter 5 also apply here. Galois introduced a number of other basic notions including, for example, simple groups and normal subgroups. In 1850, Cayley showed that every group can be represented as a permutation group, and much of the nineteenth century work dealt with this aspect of the theory. Results started to appear more quickly, Sylow produced his ground-breaking work on p-subgroups in 1872, characters and representation theory were introduced around the turn of the century, and Hall's extensions of Sylow's work appeared in the 1920s and 1930s—to mention only a few of the many major developments. Another surge began around 1950 and led in the early 1980s to the completion of the classification problem for finite simple group, hereafter referred to as CFSG, which must surely rank amongst the greatest achievements in all mathematics. A number of important corollaries have followed from this work, for example, the positive solution of the Restricted Burnside Problem (page 27).

The purpose of this book is to introduce the reader to the fine branch of mathematics called *group theory*—there is a 'great story' to tell, and we hope that it will encourage you, the reader, to develop an abiding interest in the subject, and a desire to look further and deeper into the theory.

H.E. Rose, *A Course on Finite Groups*,
Universitext,
DOI 10.1007/978-1-84882-889-6_1, © Springer-Verlag London Limited 2009

Group Examples

A group is a mathematical system (or set) with a single operation. We begin by considering two familiar examples; the first is the integers $\mathbb{Z}$. The elements are

$$\ldots, -2, -1, 0, 1, 2, 3, \ldots,$$

and the operation is standard addition '$+$'. There are a number of basic axioms from which almost all additive properties follow. The first and in some ways the most important is *closure*, or being *well-defined*; that is,

$$\text{if } a, b \in \mathbb{Z}, \quad \text{then} \quad a + b \in \mathbb{Z}.$$

This is equivalent to stating that $+$ is an operation (Appendix A). Some so-called *partial* systems have been studied, but all systems considered in this book satisfy an axiom of this type. The next property is *associativity*; that is,

$$\text{for all } a, b, c \in \mathbb{Z} \quad \text{we have} \quad (a + b) + c = a + (b + c).$$

When forming this sequence of additions, we obtain the same answer if we first add a to b, and then add the result of this addition to c, or if we first add b to c and then form the sum of a and the result of this last addition. Some algebraic systems lack this property but, in general, non-associative systems have limited uses unless some more complex rule is applied—for example, in Lie algebras—and again we shall not consider such systems in this book.

In $\mathbb{Z}$, a natural question to ask is: Does the equation

$$a + x = b \tag{1.1}$$

have a solution x? In ancient times mathematicians only 'allowed' this equation to have a solution if $b > a$, that is, if x is positive. But this is very restrictive, and in the group $\mathbb{Z}$ Equation (1.1) is always uniquely soluble, and so we need to introduce the 'zero' and 'negative' integers. The zero 0 satisfies

$$a + 0 = 0 + a = a \quad \text{for all } a \in \mathbb{Z};$$

in the sequel, we use the term *neutral element* for the entity 0; see the discussion on page 4. Further, we introduce the *negative integers* by

$$\text{for all } a \in \mathbb{Z} \quad \text{there exists a unique } c \in \mathbb{Z} \quad \text{that satisfies} \quad a + c = 0.$$

We usually write $-a$ for c (and a^{-1} for c if we are using a multiplicative notation as is the case for almost all groups discussed in this book), and we call it the *inverse* of a. It is now easy to show that (1.1) always has a unique solution. The system $\mathbb{Z}$ has one extra basic property not shared by all groups: it is *commutative*, or *Abelian*. This is given by

$$\text{for all } a, b \in \mathbb{Z} \quad \text{we have} \quad a + b = b + a,$$

the result of several additions does not depend on the order of the terms in the sum. To recap, $\mathbb{Z}$ has the four basic properties: Closure, associativity, a neutral element and inverses, and it has the extra property of Abelianness. Note also it is a countably infinite system.

For our second example, we consider another familiar system—symmetries of an equilateral triangle. Groups of symmetries provide a good range of examples, they are widely used in both mathematical and physical systems, for example, by chemists when they are studying the crystal structure of matter. The group of symmetries of a triangle has two aspects which are different from those in our first example: It is finite, and it is not Abelian, but as we shall see below it shares the four basic properties with $\mathbb{Z}$. Consider an equilateral triangle with vertices labelled A, B, C. This geometric object has a number of *symmetries*, that is, transformations (rotations and reflections) that give another copy of the original triangle. We work in the standard Euclidean plane.

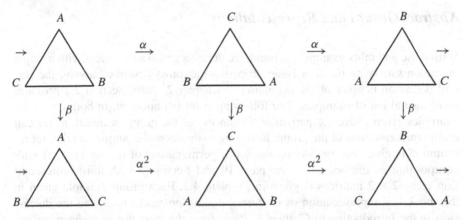

The elements of the group are the six rotations and reflections of the triangle illustrated above, and the operation is composition; that is, do one rotation or reflection, and then do another on the result of the first. We take the basic rotation α to be clockwise about the centre of the triangle by the angle $2\pi/3$, see the top row of the diagram above. Note that three applications of α (that is, α^3, a rotation by 2π) maps the triangle to itself identically, and so can be taken as the neutral element. This also shows that the inverse of α is α^2; see the bottom row in the diagram. The triangle has three reflections, they are mirror transformations about a line through a vertex and the centre of the corresponding opposite edge. One, labelled β (three times) in the diagram, is about a vertical line through the top vertex of the corresponding triangle and the centre of its base. The other two reflections can be generated as follows. If we first apply α to the top-left triangle and then apply β to the result, we obtain the middle triangle in the bottom row. This gives the second reflection of the top-left triangle, now about a line through the bottom right-hand vertex B and the centre of the opposite edge AC. Note that we would obtain the same result if we first applied β to the top-left triangle, and then applied α^2 to the result. We can obtain the third reflection if we repeat this construction but begin by applying α^2 instead of α to the top-left triangle. Incidentally, this shows that the group is not Abelian (that is, not commutative as $\beta\alpha = \alpha^2\beta \neq \alpha\beta$). The inverse of a reflection is itself: Two applications of a reflection gives the neutral element. We call a group element of this type an *involution*, and we shall see later that these elements play an important role in the theory. It is straightforward to check that this system is closed and has the associativity property (reader, try a few cases). Hence the system contains the six symmetries of the triangle: Neutral element (where a triangle is mapped to itself identically), α, α^2, β, $\alpha\beta$, and $\alpha^2\beta$, and it satisfies the four basic group axioms (closure, associativity, and possession of a neutral element and inverses for all of its elements) as in our first example. It is also finite and not Abelian. Later we shall call this system the *dihedral group* of the triangle and denote it by D_3.

Abstract Groups and Representations

With these and other examples in mind, we define a group as a system with a single operation satisfying the four basic properties (axioms) described above, the formal definition is given at the beginning of Chapter 2. Also, Section 2.2 provides a substantial list of examples. The following point is important. In both cases, the examples given above are particular 'instances' of the group in question; we call them *representations* of the group. Referring to the second example, another representation is given by considering the set of permutations of the set $\{1, 2, 3\}$ with composition as the operation; see page 19 and Section 3.1. A third representation using 2×2 matrices is given in Problem 4.2. Each group example given in this book is a representation of its corresponding *abstract group*, see the discussion in the Introduction to Chapter 3. *Right from the start this is a characteristic feature that the reader should note, and we use a corresponding nomenclature.* So when discussing a group in the abstract we call it a 'group', but when discussing a particular example or representation, say using permutations or matrices or some analogous construction, we call it a 'permutation group' or a 'matrix group' or analogous group. Similarly, the 'neutral element' which we always denote by e is the element in the abstract group satisfying the relevant axiom, and it has many instances or representations in particular groups. In the examples given in Section 2.2, it is 0, or 1, or -3, or a matrix, or a collection of maps, or the point at infinity on a curve, *et cetera*. The term 'neutral element' is not standard, nor is it 'new'; for example, it occurs in Cohn (1965), page 50. Some authors use either 'identity' or 'trivial element'; the first only really applies to an 'operation of multiplicative type', whilst the second gives the wrong impression—this element is an important one in the group, and certainly not trivial in the usual sense of this word. We have extended this nomenclature to the single element (sub)group $\langle e \rangle$ which we call the *neutral (sub)group*, see Definition 2.12. Also an 'inverse' can be an additive inverse, or a multiplicative inverse, or an inverse matrix, *et cetera*.

Classes of Groups

The class of all groups is a large one. Set-theorists call it a *proper class* as opposed to a *set*, but as we are taking the usual naive view of set theory (Appendix A) we shall treat sets and classes synonymously. We shall see that it is convenient to consider subclasses defined by some of the basic properties. For example, groups can be finite or infinite, and Abelian or non-Abelian; these distinctions are fundamental. We shall study further distinctions later, for instance, in Chapter 11 between 'soluble' and 'non-soluble' groups. Here we divide the class of all groups into four subclasses and, as we shall see, both the theory and the actual groups in each subclass have distinct characteristics.

The first subclass contains the

Finite Abelian Groups In Chapter 7, we show that groups in this class can be characterised completely, and they have a particularly simple form—that is, as 'products of cyclic groups'. So in one sense they are 'a bit boring'; but in an application, if we know *a priori* that the group or groups under discussion are in this subclass, then we can be sure that they take this simple form which can have a major influence on the result. A good example occurs in the theory of rational points on elliptic curves discussed on page 22. Amongst our four subclasses, this is the only one for which we have a complete description of all of the groups involved.

The second subclass contains the

Finite Non-Abelian Groups Most of the work in this book deals with groups in this class. For finite groups in general, there is a strong interplay between the 'group theory' and the 'number theory' of the group in question. In part this is a consequence of Lagrange's Theorem which states that the order (number of elements) of a subgroup H of a group G must divide the order of G; so the prime factorisation of the order of a finite group is an important invariant of the group. One major development from this is the Sylow theory discussed in Chapter 6 which asserts the existence of subgroups with prime power order. Another important distinction in the theory is between the so-called 'simple' and 'non-simple' groups; see the definition on page 33. The Jordan–Hölder Theorem states, roughly speaking, that all finite (and some infinite) groups can be 'built up' from simple groups using 'extensions'; this will be discussed in Chapter 9—note that one of the main aims our work is to describe all groups. A theory of extensions has been developed, but a considerable amount of work and many new ideas will be needed before it can be described as finished; see Section 9.2. On the other hand, a complete list of finite simple groups is now known, much of the development work was undertaken between 1955 and 1985 and, as noted above, it forms one of the crowning achievements of twentieth century mathematics. We give a brief introduction to this topic in Chapter 12. Hence considerable progress has been made in the theory of finite non-Abelian groups and this will be discussed in the following chapters and web appendices, but work still needs to be done.

The third subclass contains the

Infinite Abelian Groups   For infinite groups in general, number theory only plays a small role, but questions concerning cardinality can be important. The so-called finitely-generated Abelian groups are similar to those in the first subclass as they can be represented as products of cyclic groups. But many groups in this class are not finitely generated, for example, the rational numbers with addition or the positive reals with multiplication. Apart from a brief survey in Web Section 7.5 we shall not deal with these groups in this book. Much of the work develops ideas from linear algebra, and good introductions to this topic are given in Kaplansky (1969), and Fuchs (1970, 1973).

The final subclass contains the

Infinite Non-Abelian Groups   This is perhaps the least well-understood part of the theory. A number of extensions of the finite theory have been studied, but no general classification is known, and it seems unlikely that one will be found in the near future. One approach is to use topology. For example, the reals have a natural (metric) topology, and the interplay between the group theory and the topology of this system can be exploited to gain new insights. Since 1950 a number of long-standing problems have been solved, often showing that these groups are more complicated than previously thought; for example, see Problem 6.7. A good introduction is given in Kurosh (1955), also Robinson (1982) discusses a number of aspects of this part of the theory. Infinite groups with some kind of 'finiteness condition', such as being 'finitely generated' or 'finitely presented', have also been widely studied.

Summary of the Book

Below we give a brief summary of the contents of the printed Chapters 2 to 12, Appendices A to E, the Web Chapters 13 and 14, and the Web Appendices.

Chapter 2—Elementary Group Properties   The basic entities—semigroups, groups, subgroups, cosets, normal subgroups and simple groups—are defined, Lagrange's Theorem is derived, and the second section lists a number of standard examples.

Chapter 3—Group Construction and Representation   The main construction methods and group representations are discussed. Firstly, permutations are introduced, the symmetric and alternating groups are defined, and an elementary proof of the simplicity of A_n for $n > 4$ is given. Secondly, matrix groups are briefly considered, and lastly group presentations are introduced (this topic is completed in Web Section 4.7 once the First Isomorphism Theorem has been proved). Web Section 3.6 discusses some of the various representations of the alternating group A_5 to illustrate the fact that groups can have a wide range of representations.

Chapter 4—Homomorphisms   The natural maps, called Homomorphisms (and Isomorphisms when the map is a bijection), and factor groups are introduced, and the four fundamental Isomorphism and Correspondence Theorems are derived. Cyclic groups and the basic properties of the automorphism group of a group are described. There are two Web Sections 4.6 and 4.7. The first introduces the 'transfer' which provides a useful example of a 'real' homomorphism, and the second completes the work on group presentations begun in Chapter 3.

Chapter 5—Action and the Orbit-stabiliser Theorem   This chapter, the last giving the basic material, introduces 'actions' which bring together a number of useful constructions, and gives a proof of the Orbit-stabiliser Theorem. It also describes three important particular actions: the coset action, the conjugate element action leading to centralisers and the Class Equations, and the conjugate subgroup action leading to normalisers and the N/C-theorem. Web Section 5.5 extends the work on permutation theory begun in Chapter 3, and includes a discussion of 'transitive' and 'primitive' permutation groups, and Iwasawa's simplicity lemma.

Chapter 6—p-Groups and Sylow Theory   The basic theory of p-groups (where all elements have order a power of p) is discussed, and the five Sylow theorems are derived—these results form one of the most important aspects of the finite theory. There are then two sections of applications, the first gives (a) some facts about groups whose orders have a small number of factors, (b) proves the so-called Frattini Argument, and (c) introduces nilpotent groups. The second is Web Section 6.5 which gives some more substantial applications including a proof of Burnside's Normal Complement Theorem and a discussion of groups all of whose Sylow subgroups are cyclic.

Chapter 7—Products and Abelian Groups   Direct products are introduced, and two proofs of the Fundamental Theorem of Abelian Groups are presented; see page 5. The third section discusses 'semi-direct products', a variant of the direct product construction, and the groups of order 12 are described (they can all be treated as semi-direct products). Some basic facts, but no proofs, concerning infinite Abelian groups are given in Web Section 7.5. Except for some problems, this is the only point where specifically infinite groups are considered in any detail.

Chapter 8—Groups of Order 24, Three Examples   No new theory is presented in this chapter, but three groups of order 24 are discussed in some detail. The work constructs their subgroups including those of Frattini and Fitting, the subgroup lattice, series and some of their representations. Appendix C, see pages 289 to 292, gives data on the remaining twelve groups of order 24. The purpose of this chapter is to challenge the reader to think more about the objects he or she is studying, and to ask questions. For example: can the centre of a group equal its derived subgroup or its Frattini subgroup? This chapter is also intended to motivate the remaining topics, that is series, simple groups and (on the web) representation theory.

Chapters 9—Series, Jordan–Hölder Theorem and the Extension Problem
This is the first of three shorter chapters dealing with series and the normal sub-group structure of groups. In the first of these, we prove the theorem of Jordan and Hölder on composition series—this demonstrates the importance of simple groups to the theory. Secondly, we present a brief introduction to extension theory—that is the construction of complex groups using some of their subgroups as components, and we discuss one substantial example.

Chapter 10—Nilpotency Nilpotent groups lie between Abelian and soluble groups, and this second shorter chapter continues the work on these groups begun earlier. There is a surprising number of equivalent definitions which shows the importance of the notion. The second section discusses two 'special' subgroups of a group—the Frattini and Fitting subgroups; they have some remarkable properties (including being nilpotent), and extensions of the latter have proved useful in the completion of CFSG.

Chapter 11—Solubility After a brief historical introduction the last of the shorter chapters introduces the basic facts about soluble groups, and discusses a number of equivalent conditions. The most important is due to Philip Hall and extends the Sylow theory in the soluble case.

Chapter 12—Simple Groups of Order Less than 10000 This is another 'descriptive' chapter giving an account of simple groups with order less than 10000. We introduce Steiner systems—their automorphisms provide a new way to construct groups, prove the simplicity of the linear (matrix) groups $L_n(q)$, and discuss one 'classical' ($U_3(3)$, a *unitary group*) and one 'sporadic' (M_{11}, the first *Mathieu group*) group in detail. Some numerical data is also given but many proofs are omitted. Appendix E, see page 295, gives data on the groups $L_2(q)$, and an appendix at Web Section 12.6 provides more information about Steiner systems for Mathieu groups, and data on simple groups of order less than 10^6.

Appendix A—Set Theory and Appendix B—Number Theory These two appendices give the basic definitions and results for the work on sets and number theory which underlie all the material in the book.

Appendices C, D and E These appendices provide data on several aspects of the theory. The first, C, lists properties of groups of order 24 (this is an appendix to Chapter 8), D details the number of groups with order up to 520, and E provides some representations of the linear groups $L_2(q)$ (this is an appendix to Chapter 12).

Web Chapter 13—Representation and Characters A brief introduction to representation and character theory is presented sufficient for the applications given in Web Chapter 14. This theory includes the basic definitions, Schur's Lemma and Maschke's theorem, the orthogonality relations, and 'lifts', *et cetera*. This chapter is entirely theoretical except for the examples.

Web Chapter 14—Character Tables, and Theorems of Burnside and Frobenius We give three applications of the work in Web Chapter 13 which have strong connections with the earlier material. First, we construct some character tables including those for most of the groups of order 12 or less, and some others including several of order 24 discussed in Chapter 8. These character tables provide a surprising amount of information concerning the groups in question. Second, we prove Burnside's $p^r q^s$-theorem (this completes the proof of Hall's Theorem given in Chapter 11). The third section introduces the Frobenius 'kernel' and 'complement', gives a proof of Frobenius's Theorem concerning these notions, and finally it provides some applications of this theorem including a discussion of Suzuki groups.

Web Solution Appendix This includes answers, hints on solutions, and in some cases full solutions, for all of the problems given in Chapters 2 to 12, and Appendices A and B.

Developing the theory and proving results are of course important, but two other aspects are also important.

Problems Each chapter ends with a sequence of problems for the reader to try of varying difficulty partly as indicated with a star ⋆ suggesting a greater challenge. Some readers may find it difficult to decide which problems start with and which are the most important, so some of these have been marked with the symbol ♦. These are all fairly straightforward, theoretical, and contain minor results that are used in the main part of the text. Other problems ask for examples to be constructed, these have no indication mark but should also be tackled early. As noted above, hints, sketch solutions, or in some cases detailed treatments of problems, are given in Web Solution Appendix on the web site attached to this book.

'Actual groups' We are studying groups, and so it seems essential to us that the reader 'sees' and 'experiences' as many 'actual' or 'concrete' groups as possible. This will, we hope, illuminate the theory and so induce a greater understanding in general. Some parts of the text and a number of the problems are given over to this aspect including the whole of Chapter 8.

Computers in Group Theory

During the past thirty years, and more so recently, computers have become an increasingly useful tool in pure mathematics, as well as in most other branches of mathematics, many branches of the physical sciences, and beyond. In group theory, they are particularly useful for doing matrix and permutation calculations, and for producing examples. But they can also be used for more sophisticated constructions, for example, looking for subgroups of a group or constructing homomorphisms. A number of computer algebra packages, some 'free' and some commercially available, have been developed over the past decade or so, and the reader is encouraged to make use of at least one of these while reading this book. Also a number of the problems are best tackled using one of these packages.

While writing this book, we have made extensive use of the computer algebra package called GAP—Groups, Algorithms, and Programming. This package has many authors based in Aachen in Germany, St. Andrews in Scotland, and at many other sites; we would like to take this opportunity to compliment these authors on the excellence of their product. It is available free from the St. Andrews web site at

$$\texttt{http://www-gap.dcs.st-and.ac.uk/~gap}$$

We have also made some use of the commercially available package called MAGMA which incorporates many aspects of the GAP program.

One point should be borne in mind whilst working with any of these packages, and it is one that we emphasise several times in this book. In a particular calculation, the program can only deal with a specified representation of the group under discussion, say as a permutation group or as a matrix group. The package GAP is particularly good when working with permutation groups, but it also deals well with matrix groups defined over a specific field and with presentations.

Chapter 2
Elementary Group Properties

In this chapter, we introduce our main objects of study—groups. A general overview including some historical comments was given in Chapter 1. More detail on the history of the theory can be found in Wussing (1984), van der Waerden (1985), and at www-gap.dcs.st-andrews.ac.uk/~history/. Here we give the basic definitions and an extensive list of examples, introduce subgroups and cosets, normal subgroups and simple groups, and prove the first major result in the theory—Lagrange's Theorem.

2.1 Basic Definitions

We begin by defining the group concept. Maps between groups will be discussed in Chapter 4. As a preliminary to this we introduce *semigroups* as follows.

Definition 2.1 A *semigroup* is a non-empty set $X = \{\ldots, x, y, z, \ldots\}$ together with a binary operation $\odot$ (page 281) which satisfies the following two conditions (axioms):

(i) it is *closed*, or *well-defined*: for all $x, y \in X$, we can perform the operation $x \odot y$ and
$$x \odot y \in X,$$

(ii) it is *associative*: for all $x, y, z \in X$,
$$x \odot (y \odot z) = (x \odot y) \odot z.$$

Note that (i) is implied by the definition of the operation $\odot$; see the comments below Definition 2.2.

H.E. Rose, *A Course on Finite Groups*,
Universitext,
DOI 10.1007/978-1-84882-889-6_2, © Springer-Verlag London Limited 2009

Examples The following sets with operations are semigroups.

(a) The positive integers with addition.
(b) The set of all one-variable functions with domain and codomain $\mathbb{R}$, and with the operation of composition of functions.

There is an extensive theory of semigroups which is of particular interest in some branches of analysis and combinatorics. Also a number of similar systems that are not quite groups have been studied, for instance, the operation may be only partially defined, or there may be a neutral element but no inverses, *et cetera*. We shall not consider these systems; Bruck (1966) provides a good introduction.

Definition 2.2 A *group* $(G, \odot)$ is a semigroup which satisfies the following extra conditions (axioms):

(iii) G contains a unique element e which satisfies, for all $g \in G$,

$$e \odot g = g \odot e = g,$$

(iv) for each $g \in G$, there exists a unique $g' \in G$ satisfying

$$g \odot g' = g' \odot g = e.$$

The element e is called the *neutral element* of the group G, see page 4. Some authors use the term *identity* for e, and if the operation $(\odot)$ is written additively $(+)$ then it is called the *zero* and denoted by 0. There are a number of redundancies in this definition—in particular, in axioms (i), (iii) and (iv). Strictly speaking, (i) is unnecessary as it is implied by the fact that $\odot$ is an operation; see Appendix A. But we have left it in to remind the reader that closure is vitally important—this property must be checked whenever it is required to show that a particular set and product form a group. For (iii) and (iv), see Theorem 2.5.

Definition 2.3 A group $(G, \odot)$ is called *Abelian,* or occasionally *commutative*, if its operation is commutative: For all $g, h \in G$

$$g \odot h = h \odot g.$$

The term 'Abelian' commemorates the Norwegian mathematician Niels Abel who died at the age of 27 in 1829. He was working on solutions to polynomial equations, and needed to apply a condition similar to the one above; see the Introduction to Chapter 11 and van der Waerden (1985), page 88.

Examples We give four here, and an extensive list in the next section.

(a) The set $\{1, -1\}$ with the operation of standard multiplication forms a finite Abelian group which we denote by T_1 (one copy of 'two'). The neutral element is 1, and each element is self-inverse.
(b) The set of *permutations* of a set $\{1, 2, 3\}$. The elements are the six permutations of this set, and the operation is composition: Do the first permutation, then do the

second permutation on the result of the first. For example, if the first permutation maps $1 \mapsto 1, 2 \mapsto 3$ and $3 \mapsto 2$, and the second maps $1 \mapsto 3, 2 \mapsto 2$ and $3 \mapsto 1$, then their composition maps $1 \mapsto 3, 2 \mapsto 1$ and $3 \mapsto 2$. The neutral element is the permutation that moves no symbols, and the inverse of a permutation is its reverse (Section 3.1). This system forms a finite non-Abelian group which we denote by S_3 and call the *symmetric group* of the set $\{1, 2, 3\}$. Reader, why is this group not Abelian?

(c) The positive real numbers $\mathbb{R}^+$ with multiplication form an infinite Abelian group. The neutral element is 1, and the inverse of x is $1/x$.

(d) The set of all non-singular 2×2 matrices having rational number entries with the operation of matrix multiplication is an example of an infinite non-Abelian group, it is denoted by $GL_2(\mathbb{Q})$ and called the 2×2 *general linear group* over $\mathbb{Q}$. The neutral element is I_2, the 2-dimensional identity matrix, and inverses exist by definition.

The symbols $(G, \odot)$, G, H, J, or K, sometimes with primes or subscripts, will always denote groups. We use lower case Roman letters a, b, c, d, g, h, j, k, and l, again sometimes with primes or suffixes, to stand for group elements, and we use x, y and z for set elements or occasionally for group elements following the usual mathematical convention that these letters denote entities which satisfy a proposition or equation. The words 'operation', 'multiplication' and 'product' are used more or less synonymously: If $g, h \in G$ we say that we apply the operation $\odot$ to form the product $g \odot h$, or we multiply g by h to obtain $g \odot h$.

The *underlying set* of a group G is the set of elements of G stripped of its operation; where there is no confusion, this will also be denoted by G. Also, we sometimes say that a group G is *generated* by a set X, or X is a *generating set* for G, where X is a subset of the underlying set of G, and we write $G = \langle X \rangle$. This means that the collection of all products of powers (both positive and negative) of elements of X coincides with G. For example, the set $\{1\}$ is a generating set for $\mathbb{Z}$, that is, $\mathbb{Z} = \langle 1 \rangle$ because every integer can be expressed as a sum of 1s or -1s. Note that a group may have many different generating sets, and it always has at least one because the underlying set of G clearly acts as a generating set for G. This notion is defined formally in Definition 2.16. We also write $\langle e \rangle$ for the (unique) group containing the single element e (with operation $e \odot e = e$), we call it the *neutral group*. Some authors use the term 'trivial group' for $\langle e \rangle$; it is an important component of a group and certainly not 'trivial' using the normal meaning of this word, hence we shall not use this term; see also the comments on page 4.

We noted above that Definition 2.2 can be weakened considerably without affecting our objects of study. Consider

Definition 2.2′ A group $(G, \odot)$ is a semigroup, see Definition 2.1, which satisfies the following two conditions:

(iii)′ there is an element $f \in G$ with the property: $f \odot g = g$, for all $g \in G$;

(iv)′ for each $g \in G$, and with f as in (iii)′, there exists $h \in G$ satisfying $h \odot g = f$.

These conditions imply that G has at least one *left neutral element* f, and each $g \in G$ has at least one *left inverse* h relative to f.

Definition 2.2′ is equivalent to Definition 2.2; see also Problem 2.4. Note that this equivalence is useful, for when checking if the group axioms hold for a particular set and map, once closure and associativity have been established (Axioms (i) and (ii)), it is not necessary to prove that either the neutral element or the inverse operation is unique, or two-sided, because these properties follow by Theorem 2.5 below. Also, if we find *an* inverse of an element g, then we can be sure that it is the *unique* inverse of g, again by Theorem 2.5.

We begin with the following result: In all groups, the only element which equals its square is the neutral element (in algebra generally, such elements are called *idempotents*).

Lemma 2.4 *Let* $(G, \odot)$ *be a semigroup satisfying the conditions of Definition* 2.2′. *If* $a \in (G, \odot)$ *and* $a \odot a = a$, *then* $a = f$ *where* f *is given by* (iii)′.

Proof[1] By (iv)′, if $a \in G$ we can find $b \in G$ satisfying $b \odot a = f$, so by (iii)′

$$a = f \odot a = (b \odot a) \odot a = b \odot (a \odot a) = b \odot a = f,$$

by associativity, the hypothesis and (iv)′ again.   □

Theorem 2.5 *A semigroup* $(G, \odot)$ *satisfying Conditions* (iii)′ *and* (iv)′ *in Definition* 2.2′ *forms a group as given by Definition* 2.2.

Proof We need to show that f, and the inverses, apply both on the left and on the right, and are unique; that is, f as *the* neutral element, and h as *the* inverse of g. First, we show that if $a \in (G, \odot)$ and $b \odot a = f$, then $a \odot b = f$ (a left inverse is also a right inverse). We have

$$b \odot (a \odot b) = (b \odot a) \odot b = f \odot b = b.$$

by (ii), (iv)′, and (iii)′. Hence, by (ii) again

$$(a \odot b) \odot (a \odot b) = a \odot (b \odot (a \odot b)) = a \odot b.$$

By Lemma 2.4, this shows that $a \odot b = f$; the first part follows. Secondly f is a right identity. We have, using the above subresult and (ii),

$$a \odot f = a \odot (b \odot a) = (a \odot b) \odot a = f \odot a = a$$

by (iii)′. Thirdly, we show that b is unique (that is, inverses are unique). For suppose $c \odot a = f$, then we have by the above and (ii)

$$c = c \odot f = c \odot (a \odot b) = (c \odot a) \odot b = f \odot b = b,$$

---

[1] To emphasise their importance, and to aid clarity, all proofs are typeset indented.

by hypothesis and (iii)′ applied to b. Lastly, the neutral element. For if e also satisfies (iii)′, that is $e \odot a = a$ for $a \in G$, then substituting e for a we obtain $e \odot e = e$, and so, by Lemma 2.4, $e = f$. This completes the proof. $\qquad\square$

From now on, we adopt the following conventions. We write ab for $a \odot b$, e for the neutral element, and G for $(G, \odot)$ when it is clear which operation is being used. Also, the inverse of g given by (iv) in Definition 2.2 will be written in the standard notation g^{-1} (and $-g$ if we are using addition). We normally drop brackets and write xyz for either $x(yz)$, or $(xy)z$. In some cases, we do not delete the brackets if this aids clarity.

The next three results apply to all groups, and they will often be used in the sequel usually without being specifically identified. Note that no restrictions apply, a rare occurrence in the theory!

Theorem 2.6 (Cancellation) *Suppose $a, b, x, y \in G$. If $ax = bx$, or if $ya = yb$, then $a = b$.*

Proof From $ax = bx$ we obtain, by Definition 2.2 and associativity,

$$a = ae = a(xx^{-1}) = (ax)x^{-1} = (bx)x^{-1} = b(xx^{-1}) = b.$$

A similar argument applies in the second case. $\qquad\square$

Theorem 2.7 *Suppose a and b are elements of a group G.*

(i) $(ab)^{-1} = b^{-1}a^{-1}$.
(ii) $(a^{-1})^{-1} = a$.
(iii) *If G is finite, then a^{-1} equals some positive power of a.*
(iv) *If a commutes with b, then a^{-1} also commutes with b.*

Proof (i) As $(b^{-1}a^{-1})(ab) = b^{-1}(a^{-1}a)b = b^{-1}b = e$ and, by Theorem 2.5, inverses are unique and two-sided, it follows that $b^{-1}a^{-1}$ is the inverse of ab. A similar argument applies for (ii).

(iii) If G has n elements and $a \neq e$, then an integer m must exist satisfying $1 < m \leq n$ and $a^m = e$ (the powers of a cannot all be distinct in the finite case). Now $a^{m-1} = a^{-1}a^m = a^{-1}$.

(iv) If $ab = ba$ then $b = a^{-1}ba$, and so $ba^{-1} = a^{-1}b$. $\qquad\square$

Using induction we can extend (i) to prove that $(a_1 \cdots a_n)^{-1} = a_n^{-1} \cdots a_1^{-1}$.

Theorem 2.8 *If we treat the group G as a set (that is, we consider the underlying set of G) then, for all fixed $a \in G$,*

$$G = \{ag : g \in G\} = \{g^{-1} : g \in G\}.$$

Proof Use Theorems 2.6 and 2.7; see Problem 2.1. $\qquad\square$

Most elementary exponent properties apply to groups. Note that as the group in question may not be Abelian some properties do not always hold. For example, $(ab)^2$ may, or may not, equal $a^2 b^2$.

For all elements a in a group, if $n \geq 0$ we write

$$a^0 = e \quad \text{and} \quad a^{n+1} = a^n a, \quad \text{that is} \quad a^n = aa \cdots a \ (n \text{ copies of } a).$$

Also, again if $n \geq 0$ we write a^{-n} in place of $(a^{-1})^n$. By Theorem 2.7, this also equals $(a^n)^{-1}$; reader, why?

Theorem 2.9 *Suppose a is a group element, and $r, s \in \mathbb{Z}$.*

(i) $a^{r+s} = a^r a^s$.
(ii) $(a^r)^s = a^{rs}$.

Proof (i) By induction on s. Suppose s is non-negative. We have $a^{r+0} = a^r = a^r e = a^r a^0$ and, using the inductive hypothesis in the third equation,

$$a^{r+(s+1)} = a^{(r+s)+1} = a^{r+s}a = a^r a^s a = a^r a^{s+1}.$$

Now apply induction. Using this we have $a^{r-s} a^s = a^{(r-s)+s} = a^r$, hence $a^{r-s} = a^r (a^s)^{-1} = a^r a^{-s}$ and (i) follows for negative s.

(ii) Assume first that s is non-negative. As above we use induction on s, we have $(a^r)^0 = e = a^{0r}$ and

$$\left(a^r\right)^{(s+1)} = \left(a^r\right)^s a^r = a^{rs}a^r = a^{r(s+1)},$$

and this case follows by the inductive hypothesis and (i). The reader should do the remaining case using a similar method to that given in the last part of the proof of (i).   □

We shall see below that an important invariant of a group is the number of elements in its underlying set, we define this as follows.

Definition 2.10 (i) Two groups G and H are called *equal*, $G = H$, if and only if their underlying sets are equal (page 277), and they have the same operation.

(ii) The *order* of a group G is the number (or cardinality) of elements in the underlying set of G, this is denoted by $o(G)$.

Some comments on cardinality are given in Appendix A. One or two authors reserve the word 'order' for groups and use the word 'size' for sets, we shall use 'order' for both. Note that $o(G)$ can be finite or infinite, and this distinction is important; see page 5. If the order of G is finite, then the usual number-theoretic rules apply and, as we shall show later, they have a powerful controlling influence on the structure of G. If $o(G)$ is infinite, then different considerations apply and care is needed when interpreting results.

Isomorphism—A Preliminary Note

Several groups can appear to be distinct but are, in fact, identical from the group-theoretical point of view. If we have two groups G_1 and G_2 with a bijection θ between their underlying sets which preserves or transforms all group-theoretic properties of G_1 to G_2, and vice versa, then we say they are *isomorphic*, and θ is an *isomorphism* between them, symbolically this is written $G_1 \simeq G_2$. The main property is

$$(ab)\theta = a\theta \cdot b\theta, \tag{2.1}$$

for all $a, b \in G_1$; see page 68. We shall give a formal definition of this concept at the beginning of Chapter 4, but it will be convenient to be able to use this notion from now on. As an illustration, we give two examples here. If $G = H$, see Definition 2.10(i), then the identity map (page 281) clearly acts as an isomorphism. Secondly, the real numbers with addition $\mathbb{R}$, and the positive reals with multiplication $\mathbb{R}^+$, both form groups. They are isomorphic, and one isomorphism θ defined by

$$x\theta = 2^x, \quad \text{for } x \in \mathbb{R},$$

demonstrates this fact. The map $\theta : \mathbb{R} \to \mathbb{R}^+$ is a bijection (with inverse $\log_2$) and it transfers all group-theoretic properties of the first group to the second, and vice versa via (2.1). For instance, the neutral element 0 of $\mathbb{R}$ is mapped to $2^0 = 1$, the neutral element of $\mathbb{R}^+$, and if $a, b \in \mathbb{R}$ then $(a + b)\theta = 2^{a+b} = 2^a 2^b = a\theta b\theta$ which verifies (2.1) in this case.

Isomorphism Class

Consider the statement: "There are only two groups of order 6" (Problem 2.20). This is not correct as it stands because there are infinitely many distinct groups of order 6, but many are isomorphic to one another. So to be more precise, we should say that "there are exactly two *isomorphism classes* of groups of order 6". If we take the group with elements $\{0, 1, 2, 3, 4, 5\}$ and operation addition modulo 6 ($\mathbb{Z}/6\mathbb{Z}$, the cyclic group of order 6), and D_3 (page 3) as our 'standard' groups of order 6, then it is true that all groups of order 6 are isomorphic to one of these two groups. When discussing groups of a fixed size we shall often use this short-hand.

2.2 Examples

Groups are found throughout mathematics, there is hardly a branch of the subject where they do not occur, they are also widely used in many branches of the physical sciences. We give here an extensive list of examples to illustrate the range and applicability of the group concept. No proofs will be given, in most cases it is not difficult to check that the group axioms are satisfied. Note that the notation for individual groups given in this section will be used throughout the book, see pages 303 and 304.

Number Systems

Our first examples are the standard number systems. The integers $\mathbb{Z}$, the rational numbers $\mathbb{Q}$, the real numbers $\mathbb{R}$, and the complex numbers $\mathbb{C}$, with the operation of standard addition in each case, all form infinite Abelian groups. The non-zero rational numbers $\mathbb{Q}^*$, non-zero real numbers $\mathbb{R}^*$, and non-zero complex numbers $\mathbb{C}^*$, with the operation of multiplication in each case, also form infinite Abelian groups distinct from the above. In general, for a ring or field F we let F^* denote the multiplicative group of the non-zero elements of F. Further, the positive rationals $\mathbb{Q}^+$ with multiplication form a group (which is a subgroup of $\mathbb{Q}^*$; see Section 2.3), with a similar construction for $\mathbb{R}^+$. Note that neither the non-zero integers with multiplication nor the positive integers with multiplication form groups as inverses do not exist.

Modular Arithmetic

Our second collection of examples are finite groups from number theory. If $m > 0$, the *congruence*

$$a \equiv b \ (\mathrm{mod}\ m)$$

stands for: a and b have the same remainder after division by m (in symbols, $m \mid b - a$). This notation was first introduced by C.F. Gauss in 1801 in his famous number theory text called 'Disquisitiones arithmeticae'. Let $\mathbb{Z}/m\mathbb{Z}$ denote the set $\{0, 1, \ldots, m - 1\}$. If $a, b \in \mathbb{Z}/m\mathbb{Z}$, the operation $+_m$ is given by

$$a +_m b = a + b, \quad \text{if } a + b < m, \quad \text{and}$$

$$a +_m b = a + b - m, \quad \text{if } a + b \geq m,$$

(so $a +_m b \equiv a + b \ (\mathrm{mod}\ m)$, this is called *addition modulo m*). The set $\mathbb{Z}/m\mathbb{Z}$ with the operation $+_m$ is an Abelian group of order m, the notation $\mathbb{Z}/m\mathbb{Z}$ which relates to cosets and factor groups will be explained in Chapter 4. This implies that at least one group of order m exists for each positive integer m; in some cases this is essentially the only group of this order (that is, up to isomorphism); for example, when $m = 13$ or 15, see Appendix D.

If m is a prime number p, then $(\mathbb{Z}/p\mathbb{Z})^* = (\mathbb{Z}/p\mathbb{Z})\backslash 0$ with multiplication modulo p, defined similarly to addition modulo p, forms another finite Abelian group. Inverses exist by the Euclidean Algorithm (Theorem B.2 in Appendix B). Also note that the group $T_1 = \{-1, 1\}$ (page 42) is isomorphic both to the group $\mathbb{Z}/2\mathbb{Z}$, and to the group $(\mathbb{Z}/3\mathbb{Z})^*$.

Product Groups

Given groups $G_1, \ldots, G_n$, we can form a new group by taking all (ordered) n-tuples of the form $(g_1, \ldots, g_n)$, where $g_i \in G_i$ for $i = 1, \ldots, n$, as the new elements, and defining the new operation component-wise using the operations of each G_i in turn. In many cases, several different operations can be defined; see Chapter 7. For example, suppose $n = 2$ and $G_1 = G_2 = T_1$. The elements of the product group are

$$(1, 1), \quad (1, -1), \quad (-1, 1), \quad (-1, -1),$$

and the operation is given by $(a, b)(c, d) = (ac, bd)$. This group is called the 4-*group* and is denoted by T_2, it is a product of two copies of T_1; in Chapter 7, we use the standard notation $C_2 \times C_2$ for this group. Some authors use K (for Klein group) or V (for 'Viergruppe' the German word for '4-group' or 'fours group') for this group. Note that the square of every element in T_2 is the neutral element $(1, 1)$, and it is an example of an *Elementary Abelian Group* as defined in Problem 4.18.

Matrix Groups

Matrix groups form one of the most important collections in the theory. Let F be a field (for instance, the rational numbers $\mathbb{Q}$) and let $m \geq 1$. The set of non-singular $m \times m$ matrices with entries from F and operation matrix multiplication forms a non-Abelian group (if $m > 1$) called the $m \times m$ *general linear group* over F, it is denoted by $GL_m(F)$. The group axioms can be shown to hold using some elementary matrix algebra; the matrices are non-singular, and so inverses exist by definition. See Section 3.3 for further details.

As we shall show later, subgroups (Section 2.3) of these matrix groups provide a further wide range of examples. For instance, (a) by considering only those matrices with determinant 1 in $GL_m(F)$ we obtain the $m \times m$ *special linear group* denoted by $SL_m(F)$, or (b) by considering those matrices that have zeros at all entries below the main diagonal we obtain the group of $m \times m$ upper triangular matrices denoted by $UT_m(F)$; see Section 3.3. Also many examples can be obtained by choosing different fields F. So if F is finite, these matrix groups provide a variety of examples of finite non-Abelian groups. For instance, $SL_2(\mathbb{F}_4)$ (which we usually write as $SL_2(4)$, the group of all 2×2 matrices A with $\det A = 1$ and entries in the four element field $\mathbb{F}_4$—see page 254) is an important example of a *simple* group; as given by Definition 2.33. Many other simple groups can be defined using similar constructions; see Sections 12.2 and 12.3.

Symmetries of Geometric Objects

The symmetry properties of geometric objects provide a number of group examples. In Chapter 1 (page 3), we discussed the symmetries of an equilateral triangle, the group in question being called the *dihedral group* of the triangle denoted by D_3. The elements of the group are the rotations and reflections that give the same geometric figure, and the operation is composition. A similar construction can be carried out for a regular polygon with n sides: the clockwise rotations about the centre are now by $2\pi/n$, and 'reflection' or 'turning over' is as before. This group is denoted by D_n and again called *dihedral*. For example, D_4 is the group of symmetries of the square, it has order 8. (Note that a few authors write D_{2n} for D_n, see page 303.) Some other regular geometric objects have non-neutral (rotational) symmetry groups, for example, the tetrahedron (A_4, see Problem 3.10), the octahedron (S_4, see page 170) and the dodecahedron (A_5, see Section 3.2 and Web Section 3.6).

Also under this heading is the topic of *sphere packing* in various dimensions. Consider a large container filled with identical balls, some will touch adjacent balls and some will not. In dimension 2, where we have identical discs, a regular pattern forms and the set of disc centres gives a lattice (of equilateral triangles), and we can consider the symmetries of this lattice just as we have done for the triangle. In dimension 3, no such regular pattern forms where all adjacent balls touch. In this case, there are infinitely many ways to fit twelve balls around a central ball all touching it (there is always some room to spare), but thirteen never quite fit. In dimensions 8 and 24, 'regular touching' patterns do again form, the lattices given by the centres of the 'spheres' have some remarkable properties and give rise to some remarkable groups. For further details, see Conway and Sloane (1993). As a preliminary to this you should consider the following. The *kissing number* for these lattices is the maximum number of spheres that can fit around a central sphere S so that every sphere touches (kisses) S. In dimension 2, the kissing number is well-known to be six, and in dimension 3 it is, as noted above, twelve with some room to spare. But in dimension 8 it is 240, and in dimension 24 it is 196560 ! The first of these lattices has connections with the Mathieu group M_{24}, and the second with the sporadic group called the 'Monster' or 'Friendly Giant', see Chapter 12, the ATLAS (1985), and the reference quoted above.

Permutations

Permutations play a vital role in group theory, especially in the early development. If X is a set and S_X denotes the collection of all permutations on X (that is, bijections of X to itself), then this collection forms a group under the operation of composition called the *symmetric group* on X. If X is finite with n elements, we usually take X to be the set $\{1, 2, \ldots, n\}$ and write S_n for S_X. See page 12 for the case $n = 3$. Note that S_n is non-Abelian if $n > 2$, and has order $n!$ (count all possible maps). As with many other groups, the symmetric groups have a number of important subgroups, that is, subsets that form groups; see Definition 2.11. For example, the *alternating group* A_n which is the group contained in S_n of all even permutations (a permutation is *even* if it can be expressed as an even number of interchanges of pairs of symbols; see Section 3.1)

Examples from Analysis

Some classes of functions form groups. For example, let Z denote the set of all continuous, strictly-increasing functions f which map $[0, 1]$ onto $[0, 1]$, and satisfy $f(0) = 0$ and $f(1) = 1$. This set Z forms a group if the operation is taken to be composition of functions (the identity function f_0, where $f_0(x) = x$ for all x, acts as the neutral element, and inverses exist as the functions f are continuous and strictly monotonic). We can construct further groups (subgroups) inside this one, for instance, we could consider only those functions in Z which are differentiable. These are examples of 'topological groups', see page 6.

Free Groups and Presentations

This construction provides another way to introduce groups, it will be discussed in more detail in Section 3.4 and Web Section 4.7. Consider an *alphabet* of letters $A = \{a, a', b, b', \ldots\}$. The letter a' is going to act as the inverse of a, *et cetera*, see page 57. A *word* $c_1 c_2 \cdots c_k$ consists of a finite string of letters c_i from the alphabet A, for example,

$$aabb'a', \ b, \ \text{or} \ ababa$$

are words. The set of words with the operation of concatenation forms a semigroup; to obtain a group we proceed as follows. We define a *reduced word* as a word in which all pairs of consecutive letters $aa', a'a, bb', \ldots$ do not occur or have been removed, for example, $aabb'a'$ reduces to a, whilst b and $aba'b'a$ are reduced. As a' will act as the inverse of a, *et cetera*, each of these removals corresponds to the use of axiom (iv) in Definition 2.2. The operation of the group is as for the semigroup, that is concatenation—write one reduced word and then write the second reduced word immediately following the first, except that the resulting word must be reduced by removing all pairs $aa', a'a, bb', \ldots$ if they are formed by the concatenation, or by previous removals. For instance,

$$\text{the product of } ac'b \text{ and } b'ca'c \text{ is } c.$$

The *empty word*—that is the word with no symbols from A which is written as e where $e \notin A$—acts as the neutral element of the group, and inverses are constructed as in the example above—for instance, the inverse of $aab'cbc'$ is $cb'c'ba'a'$. The group is denoted by $\langle a, b, c, \ldots \rangle$ (in this notation it is assumed that the letters $e, a', b', \ldots$ are also present), and the letters $a, b, c, \ldots$ are the *generators*. It is called *free* because there are no constraints on possible words other than those ensuring the group properties hold; note that all free groups are necessarily infinite. A free group with just one generator a, say, is called an *infinite cyclic group*, it is isomorphic to $\mathbb{Z}$ and so we denote it either by $\langle a \rangle$ or by $\mathbb{Z}$. Non-free groups have more conditions called *relations*, or sometimes *defining relations*. For example, the finite cyclic group C_n of order n can be treated as the infinite cyclic group $\mathbb{Z} \simeq \langle a \rangle$ with the extra relation

$$a^n = e.$$

In this group, each time a^n occurs it is replaced by the neutral element e in the same way that terms of the form aa' or $a'a$ are replaced by e. In Section 3.4, we shall see that this method for constructing groups has a number of advantages, but also some disadvantages. For instance, in a few cases it may be difficult, or sometimes impossible, to determine the group order.

Elliptic Curves

The collection of solutions of some equations can be formed into groups. For example, consider the set of rational solutions of the equation

$$y^2 = x^3 + k \quad \text{where } k \in \mathbb{Z}\backslash\{0\}. \tag{2.2}$$

Suppose $P_1 = (x_1, y_1)$ and $P_2 = (x_2, y_2)$ lie on the curve. The straight line $P_1 P_2$ passing through the points P_1 and P_2 meets the curve in exactly one further point $P_3 = (x_3, y_3)$, say, and the pair (x_3, y_3) forms a new solution of (2.2), and if $x_1, \ldots, y_2 \in \mathbb{Q}$ then also $x_3, y_3 \in \mathbb{Q}$. If $P_1 = P_2$, then the line is the tangent to the curve at P_1; and the whole procedure is called the *chord–tangent process*.

Points on the curve with rational coordinates form the elements of a group, and the operation is defined using the chord–tangent process. It is closed because, given rational points P_1 and P_2 on C, a rational point P_3 on C always exists, and we set $P_1 + P_2 = -P_3$. The neutral element is the 'point at infinity' on the curve in the y-direction. (To set this procedure up properly we use homogeneous coordinates $(x : y : z)$, where

$$(tx : ty : tz) = (x : y : z) \quad \text{for all } t \in \mathbb{Q}^*.$$

The usual notation for a point (x, y) is identified with $(x : y : 1)$, and the point $(x : y : 0)$ lies on the 'line at infinity'. Equation (2.2) becomes $y^2 z = x^3 + kz^3$. The point $(0 : 1 : 0)$, the neutral element of the group, clearly lies on the curve, and is the 'point at infinity' in the y-axis direction. We set $P_1 + P_2$ to equal 'minus' P_3 to obtain a valid inverse operation, so using the standard two variable (affine) coordinates, the inverse of the point $P_1 = (x_1, y_1)$ is $-P_1 = (x_1, -y_1)$.) In this 'projective geometry' all vertical lines 'meet' at the point at infinity $(0 : 1 : 0)$, and some results from algebraic geometry are needed to prove associativity. These groups can be finite or infinite, and they are Abelian because the line through the points P_1 and P_2 is clearly the same as the line through P_2 and P_1. See for example Rose (1999), Chapters 15 and 16, for further details.

Examples from Topology

The basic structure of a topological space is best described using groups. For example, the *fundamental group* of a space, which was first defined by H. Poincaré over a century ago, is constructed as follows:

Fix a point P in a path-wise connected topological space T, and consider the set of all continuous closed and directed loops from P to P. Call two loops 'equivalent' if one can be continuously deformed into the other (topologists call them 'homotopic'), the group operation is composition. The neutral element is the set of loops that can be continuously contracted to the point P, and the inverse of the loop L is L with its direction reversed. For instance, the fundamental group of the real plane $\mathbb{R}^2$ with the origin removed is the infinite cyclic group (all loops through P that do not enclose the origin can be contracted to P, but, for example, a loop that passes around the origin clockwise four times, say, cannot be contracted and 'equals' four times a loop which passes around the origin clockwise only once).

Although the definition appears to depend on the point $P \in T$, it can be shown that fundamental groups for different points P are isomorphic; that is, there exists a fundamental group for the space T. Further details can be found in a standard book on general topology, for example, Kelley (1955) or Willard (1970). Algebraic topology is another area where groups—homology and cohomology groups-provide insights into the structure of topological spaces, see for example Benson (1991).

Examples from the Physical Sciences

Particle physicists make extensive use of group theory. Many of the essential properties of the basic constituents of matter are best described using the language and properties of groups. At least one elementary particle was discovered using the abstract theory. A collection of particles was associated with a particular class of groups, and it was realised that there was one more group (that is, 28) than known particles in this collection (at the time, 27); this led the experimental workers to look for the 'missing' particle (called Ω^-), and it was duly found a few years later! An excellent 'down-to-earth' account of this topic is given in Close (2006), and Williams (1994) provides a good technical introduction.

Some chemists use groups to describe the structure of molecules. A notable example was given in 1985 when a crystalline form of carbon, called Carbon60, was discovered by the chemists Kroto, Curl and Smalley;[2] its structure is closely related to that of a dodecahedron, and so also to the alternating group A_5; the subsection on symmetries above, Section 3.2, and Web Section 3.6 all give some details.

---

[2]They were awarded the Nobel prize in chemistry for this work.

2.3 Subgroups, Cosets and Lagrange's Theorem

Most groups contain a number of smaller groups using the same operation, we shall consider these now.

Definition 2.11 A *subgroup* H of a group G is a non-empty subset of G which forms a group under the operation of G.

We write $H \leq G$ when H is a subgroup of G. For example, if G is the group $\mathbb{Q}$, then $\mathbb{Z}$ is a subgroup, that is, $\mathbb{Z} \leq \mathbb{Q}$. But note that $\mathbb{Q}^+$ is not a subgroup of $\mathbb{Q}$ even though the underlying set in the first group is contained in the second; reader, why?

Definition 2.12 (i) A subgroup J of a group G is called *proper* if $J \neq G$, this is denoted by $J < G$.

(ii) The singleton set $\{e\}$ forms a subgroup of all groups, it is called the *neutral subgroup* and is denoted by $\langle e \rangle$.

(iii) A subgroup H of a group G is called *maximal* in G if it is a *proper* subgroup of G, and whenever a subgroup J exists satisfying $H \leq J \leq G$, then either $J = H$ or $J = G$, so no subgroup lies strictly between H and G.

Notes (a) The neutral subgroup $\langle e \rangle$ is sometimes called the identity, trivial, or unit, subgroup; see page 4.

(b) Clearly $G \leq G$; so all groups with more than one element have at least two subgroups; some have no more, see Theorem 2.34.

(c) Maximal subgroups are not necessarily 'large'. For an extreme example, consider the alternating group A_{13} which has order 3113510400, remarkably it possesses a maximal subgroup of order 78. Also arbitrarily large groups with maximal subgroups of order 2 exist—Problems 3.20 and Corollary 6.12.

(d) There are connections between maximal subgroups and generators, see Problem 2.13 and Section 10.2, and reasoning with maximal subgroups is used in several proofs, for example, in that for the Frattini Argument (Lemma 6.14).

The next result gives conditions for a group subset to be a subgroup.

Theorem 2.13 *If H is a subset of G, then $H \leq G$ if, and only if,*

(a) *H is non-empty, and*
(b) *whenever $a, b \in H$, we also have $a^{-1}b \in H$.*

Proof Clearly (a) and (b) are valid if $H \leq G$ (Definitions 2.2 and 2.11). Conversely, suppose (a) and (b) hold for a subset H of G. By (a), there is at least one element $a \in H$, and so, by (b), $e = a^{-1}a \in H$. Applying (b) again, we have, as $a, e \in H$, $a^{-1} = a^{-1}e \in H$, and so H is closed under inverses. Thirdly, if $a, b \in H$, then $a^{-1} \in H$, and so together these give

$ab = (a^{-1})^{-1}b \in H$ by Theorem 2.7(ii), that is, H is closed under the operation of G. Finally, we note that associativity holds in H because it holds in G;
the result follows.   □

There is also a 'right-hand version' of this result where in (b) '$a^{-1}b \in H$' is substi-
tuted by '$ab^{-1} \in H$'. In practice, it is often better to replace (b) by:

(b1) if $a, b \in H$ then $ab \in H$, and
(b2) if $a \in H$ then $a^{-1} \in H$.

For example, suppose $G = GL_2(\mathbb{Q})$ and H is the subset of these matrices with determinant 1. The identity matrix I_2 belongs to H, and so H is not empty. Also if
$A, B \in H$, then $\det A = \det B = 1$ and so, as $\det(AB) = \det(A)\det(B) = 1$, we deduce $AB \in H$. Finally, if $C \in H$, then $1 = \det C = \det C^{-1}$, and so $C^{-1} \in H$; this
gives $H \leq G$. We use the notation $SL_2(\mathbb{Q})$ for H, see Section 3.3. Some further
examples are given in Problem 2.10.

We consider now some set-theoretic operations on subgroups.

Corollary 2.14 (i) *If $H \leq J$ and $J \leq G$, then $H \leq G$, that is the subgroup relation
is transitive.*
(ii) *If $H, J \leq G$ and $H \subseteq J$, then $H \leq J$.*

Proof Both of these results follow from Theorem 2.13, see Problem 2.5.   □

Intersections of subgroups are always subgroups (but unions are usually not subgroups because closure fails). See the note concerning subgroup lattices on page 32.

Theorem 2.15 *Suppose I is a non-empty index set. If $H_i \leq G$, for each $i \in I$, and
$J = \bigcap_{i \in I} H_i$, then $J \leq G$.*

Proof As $e \in H_i$ for all $i \in I$, we have $e \in J$, so J is not empty. Secondly, if
$a, b \in J$, then $a, b \in H_i$ for all $i \in I$, but each $H_i \leq G$ so, by Theorem 2.13,
$a^{-1}b \in H_i$, for all $i \in I$, which shows that $a^{-1}b \in J$. Now use Theorem 2.13
again.   □

In Section 2.1 (page 13), we introduced the notion of a *generating set* for a group,
this can be formally defined by

Definition 2.16 A subset X of the underlying set of a group G is said to *generate*
G if the intersection of all subgroups of G that contain X coincides with G, or
to put this another way, the only subgroup of G that contains X is G itself. This
intersection is denoted by $\langle X \rangle$.

Theorem 2.17 *Suppose X is a non-empty subset of the underlying set of the
group G. The set X generates G if and only if the set of all products of powers
(positive and negative) of elements of X equals G.*

Proof By Theorem 2.15 and Definition 2.16, $\langle X \rangle \le G$. Suppose $\langle X \rangle = G$. Let Z denote the set of all powers of products of elements of X; $Z \subseteq G$). The set Z is non-empty as X is non-empty, and so $Z \le G$ by Theorem 2.13 and the definition of Z. Also $X \subseteq Z$, and so Z is one of the subgroups used in the formation of the intersection $\langle X \rangle$; hence $\langle X \rangle \le Z \le G$. Therefore, as X generates G (by supposition), we have $Z = G$.

For the converse suppose $Z = G$. Now for given H, if $X \subseteq H$ and $H \le G$, then $Z \le H$, again by Theorem 2.13 and the definition of Z. This holds for all such H, and so it holds for $\langle X \rangle$ by Theorem 2.15; that is, $Z \le \langle X \rangle$. But by our supposition $Z = G$, and so $\langle X \rangle = G$, and the result is proved.   □

We set $\langle X \rangle = \langle e \rangle$, if X is empty.

Now we consider group elements in more detail. For $g \in G$ we write $\langle g \rangle$ for the set of powers of $g \in G$, that is $\langle g \rangle = \{g^t : t \in \mathbb{Z}\}$; see Section 4.3. We have

Theorem 2.18 *If $g \in G$ then $\langle g \rangle \le G$.*

Proof The set $\langle g \rangle$ is clearly not empty, and if $m, n \in \mathbb{Z}$, then $g^m, g^n \in \langle g \rangle$, and $(g^m)^{-1} g^n = g^{n-m} \in \langle g \rangle$. Result follows by Theorems 2.9 and 2.13.   □

We say that g is a *generator* of the subgroup $\langle g \rangle$ of G (page 13). This result ensures that almost all groups have at least some non-neutral proper subgroups, see Theorem 2.34 for the exceptions.

Examples (a) Let $G = \mathbb{Z}$ and $g = 7$, then $\langle 7 \rangle$ is the proper subgroup of $\mathbb{Z}$ consisting of the set of integers divisible by 7.

(b) Secondly, let $G = (\mathbb{Z}/7\mathbb{Z})^*$ and $g = 3$. In this case, the subgroup $\langle 3 \rangle$ is G itself because the powers of 3 modulo 7 generate the whole group; the reader should check this and also consider the case $g = 2$.

Theorem 2.18 and these examples suggest the following

Definition 2.19 Let $g \in G$.

(i)   The subgroup $\langle g \rangle$ given by Theorem 2.18 is called *cyclic*.
(ii) The *order* of g, denoted by $o(g)$, is defined by $o(g) = o(\langle g \rangle)$; that is, $o(g)$ equals the order of the cyclic subgroup generated by g in G.
(iii) An element of order 2 is called an *involution*.
(iv) The *exponent*, if it exists, of a group G is the least common multiple of the orders of all of the elements of G; that is, the least positive integer m with the property: $g^m = e$ for *all* $g \in G$.

Notes (a) All parts of this definition are relative to a fixed group G.

(b) Orders can be finite or infinite, and if the orders of two elements are finite it does not follow that the order of their product is finite (Problem 2.7).

(c) If G is finite, it has an exponent which is not greater than $o(G)$. The group $\mathbb{Q}$ is an example of an infinite group with no exponent. In 1902, W. Burnside (1852–1927) conjectured that a group G with finite generating set and finite exponent must be finite, and this is true if G is Abelian. But it can be false if G is not Abelian as was shown by Adian and Novikov in 1968 for a group with an exponent larger than 665; see Vaughan-Lee (1993).

(d) Elements of order 2 are called *involutions* to signal the fact that they play a unique role in the theory, particularly to CFSG. (It is the subgroups called *centralisers* of these involutions, see Section 5.2, that play this vital role.) Also, apart from the neutral element e they are the only group elements which equal their own inverses. Further properties are given in Problems 2.28, 3.1(iv) and 3.20, and in the note about Coxeter Groups on page 64.

We illustrate these concepts with the following result.

Corollary 2.20 *If a group G has exponent 2, then it is Abelian.*

Proof Suppose $a, b \in G$, then $ab \in G$ and $e = (ab)^2 = abab$. Multiplying on the left by a and on the right by b, we obtain

$$ab = aeb = a(abab)b = a^2bab^2 = ba$$

as both a and b have order 2. This holds for all $a, b \in G$. $\qquad\qquad\square$

Given a prime p, an Abelian group all of whose non-neutral elements have order p is called an *Elementary Abelian p-group*. We shall see later (Problem 4.18) that these groups can be treated as vector spaces defined over a p-element field. The corollary above shows that all groups of exponent 2 are of this type, this is not true for primes $p > 2$; an example is given in Problem 6.5.

Cosets and Lagrange's Theorem

For our next results, we require some new notation. If X and Y are non-empty *subsets* of a group G, then we write XY for the subset

$$XY = \{xy : x \in X \text{ and } y \in Y\} \subseteq G.$$

If X is the singleton set $\{x\}$, then we write xY for $\{x\}Y$ (and Yx for $Y\{x\}$). Note that if $X, Y, Z \subseteq G$ then, by associativity,

(i) $X(YZ) = (XY)Z$, and
(ii) $XY = YX$, if G is Abelian.

Definition 2.21 For $H \leq G$ and $g \in G$, the set $gH = \{gh : h \in H\}$ is called a *left coset* of H in G, and the set Hg is called a *right coset* of H in G.

Cosets play an important role in the theory, here they lead to our first major result—Lagrange's Theorem, and in Chapter 4 they form part of the important ideas associated with factor groups. One of the origins of this work is Gauss's development of modular arithmetic undertaken two centuries ago (page 18): if $G = \mathbb{Z}$ and $H = n\mathbb{Z}$, then the cosets are

$$kH = \{k + nz : z \in \mathbb{Z}\} \quad \text{for } k = 0, \ldots, n-1;$$

the coset kH equals the set of integers congruent to k modulo n.

When referring to the set T of cosets of H in G, we often write $T = \{gH : g \in G\}$. Here we are using the convention that in an un-ordered set duplication is ignored, for instance, the set $\{\ldots, a, a, \ldots, a, \ldots\}$ is the same as $\{\ldots, a, \ldots\}$. If we did not use this convention in the coset case, we would need to specify a unique g in each coset gH, and this would cause problems.

We begin with some basic lemmas. The first will be used often in the following pages, it characterises the coset representatives.

Lemma 2.22 *If $H \leq G$ and $a, b \in G$, then*

$$aH = bH \quad \text{if and only if} \quad a \in bH \quad \text{if and only if} \quad b^{-1}a \in H.$$

There is an exactly similar result for right cosets; the reader should write it out and redo the following proof in this second case.

Proof Suppose firstly $aH = bH$. As H is a subgroup, $e \in H$ and so $a = ae \in aH = bH$. Secondly, suppose $a \in bH$, then there exists $h \in H$ satisfying $a = bh$, and so $b^{-1}a = h \in H$. Lastly, suppose $b^{-1}a \in H$. As above this gives $a = bh$ for some $h \in H$, and hence

$$ah_1 = bhh_1 \in bH \quad \text{for all } h_1 \in H;$$

that is, $aH \subseteq bH$. For the converse inclusion, as $H \leq G$, we have by Theorem 2.13, $a^{-1}b = (b^{-1}a)^{-1} \in H$, and so we can repeat the previous argument with a and b interchanged, the equation $aH = bH$ follows.   □

To derive Lagrange's Theorem, we require the following three lemmas, the first shows that cosets are either disjoint or identical.

Lemma 2.23 *If $H \leq G$, then the underlying set of G can be expressed as a disjoint union of the collection of all left cosets of H in G. There is an exactly similar result for right cosets.*

Proof Clearly, each element $g \in G$ belongs to a left coset because $g \in gH$. Suppose further $g \in aH$ and $g \in bH$, then by Lemma 2.22, $aH = gH = bH$, and the lemma follows. The right coset version follows similarly.   □

Lemma 2.24 *If $H \leq G$ and $g \in G$, then $o(H) = o(gH) = o(Hg)$.*

Proof We give the proof for left cosets, the right coset result is proved similarly. To establish this lemma, we construct a bijection between the sets involved. Let ϕ be the map from H to gH defined by

$$h\phi = gh;$$

see note on 'left or right' on page 68. If $h\phi = h'\phi$ then $gh = gh'$, and by cancellation (Theorem 2.6) this gives $h = h'$, and so ϕ is injective, hence it is bijective because it is clearly surjective.   □

Lemma 2.25 *If $H \leq G$, then the number (cardinality) of left cosets equals the number of right cosets.*

Proof As in the previous proof, we construct a bijection between the sets. Let θ be the map from the set of left cosets to the set of right cosets given by

$$(gH)\theta = Hg^{-1}.$$

This is well-defined; for if $gH = g_1 H$, then $Hg^{-1} = Hg_1^{-1}$ (use left and right versions of Lemma 2.22 and closure under inverses). It is also clearly surjective because each element of G is the inverse of some element in G (Theorem 2.8). To prove injectivity, suppose

$$(gH)\theta = (g_1 H)\theta, \quad \text{that is} \quad Hg^{-1} = Hg_1^{-1}.$$

Using the right-hand version of Lemma 2.22, this gives $g_1^{-1} \in Hg^{-1}$, and hence $g_1^{-1} = hg^{-1}$ for some $h \in H$. Therefore, $g_1 = gh^{-1} \in gH$ (as H is a subgroup), and so $g_1 H = gH$ by Lemma 2.22 again.   □

Having established these lemmas, we can now derive Lagrange's Theorem. Much of the early work in algebra was concerned with properties of polynomials defined over the rational numbers. J.-L. Lagrange (1736–1813), an Italian mathematician working in France, studied these polynomials as well as in many other topics in mathematics. He postulated a result similar to that given in the last part of Problem 5.1 which relies on what we now call 'Lagrange's Theorem'; and it is for this reason that the following result is so named. In fact, Galois gave the first proof of the theorem for permutation groups in 1832, and it was probably C. Jordan (1838–1932) who gave the first proof for general groups some thirty years later.

We begin by making the following

Definition 2.26 Let $H \leq G$. The number (cardinality) of left cosets of H in G is called the *index* of H in G, it is denoted by $[G : H]$.

By Lemma 2.25, this equals the number of right cosets of H in G.

Theorem 2.27 (Lagrange's Theorem) *If $H \leq G$ then $o(G) = o(H)[G : H]$.*

Proof By Lemma 2.23, the underlying set of G is a disjoint union of $[G : H]$ cosets, and by Lemma 2.24, each of these cosets has the same cardinality (number of elements), that is $o(H)$, the theorem follows.   □

This result is particularly useful in the finite case where it shows that H can only be a subgroup of G if $o(H) \mid o(G)$; that is, the prime factorisation of the order of a group G is an important invariant of G. For instance, a group of order 30 cannot have subgroups of order $4, 7, 8, 9, 11, \ldots, 29$. Also, it cannot have elements of order $4, 7, \ldots$, see Definition 2.19. In the infinite case, the theorem shows that either the order of the subgroup, or the index (or both), must be infinite. Note that there exist infinite groups all of whose proper subgroups are finite, see Problem 6.7.

2.4 Normal Subgroups

The last topic in this chapter concerns a special type of subgroup in which left and right cosets are equal, they play a vital role in the theory. These subgroups were first defined by Galois in the 1820s when he was working on the solution of polynomial equations by radicals; see the Introduction to Chapter 11.

Definition 2.28 (i) Let $K \leq G$. The subgroup K is called *normal* in G if, and only if,

$$gK = Kg \quad \text{for all } g \in G.$$

This is denoted by $K \lhd G$.

(ii) If $g, h \in G$, $h^{-1}gh$ is called the *conjugate* of g by h in the group G.

(iii) For a fixed element $g \in G$, the set $\{h^{-1}gh : h \in G\}$, that is, the set of conjugates of g in G, is called the *conjugacy class* of g in G; see also Definition 5.17.

Notes (a) The subgroups $\langle e \rangle$ and G are normal in G for all groups G.

(b) If G is Abelian, all subgroups are normal, all conjugates of g equal g, and so the conjugacy class of g in G is $\{g\}$.

(c) We reserve the symbol 'K', possibly with primes or subscripts, to denote a normal subgroup, but other symbols will occasionally be used where necessary. In Chapter 4, we discuss the connection between normal subgroups and *kernels* of homomorphisms—K for kernels and so also for normal subgroups.

(d) Conditions stronger than normality are useful at times. The first is *characteristic*, for a definition and basic properties see Problem 4.22; the main point is that characteristic is a transitive property whereas normality is not. Some authors use an even stronger property called *fully invariant* which is defined similarly to characteristic except that in the definition on page 89 the word 'automorphism' is replaced by 'endomorphism', see Definition 4.2.

The following theorem gives two conditions for normality, see note below the statement of Lemma 4.6.

Theorem 2.29 (i) *If $K \leq G$, then the following conditions are equivalent*:

(ia) $K \triangleleft G$;
(ib) *for all $g \in G$, $g^{-1}Kg \subseteq K$*;
(ic) *for all $g \in G$ and all $k \in K$, $g^{-1}kg \in K$.*

(ii) *Suppose $K \triangleleft G$. If $k \in K$, then all conjugates of k in G belong to K, and K is the union of a collection of the conjugacy classes of G.*

Proof Note first that both parts of (ii) follow immediately from (i). Suppose (ia) holds, so if $g \in G$, $gK = Kg$ by definition. Hence, for *all* $k \in K$, we can find $k' \in K$ to satisfy

$$gk' = kg, \quad \text{that is} \quad g^{-1}kg = k' \in K,$$

which gives (ib). Secondly, note that (ic) follows immediately from (ib) (as $g^{-1}kg \in g^{-1}Kg$). Finally, suppose (ic) holds. So if $g \in G$ and $k \in K$, we can find $k' \in K$ to satisfy

$$g^{-1}kg = k', \quad \text{which gives} \quad kg = gk' \quad \text{and so} \quad Kg \subseteq gK,$$

as this argument holds for all $k \in K$. For the converse, we have $gkg^{-1} = (g^{-1})^{-1}kg^{-1} \in K$, and so we can find $k'' \in K$ to satisfy $gkg^{-1} = k''$ or $gk = k''g$. This gives the reverse inclusion and (ia) follows. □

Notes To prove that $K \triangleleft G$ it is necessary to prove *both* $K \leq G$ *and* K is normal in G. Secondly, normality is not transitive (*cf.* Corollary 2.14); that is, if $K \triangleleft G$ and $H \triangleleft K$, it does not follow that $H \triangleleft G$; see Problem 2.19(iii) for an example. On the other hand, if $K \triangleleft G$ and $K \leq H \leq G$, then $K \triangleleft H$; see Problem 2.14. A stronger property called *characteristic* which is transitive was mentioned in (d) opposite.

Our first application of the normal subgroup concept answers the question: When is the product HJ of two subgroups H and J itself a subgroup? Note that in general HJ is not a subgroup because it is not closed under the group operation. We use the notation $H \vee J$ (or sometimes $\langle H, J \rangle$)—the *join* of H and J—for the group generated by the elements of both H and J (Definition 2.16). Clearly $HJ \subseteq H \vee J$, we have

Theorem 2.30 *Suppose $H, J \leq G$.*

(i) *If either H or J is a normal subgroup of G, then $HJ \leq G$ and $H \vee J = HJ = JH$.*
(ii) *If both $H \triangleleft G$ and $J \triangleleft G$, then $HJ \triangleleft G$.*

Proof (i) Suppose $h_i \in H$, $j_i \in J$, $i = 1, 2$, and $H \lhd G$ (the proof is similar if $J \lhd G$). Then $j_1^{-1}(h_1^{-1}h_2)j_1 = h^*$ for some $h^* \in H$ (as $h_1^{-1}h_2 \in H$, $j_1 \in J$ and $H \lhd G$). Hence

$$(h_1 j_1)^{-1}(h_2 j_2) = j_1^{-1}h_1^{-1}h_2 j_1 j_1^{-1} j_2 = h^* j_1^{-1} j_2 \in HJ,$$

and, as HJ is clearly not empty, $HJ \leq G$ follows by Theorem 2.13. A similar argument shows that a product of terms each of the form hj, for $h \in H$ and $j \in J$, is itself of this form; that is, $HJ = H \vee J$. The last equation in (i) follows because $H \vee J = J \vee H$, or we can show directly as above that $HJ \subseteq JH$ and $JH \subseteq HJ$.

(ii) By (i), we only need to check normality. If $g \in G$, $h \in H$ and $j \in J$, we have $g^{-1}hjg = g^{-1}hgg^{-1}jg \in HJ$ by hypothesis, the result follows. $\square$

Subgroup Lattices

Using Corollary 2.14 and Theorems 2.15 and 2.30, the collection of subgroups of a group forms a (complete) lattice L, that is, a non-empty partially ordered set (Definition A.5 in Appendix A) in which every subset has a greatest lower bound and a least upper bound in L. Note that both the intersection and the join of two subgroups of a group G are themselves subgroups of G. Some examples are given in Chapter 8. The structure of this lattice can have an important bearing on the group in question. The first major result (Ore 1938) states that the lattice of a finite group G is distributive (that is, $H \vee (J \cap K) = (H \vee J) \cap (H \vee K)$, *et cetera*.) if and only if G is cyclic. It should be noted that non-isomorphic groups can have identical subgroup lattices; see Problem 6.4. For a detailed account of this aspect of the theory, the reader should consult Schmidt (1994).

The *centre* of a group is an important example of a normal subgroup, it is given by

Lemma 2.31 *In a group G, the set*

$$J = \{a \in G : ag = ga \text{ for all } g \in G\}$$

forms a normal Abelian subgroup of G.

Proof Suppose $a \in J$. For all $g \in G$, we have $eg = ge$, $ag = ga$ implies $a^{-1}g^{-1} = g^{-1}a^{-1}$ by Theorem 2.7, and if $ag = ga$ and $bg = gb$ then $abg = agb = gab$; and so $J \leq G$ by Theorem 2.13. Also J is Abelian by definition. Lastly, note that $ag = ga$ implies $g^{-1}ag = a$, and so $J \lhd G$ by Theorem 2.29. $\square$

Definition 2.32 The subgroup J of G given in Lemma 2.31 is called the *centre* of G, it is denoted by $Z(G)$.

Notes The notation $Z(G)$ is used because German authors call this subgroup the *Zentrum*. The centre of a group G gives some important information about G. Clearly, $Z(G) = G$ if and only if G is Abelian. On the other hand, some groups are *centreless*, that is, $Z(G) = \langle e \rangle$; examples are D_3 and S_4, see Problem 2.26. A centreless group can in some ways be treated as the opposite of an Abelian group.

We end this chapter by introducing *simple groups*. We shall show later they can be treated as the basic 'building blocks' for the construction of all finite and some infinite groups; see Chapter 9.

Definition 2.33 A group G is called *simple* if it contains no proper non-neutral normal subgroup.

The term 'simple' is perhaps not well-chosen because some simple groups are very complicated! But as noted above they can be used as the basic constituents of all groups; of course, they are all centreless. A full list of finite simple groups is now known; see the ATLAS (1985) and Chapter 12. Many simple groups are given by Lagrange's Theorem for we have

Theorem 2.34 *If $o(G)$ is a prime number, then the group G is simple and cyclic.*

For the converse see Theorem 9.6.

Proof By Lagrange's Theorem (Theorem 2.27), the order of a subgroup of G divides $o(G)$. But in this case, the only positive divisors of the integer $o(G)$ are 1 and p, hence G has no proper non-neutral subgroups at all, and so clearly no proper non-neutral normal subgroups. Also by Theorem 2.18 and Definition 2.19, every element has order 1 or p. There is only one element, e, of order 1 (Lemma 2.4). Hence G has $p - 1$ elements of order p; let a be one of them. As a has p distinct powers (including $p^0 = e$), it follows that all elements of G equal powers of a, and so G is cyclic.   □

In fact, 'most' simple groups (counted by the size of their orders) are of this type, that is Abelian (and cyclic). For example, there are 173 (isomorphism classes of) simple groups with order less than 1000 but only five are non-Abelian. The construction of non-Abelian simple groups is a much more difficult task, in the next chapter we introduce the first groups of this type—*alternating groups*, and more will be discussed in Chapter 12. These include a number of infinite classes of matrix groups, especially the linear groups $L_n(q)$ and the unitary groups $U_n(q)$, and also 26 (!) so called *sporadic groups*. These groups range in size from 7920 (Mathieu group M_{11}) to about 10^{84} (Friendly giant M) and they have a wide variety of constructions. The existence of these non-Abelian simple groups is surely one of the most interesting and challenging aspects of the theory.

2.5 Problems

A number of the problems given below have important applications in the sequel. For an explanation of the symbols $\star$ and $\blacklozenge$, see page 9.

Problem 2.1 (i) Write out a proof of Theorem 2.8.

  (ii) Using induction on n, prove the *generalised associativity law* for groups: If $g_1, \ldots, g_n \in G$, then all expressions formed by inserting or deleting brackets (in corresponding pairs) in the term $g_1 \odot \cdots \odot g_n$ are equal.

Problem 2.2 Show that the following sets with operations form groups, and indicate which are Abelian.

(i) $\mathbb{Z}/7\mathbb{Z}$ with addition modulo 7.
(ii) $(\mathbb{Z}/7\mathbb{Z})^*$ with multiplication modulo 7.
(iii) The set $\mathbb{Q}$ with the operation $*$ where, for $a, b \in \mathbb{Q}$, we have $a * b = a + b + 3$.
(iv) $GL_2(\mathbb{Q})$ with matrix multiplication, see also Problem 2.10 below.
(v)   The set of powers of products Q (that is, the group generated by) of the complex matrices $A = \left(\begin{smallmatrix} 0 & 1 \\ -1 & 0 \end{smallmatrix}\right)$ and $B = \left(\begin{smallmatrix} 0 & i \\ i & 0 \end{smallmatrix}\right)$ with the operation of matrix multiplication. What is the order of this group?
(vi)   Let $\overline{\mathbb{R}} = \mathbb{R} \cup \{\infty\}$ where the symbol ∞ satisfies the usual naive rules: $1/0 = \infty$, $1/\infty = 0$, $\infty/\infty = 1$ and $1 - \infty = \infty = \infty - 1$. Define six functions mapping $\overline{\mathbb{R}}$ onto itself by:

$$f_1(x) = x, \qquad f_2(x) = \frac{1}{x}, \qquad f_3(x) = 1 - x,$$

$$f_4(x) = \frac{1}{1-x}, \qquad f_5(x) = \frac{x}{x-1}, \qquad f_6(x) = \frac{x-1}{x}.$$

Show that this set forms a finite group under the operation of composition.
(vii) Let R denote the real plane $\mathbb{R}^2$, let d denote the standard distance function (metric) on R, and let Θ denote the set of bijective maps of R to itself which preserve distance—if $x, y \in R$ and $\theta \in \Theta$, then $d(x, y) = d(\theta(x), \theta(y))$. A function of this type is called an *isometry*; rotation by $\pi/3$ about the origin is an example. Show that Θ with the operation of composition forms a group.

  The reader needs to be convinced that all the sets with operations described in Section 2.2 are, in fact, groups.

Problem 2.3 Why are the following sets with operations not groups?

(i)   The integers $\mathbb{Z}$ with subtraction.
(ii)   The set of odd integers with addition.
(iii) The set $\left\{\left(\begin{smallmatrix} a & r \\ 0 & b \end{smallmatrix}\right)\right\} \cup \left\{\left(\begin{smallmatrix} c & 0 \\ s & d \end{smallmatrix}\right)\right\}$ where $a, b, \ldots, s \in \mathbb{R}$ and $ab = 1 = cd$, with matrix multiplication.
(iv)   The rational numbers $\mathbb{Q}$ with multiplication.

Problem 2.4 (i) Let S be a semigroup with cancellation, so it has closure under its operation which is associative, and for all $a, b \in S$, we can find $x, y \in S$ to solve the equations

$$ax = b \quad \text{and} \quad ya = b.$$

Show that S forms a group.

(ii)* If T is a semigroup, and for all $a \in T$ there is a unique $a^* \in T$ satisfying

$$aa^*a = a,$$

prove that T is a group.

Problem ♦ 2.5 If $H, J \leq G$ and in (iii) p is a prime, show that

(i)   If H is a *subset* of J, then $H \leq J$.
(ii) $H \cap J = H$ if, and only if, $H \leq J$.
(iii) If $o(H) = o(J) = p$, then either $H = J$ or $H \cap J = \langle e \rangle$.

Problem 2.6 Prove that if G is a group and $S \leq G$, then $SS = S$. Conversely, if T is a non-empty finite subset of G and $TT = T$, prove that $T \leq G$. Is this true if T is infinite?

Problem ♦ 2.7 (Order Function) Let $g, h \in G$. Prove the following properties of the order function.

(i) $o(gh) = o(hg)$.
(ii)   If $o(g) = n$ and $m \in \mathbb{Z}$, then $o(g^m) = n/(m, n)$; see page 284.
(iii) If $o(g) = m$ and $(m, n) = 1$, there exists $h \in G$ satisfying $h^n = g$.
(iv)   If $o(g) = m$, $o(h) = n$, and g and h commute, then $o(gh) = LCM(m, n)$; see part (vii).
(v)   If G is finite and $g \in G$, then $g^{o(G)} = e$, and $o(g) \mid ex(G)$ where ex denotes the exponent of G, see Definition 2.19.
(vi)   Suppose $g \in G$ and $o(g) = mn$ where $(m, n) = 1$. Show how to find unique $a, b \in G$ to satisfy $ab = g = ba$, $o(a) = m$ and $o(b) = n$.
(vii) In (iv), if we drop the commutativity condition show that $o(gh)$ can be infinite. (Hint. Try $G = GL_2(\mathbb{Q})$.)

Problem 2.8 (i) Suppose G is a finite group and $o(G)$ is even. Is the number of elements of order 2 in G odd—does a group of even order always contain an involution? See also Cauchy's Theorem (Theorem 6.2).

(ii) Using the group $(\mathbb{Z}/p\mathbb{Z})^*$ where p is prime, see the definition on page 18, give a proof of Fermat's Theorem:

$$a^{p-1} \equiv 1 \quad (\text{mod } p) \quad \text{if } (a, p) = 1.$$

(iii) Using the same group as in (ii), show that $(p - 1)! \equiv -1 \pmod{p}$, a result sometimes (wrongly) known as Wilson's Theorem. You are given: If $p > 2$, then $(\mathbb{Z}/p\mathbb{Z})^*$ contains exactly two elements of order at most 2 (Theorem B.13 in Appendix B).

Problem 2.9 (Multiplication Tables) Given a group G of order n with elements $g_1, \ldots, g_n$ where $g_1 = e$, we can form a square array or table, with n rows and n columns, whose (i, j)th entry is the product $g_i g_j$. Show that each row and each column of this table is a permutation of the elements $g_1, \ldots, g_n$. What can you say about the first row and first column?

Is the converse true? That is, if we have a square array of elements such that each row and each column is a permutation of some fixed set, and the first row and column have the property mentioned above, does the corresponding array always form the multiplication table of a group?

Problem 2.10 Show that the following subsets are subgroups of the corresponding groups, and determine whether they are normal.

(i)   The set $\{1, -1\}$ in $\mathbb{R}^*$.
(ii) The set of permutations on $Y = \{1, \ldots, 6\}$ which leave 3 fixed in S_6, the set of all permutations on Y; see Section 3.2.
(iii) The subsets of $GL_2(\mathbb{Q})$ of matrices (a) which have determinant 1, and (b) which are upper triangular (that is, the bottom left-hand entry is zero); see page 19 and Section 3.3.
(iv) The set of complex numbers with absolute value 1 in $\mathbb{C}^*$.
(v)   The set of differentiable functions in the group Z described in the subsection on groups in analysis on page 21.

Problem 2.11 (i) Show that a finite subgroup of the multiplicative group of the complex numbers $\mathbb{C}^*$ is cyclic. (Hint. Consider roots of unity.)

(ii) Find as many subgroups as you can of the additive group of the rational numbers $\mathbb{Q}$; see Web Section 7.5.

Problem 2.12 List the left and right cosets of the subgroups given in Problem 2.10; note that the last part is not easy!

Problem 2.13 (i) Can a subset of a group G be the left coset of two distinct subgroups of G?

(ii) If G is finite and has a unique maximal subgroup H, show that it is cyclic. (Hint. Consider an element in $G \backslash H$.)

Problem ♦ 2.14 (Normality Properties) Prove the following statements—all widely used in the sequel.

(i)   If $K \lhd G$ and $K \le H \le G$, then $K \lhd H$.
(ii) A subgroup of the centre of a group G is normal in G.
(iii) If $K_i \lhd G$ for $i = 1, 2, \ldots, n$, then $\bigcap_{i=1}^n K_i \lhd G$.
(iv) If $H, J, K \le G$ and $K \lhd J$, then $K \cap H \lhd J \cap H$.

Problem ♦ 2.15 (i) Show that if $[G : H]$ is finite and $H \leq J \leq G$, then

$$[G : H] = [G : J][J : H].$$

(ii) Prove that if $H, J \leq G$ with $[G : H] = m$ and $[G : J] = n$, then $[G : H \cap J] \geq LCM(m, n)$, and equality occurs if, and only if, m and n are coprime.

Problem ♦ 2.16 (Derived Subgroup) If $g, h \in G$ we set $[g, h] = g^{-1}h^{-1}gh$, it is called the *commutator* of g and h. Also the subgroup of G generated by the set of all products of powers of the commutators of G is called the *derived* (or *commutator*) *subgroup* of G, and it is denoted by G'. In some cases, the set of commutators of a group does, in fact, form a subgroup of the group, but not always; for an example, see Rotman (1994), page 34. More generally, if $H, J \leq G$, we let $[H, J]$ denote the subgroup generated by all commutators of the form $[h, j]$ where $h \in H$ and $j \in J$; so, for example, $[G, G] = G'$. See also Problem 4.6(ii), and Section 11.1, especially page 234.

(i)   Show that $G' \lhd G$.
(ii) Find G' when G is (a) $\mathbb{Z}$, (b) D_3, and (c) Q, see Problem 2.2(v).
(iii) Prove that if $J \leq G$ and $J \supseteq G'$, then $J \lhd G$—an important fact with many applications.
(iv) Show that if $K \lhd G$ and $K \cap G' = \langle e \rangle$, then $K \leq Z(G)$, and so in particular K is Abelian.
(v)   Finally, prove that if $K \lhd G$ and $J = [K, G]$, then $J \leq K$ and $J \lhd G$.

Problem 2.17 (Commutator Identities) Prove the following identities where $[a, b, c] = [[a, b], c]$ for a, b and c in the same group G. Identity (iv) is called the *Hall–Witt Identity*.

(i) $[b, a] = [a, b]^{-1}$,
(ii)* If $a, b \in G$, and both a and b commute with $[a, b]$, show that

$$[a^r, b^s] = [a, b]^{rs} \quad \text{for } r, s \in \mathbb{Z},$$

$$(ab)^t = a^t b^t [b, a]^{t(t-1)/2} \quad \text{if } t \geq 0.$$

   (Use induction on t, (i), and the given relationship between G' and $Z(G)$.)
(iii) $[ab, c] = (b^{-1}[a, c]b)[b, c]$ and $[a, bc] = [a, c](c^{-1}[a, b]c)$,
(iv) $b^{-1}[a, b^{-1}, c]bc^{-1}[b, c^{-1}, a]ca^{-1}[c, a^{-1}, b]a = e$.
(v)   If $a_1, \ldots, a_m, b_1, \ldots, b_n \in G$ and $H = \langle a_1, \ldots, b_n \rangle$, then we can express $[a_1 \ldots a_m, b_1 \ldots b_n]$ as a product of conjugates of $[a_i, b_j]$ by some $c_{ij} \in H$.
(vi) If $H, J \leq G$ where $G = \langle H, J \rangle$, then $[H, J] \lhd G$.

Problem ♦ 2.18 Suppose $A, B, C \leq G$ and $A \leq B$. Show that

(i) $B \cap (AC) = A(B \cap C)$,
(ii) if $G = AC$ then $B = A(B \cap C)$,
(iii) if $AC = BC$ and $A \cap C = B \cap C$, then $A = B$.

(Note that AC and/or BC may not be subgroups of G, also (i) and (ii) are sometimes known as *Dedekind's Modular Laws*.)

(iv) Now suppose $A, B, C, D \leq G$ where also $AB, CD \leq G$. Show that if $A \leq D$ and $C \leq B$ then

$$AB \cap CD = AC(B \cap D).$$

Problem ♦ 2.19 Let $H, J \leq G$. Prove the following results.

(i)   If $[G : H] = 2$, then (a) $H \lhd G$, and (b) $a^2 \in H$ for all $a \in G$—facts we use many times.

(ii) If G is finite and $o(H) > o(G)/2$, then $H = G$—no finite group can have a proper subgroup of order larger than half the group order. Further, if G is also simple and $J \leq G$, then $o(J) \leq o(G)/3$. For large simple groups, the denominator 3 can be replaced by a much bigger integer; see example below Theorem 5.15, page 101.

(iii) Show that normality is not transitive (that is if $H \lhd J$ and $J \lhd G$, it does not follow that $H \lhd G$); one example occurs in D_4 using (i).

(iv) If H and J are proper subgroups of G, prove that there exists $g \in G$ which does not belong to either H or J.

(v)   Show that $HJ \leq G$, if $HJ = JH$; cf. Theorem 2.30(i).

Problem 2.20 Using Corollary 2.20, Lagrange's Theorem (Theorem 2.27) and Problem 2.5, show that up to isomorphism there are only two groups of order 4, and only two groups of order 6—that is, there are exactly two isomorphism classes of groups of order 4, and also exactly two of order 6. (Hint. For order 6, show that the group always contains an element of order 3.) See Problem 4.2(i), different methods to prove these facts are given in Chapters 5 and 6.

Problem 2.21 Let G be a group. If $H_i \leq G$ and $[G : H_i]$ is finite for $i = 1, \ldots, n$, show that

$$\left[G : \bigcap_{i=1}^{n} H_i \right] \leq \prod_{i=1}^{n} [G : H_i].$$

(Hint. Derive the result for $n = 2$ first.)

Problem 2.22 (Poincaré's Theorem) Prove that the intersection of a finite number of subgroups of G, each with finite index, is itself a subgroup of G with finite index.

Problem ♦ 2.23 If $H \leq G$ and $g \in G$, then $g^{-1}Hg$ is called a *conjugate subgroup* of H (Definition 2.28). Prove the following statements:

(i) $g^{-1}Hg \leq G$,
(ii) $o(g^{-1}Hg) = o(H)$,
(iii) $g^{-1}Hg = \{ j \in G : gjg^{-1} \in H \}$.

Problem ◆ 2.24 (Core of a Subgroup) If $H \leq G$, the *core* of H in G, core(H), is defined by

$$\text{core}(H) = \bigcap_{g \in G} g^{-1} H g,$$

see Section 5.2. Show that

(i)   core(H) ◁ G,
(ii) core(H) is the join of all normal subgroups of G which are contained in H,
(iii) core(H) is the unique largest normal subgroup of G contained in H.

Problem 2.25 (Normal Closure of a Subgroup) If $H \leq G$, then the *normal closure* H^* of H is defined as the intersection of all normal subgroups of G which contain H. Show that

(i) $H^* \triangleleft G$,
(ii) $H^* = \langle g^{-1} H g : g \in G \rangle$,
(iii) H^* is the smallest normal subgroup of G containing H.

Problem 2.26 Find the centres of the following groups.

(i)   Integers $\mathbb{Z}$,
(ii) Dihedral group D_4,
(iii) Dihedral group D_5,
(iv) 2×2 General linear group $GL_2(\mathbb{Q})$, and
(v)   Permutation group S_3.

Problem 2.27 Prove that if $H, J \leq G$, then

$$o(HJ)o(H \cap J) = o(H)o(J).$$

One method is as follows. Define a map $\theta : H \times J \to HJ$ by $(h, j)\theta = hj$. Show that if $g = hj$ where $h \in H$ and $j \in J$, then

$$g\theta^{-1} = \{(ha, a^{-1}j) : a \in H \cap J\},$$

by proving inclusion both ways round. Further, show that if $(ha, a^{-1}j) = (hb, b^{-1}j)$ then $a = b$, and so $o(g\theta^{-1}) = o(H \cap J)$. Lastly, count ordered pairs using the property $o(H \times J) = o(H)o(J)$.

Note that (a) HJ need not be a subgroup of G, and (b) a second proof of this result is given in Theorem 5.8.

Problem ◆ 2.28 Suppose G is a finite simple group of even order. Using Problem 2.8, show that G is generated by its involutions. (Hint. Note that an involution is self-inverse.) By the Feit–Thompson Theorem (Chapters 11 and 12), this shows that all finite non-Abelian simple groups are generated by a set of their involutions.

Problem 2.29 (Double Cosets) Suppose $H, J \leq G$ and $a \in G$. The set $HaJ = \{haj : h \in H, j \in J\}$ is called the *double coset* of a with respect to H and J. Show that

(i)   Each element of G belongs to exactly one double coset.
(ii) G is the disjoint union of its double cosets.
(iii) Each double coset (with respect to H and J) is a union of right cosets of H, and a union of left cosets of J.
(iv) The number of right cosets of H in the double coset HaJ is $[J : J \cap a^{-1}Ha]$. Hence

$$[G : H] = \sum_{c \in C} [J : c^{-1}Hc]$$

provided this sum is finite, where C is a set of double coset representatives for H and J.
(v)   Using the notation set up in Section 3.1, if $G = S_4$, $a = (1, 2)$, $H = \langle (1, 2, 3) \rangle$, $J_1 = \langle (1, 2, 3, 4) \rangle$ and $J_2 = \langle (1, 4)(2, 3) \rangle$, write out the double cosets HaJ_i for $i = 1, 2$.

Problem 2.30 (i) Suppose $J_1 \leq J_2 \leq \cdots \leq G$, that is we have an infinite sequence of subgroups of G. Let $J = \bigcup_{i=1}^{\infty} J_i$. Show that $J \leq G$. Note that in general a union of subgroups is not itself a subgroup.

  (ii) In (i), if J_i is simple for infinitely many i, show that J is also simple.

Problem 2.31 (Project) Whilst reading this book, list all those theorems which apply without restriction or caveat, for example, one of the first is Cancellation (Theorem 2.6).

Chapter 3
Group Construction and Representation

Bertrand Russell defined the integer '3' as that property common to all sets having three elements, with similar definitions for other positive integers. The abstract entity '3' has many *representations* in the myriad of sets with three elements.[1] So it also is with groups. For example, the *alternating group* A_5 to be introduced in Section 3.2 is defined as the group of even permutations on a five-element set—the first *representation* of this group we give is expressed in terms of permutations. But it has several other representations (as a matrix group or a symmetry group, *et cetera*, see Web Section 3.6). The point being that when we discuss an individual group, we almost always discuss a particular *representation* of the group, as a matrix, or permutation, or other type of group, and the corresponding 'abstract' group is that entity common to all of these representations—this is an important point to bear in mind when discussing individual groups. Also these ideas have led to a branch of the subject called "group representation theory" which we shall introduce in Web Chapter 13.

There is one type of representation which comes close to the 'abstract' group, that is a 'group presentation' which was introduced briefly on page 21. The group is defined on an 'alphabet', the elements are the 'words' in this alphabet, and the operation is defined using concatenation. For instance, one presentation of the dihedral group D_3 described on page 3 is

$$\langle a, b \mid a^3 = b^2 = (ab)^2 = e \rangle.$$

In this representation, the elements of D_3 are given by the products of powers of a and b subject to the conditions given above. Assuming that we can apply the usual elementary group rules, it is easy to show that this system only has six members: $e, a, a^2, b, ab,$ and a^2b, and that they 'correspond' to the six symmetries of the triangle described on page 3. We shall consider this topic in more detail in Section 3.4.

---

[1] Nowadays a more inductive definition is given using the integer zero and the successor function.

H.E. Rose, *A Course on Finite Groups*,
Universitext,
DOI 10.1007/978-1-84882-889-6_3, © Springer-Verlag London Limited 2009

In this chapter, we discuss three methods for defining groups, that is, three types of group construction or representation. For another, see Section 12.1. They are defined using (a) permutations, (b) matrices, or (c) generators and relations, that is, presentations, and they all have important roles to play in the theory. We begin by describing the basic properties of permutations, and the *symmetric* and *alternating groups*. Historically, the development of the theory started with them and A. Cayley (1821–1895) in 1850 showed that every group can be represented as a permutation group (Theorem 4.7). And it is for this reason that some authors describe group theory as 'the science of symmetry'; see Weyl (1952). Next we give the basic properties of matrix groups, many of the more 'interesting' groups in the theory, especially many simple groups, arise first as matrix groups; see Chapter 12. Lastly, we give the formal definition of a group presentation which will be further discussed and verified in Web Section 4.7.

3.1 Permutations

We begin by developing the basic properties of permutations. Remember that we always read from left to right. Most of the work in this section first appeared in print in a series of papers published in the 1840s by the French mathematician A. Cauchy (1789–1857); see Section 6.1. The notation introduced below is also due to him.

Definition 3.1 A *permutation* σ on a set X is a bijection of X to itself.

As we shall normally be using finite sets X, it is convenient, but not essential, to take $X = \{1, 2, \ldots, n\}$ when $o(X) = n$. Apart from their natural ordering, no arithmetical properties of the integers 1 to n are used, they are just easily recognised labels for the elements of a set with n elements. We use two notations. First we have the 'matrix' form:

$$\sigma = \begin{pmatrix} 1 & 2 & \ldots & n \\ a_1 & a_2 & \ldots & a_n \end{pmatrix}, \tag{3.1}$$

where $a_i \in X$ and $i\sigma = a_i$, for $i = 1, \ldots, n$; see the note on 'left and right' on page 68. Each element in the second row of (3.1) is the result of applying the permutation σ to the element in the first row directly above it: $i \mapsto a_i$, $i = 1, \ldots, n$, so no two elements in the second row are equal (some authors just print the second row taking our first row as read). The order of the columns is unimportant, but we usually write them as in (3.1).

Using the fact that composition of two bijections is a bijection (Appendix A), we define the 'product' of two permutations by

Definition 3.2 Let σ and τ be permutations on the same set X, where, for $i \in X$, $i\sigma = a_i$ and $i\tau = b_i$. The *product* $\sigma\tau$ is given by

$$i(\sigma\tau) = (i\sigma)\tau = b_{a_i} \quad \text{for all } i \in X.$$

Using the matrix form (3.1) above, we can rewrite this as

$$\sigma\tau = \begin{pmatrix} 1 & 2 & \cdots & n \\ a_1 & a_2 & \cdots & a_n \end{pmatrix} \begin{pmatrix} 1 & 2 & \cdots & n \\ b_1 & b_2 & \cdots & b_n \end{pmatrix}$$

$$= \begin{pmatrix} 1 & 2 & \cdots & n \\ a_1 & a_2 & \cdots & a_n \end{pmatrix} \begin{pmatrix} a_1 & a_2 & \cdots & a_n \\ b_{a_1} & b_{a_2} & \cdots & b_{a_n} \end{pmatrix}$$

$$= \begin{pmatrix} 1 & 2 & \cdots & n \\ b_{a_1} & b_{a_2} & \cdots & b_{a_n} \end{pmatrix},$$

where the second matrix in the second line is the same as the second matrix in the first line except that its columns have been permuted by σ. This does not affect the result.

In the next section, we show that this product generates a number of new groups. The neutral element is the permutation that moves no element of X (the identity map ι on X), and the inverse of a permutation is its reverse, that is, if σ is given by (3.1), then

$$\sigma^{-1} = \begin{pmatrix} a_1 & a_2 & \cdots & a_n \\ 1 & 2 & \cdots & n \end{pmatrix}.$$

The second notation for permutations uses *cycles*. We begin with an example. Let

$$\sigma_1 = \begin{pmatrix} 1 & 2 & 3 & 4 & 5 & 6 & 7 & 8 & 9 \\ 7 & 3 & 1 & 8 & 5 & 2 & 6 & 9 & 4 \end{pmatrix}.$$

This permutation maps $1 \mapsto 7, 7 \mapsto 6, 6 \mapsto 2, 2 \mapsto 3$ and $3 \mapsto 1$ so forming a *cycle* with five entries $(1, 7, 6, 2, 3)$. Alternatively, we can write

$$1\sigma_1 = 7, \quad 1\sigma_1^2 = 6, \quad 1\sigma_1^3 = 2, \quad 1\sigma_1^4 = 3, \quad \text{and} \quad 1\sigma_1^5 = 1.$$

As the symbol 4 has not been used so far, we can start again: σ_1 maps $4 \mapsto 8, 8 \mapsto 9$ and $9 \mapsto 4$, giving another cycle with three entries $(4, 8, 9)$ in this case, which is disjoint from the first cycle. Finally, 5 is the only symbol in X which has not so far been used, and $5\sigma_1 = 5$ giving a third cycle with only one element. So we can treat σ_1 as the 'product' of these three cycles, that is, we write

$$\sigma_1 = (1, 7, 6, 2, 3)(4, 8, 9)(5)$$

(although sometimes single cycles, like (5), are taken as read and not printed). Note that the cycles are disjoint from one another, and the order in which they are written does not affect the final outcome.

This example is typical as we show below.

Definition 3.3 Let σ be a permutation on the finite set X and let $x \in X$. The ordered k-tuple

$$\left(x, x\sigma, x\sigma^2, \ldots, x\sigma^{k-1}\right),$$

where k is the smallest positive integer with the property $x = x\sigma^k$ is called the *cycle of length k*, or the *k-cycle*, containing x. The integer k is called the *length* of the cycle.[2] Some authors use the term 'transposition' for a 2-cycle, say (x, y), that is, a permutation which maps $x \mapsto y$ and $y \mapsto x$.

Notes We use the same notation for an ordered k-tuple and a cycle, no confusion should arise. No two elements of a cycle are equal and $1 \leq k \leq o(X)$. Also compared with our first 'matrix' notation, cycle maps are horizontal left to right except that the last entry is mapped to the first, see footnote.

The following result gives the essential facts about cycles.

Theorem 3.4 *Suppose σ is a permutation of the set $X = \{1, 2, \ldots, n\}$.*

(i)   *If τ is the cycle $(a_1, a_2, \ldots, a_n)$, then $\tau^{-1} = (a_n, a_{n-1}, \ldots, a_1)$.*
(ii) *The permutation σ can be expressed as a product of cycles $\tau_1, \tau_2, \ldots, \tau_k$, where $k \geq 1$. They are disjoint and commute in pairs.*
(iii) *The representation of σ given in* (ii) *is unique except for the order in which the cycles τ_i appear in the product.*

Proof (i) This follows immediately from the definition.

(ii) We repeat the argument given in the example above. The sequence

$$1, \ 1\sigma, \ 1\sigma^2, \ \ldots$$

forms a cycle C_1 of length k_1, where k_1 is the least positive integer satisfying $1\sigma^{k_1} = 1$. If $C_1 = X$, the result follows, for then σ forms a single cycle. If not, let a_1 be the least positive integer in X not used in C_1, and consider the cycle $C_2 = (a_1, a_1\sigma, a_1\sigma^2, \ldots)$ of length k_2, where k_2 is defined in a similar way to k_1. Now

$$C_1 \cap C_2 = \emptyset. \tag{3.2}$$

For if not, positive integers m and n exist satisfying $1\sigma^r = a_1\sigma^s$, which gives $a_1 = 1\sigma^{r-s}$ contrary to our assumption that $a_1 \notin C_1$. We can continue this process forming $C_3, C_4, \ldots$ until all of X is used up, and then $\sigma = C_1 C_2 \cdots$. By (3.2), each C_i is disjoint from the other cycles, and they commute for the same reason.

(iii) The result clearly holds for 1-cycles. Suppose $\sigma = \tau_1 \cdots \tau_r = \upsilon_1 \cdots \upsilon_s$ where $\tau_1, \ldots, \tau_r$ ($\upsilon_1, \ldots, \upsilon_s$, respectively) are disjoint cycles each of which

---

[2]For ease of typesetting, cycles are printed linearly, but perhaps they should be printed in a circle as this would more truly represent them.

moves at least two entries. Let $1 \leq i \leq n$, and suppose both τ_k and ν_l move i, so $i\tau_k = i\sigma = i\nu_l$. Then $i\tau_k^t = i\sigma^t = i\nu_l^t$ for all t by assumption. Using Problem 3.4(iii), this shows that $\tau_k = \nu_l$, and so $\tau_1 \cdots \tau_{k-1}\tau_{k+1} \cdots \tau_r = \nu_1 \cdots \nu_{l-1}\nu_{l+1} \cdots \nu_s$. We can repeat this argument, so use induction.   □

Note that in this proof $k_1 + k_2 + \cdots = n = o(X)$.

We need to prove the following two results before we can introduce the next topic—*even* and *odd permutations*.

Lemma 3.5 *Every permutation on a finite set can be expressed as a product of 2-cycles.*

Proof By Theorem 3.4, every permutation σ can be expressed as a product of cycles. Hence we need to show that every cycle equals a product of 2-cycles. But this follows by relabelling from the identity

$$(1, 2, \ldots, n) = (1, 2)(1, 3) \cdots (1, n),$$

which the reader should check noting we read from left to right.   □

We can extend this result to: Every permutation on $X = \{1, 2, \ldots, n\}$ can be expressed in terms of the 2-cycles $(1, 2), (1, 3), \ldots, (1, n)$; see Problem 3.1.

Cyclic structure is preserved by conjugation, see Definition 2.28. We have

Theorem 3.6 *Suppose σ and τ are permutations on $X = \{1, 2, \ldots, n\}$. The permutations σ and τ are conjugate (a permutation α on X exists satisfying $\tau = \alpha^{-1}\sigma\alpha$) if and only if σ and τ have the same cyclic structure, in which case τ can be obtained by applying α to the symbols of σ.*

First, we consider an example. Let $X = \{1, \ldots, 6\}$, and let

$$\sigma = (1, 5)(3)(2, 6, 4), \quad \tau = (2, 3)(6)(4, 5, 1) \quad \text{and} \quad \alpha = (1, 2, 4)(3, 6, 5).$$

We have $2\alpha^{-1} = 1$, $1\sigma = 5$ and $5\alpha = 3$, and so $2\alpha^{-1}\sigma\alpha = 3$. This agrees with τ which also maps 2 to 3. Repeating this for the other members of X shows that $\alpha^{-1}\sigma\alpha = \tau$. Also as $1\alpha = 2, \ldots, 6\alpha = 5$, we have

$$(1\alpha, 5\alpha)(3\alpha)(2\alpha, 6\alpha, 4\alpha) = (2, 3)(6)(4, 5, 1) = \tau,$$

that is, if we replace the entries in σ by their images under α, we obtain τ. Conversely, note that if we construct a 'permutation matrix' whose top row is the entries of σ, and whose bottom row is the entries of the cycle τ, we obtain α, *viz.*:

$$\alpha = \begin{pmatrix} 1 & 5 & 3 & 2 & 6 & 4 \\ 2 & 3 & 6 & 4 & 5 & 1 \end{pmatrix},$$

which is a matrix form of α. This only works because σ and τ have the same cyclic structure—a product of a 2-cycle, a 1-cycle and a 3-cycle.

Proof Suppose first σ and τ have the same cyclic structure with cycle lengths corresponding:

$$\sigma = (\dots) \cdots (\dots, l, m, \dots) \cdots (\dots),$$
$$\tau = (\dots) \cdots (\dots, l', m', \dots) \cdots (\dots).$$

Form α (using its 'matrix' form, see (3.1)) by taking the entries of σ as its top row and the entries of τ as its bottom row:

$$\alpha = \begin{pmatrix} \dots & l & m & \dots \\ \dots & l' & m' & \dots \end{pmatrix}. \tag{3.3}$$

Now as $l\sigma = m$, $l\alpha = l'$ so $l'\alpha^{-1} = l$, and $m\alpha = m'$, we have

$$m\alpha = m' \quad \text{so} \quad l\sigma\alpha = m', \quad \text{and so} \quad l'\alpha^{-1}\sigma\alpha = m'.$$

But $l'\tau = m'$, therefore we can deduce $\tau = \alpha^{-1}\sigma\alpha$, that is, σ and τ are conjugate, because the above calculation can be applied to all corresponding consecutive pairs of entries in σ and τ (if l is the last entry in a cycle then m is the first; see footnote on page 44).

For the converse, suppose σ has the form

$$\sigma = (\dots) \cdots (\dots, j, k, \dots) \cdots (\dots),$$

as above. Further, suppose $j\alpha = r$ and $k\alpha = s$, then as $j\sigma = k$ we have

$$k\alpha = s \quad \text{so} \quad j\sigma\alpha = s, \quad \text{and so} \quad r\alpha^{-1}\sigma\alpha = s.$$

Now as α and σ are permutations (bijections on X), so is $\alpha^{-1}\sigma\alpha$. Therefore, if we define τ by

$$\tau = \alpha^{-1}\sigma\alpha,$$

the conjugate of σ by α, then it has the same cyclic structure as σ because the replacements $l \mapsto r$ and $m \mapsto s$ are themselves bijective, and the cyclic structure of σ is unaltered by this procedure. □

Returning to the example on page 45, we note that to obtain the given α we had to write σ and τ in the 'right' way. The cycles $(4, 5, 1)$, $(1, 4, 5)$ and $(5, 1, 4)$ are identical, and so a number of different α can be constructed using the process given in the proof of the theorem. This is to be expected because there is likely to be a number of different solutions α to the equation $\alpha\tau = \sigma\alpha$. For example, we can use

$$\alpha' = \begin{pmatrix} 1 & 5 & 3 & 2 & 6 & 4 \\ 2 & 3 & 6 & 1 & 4 & 5 \end{pmatrix} = (1, 2)(3, 6, 4, 5).$$

The reader should check that this new α' also gives the correct answer, and write out the remaining possible solutions (Problem 3.2).

Even and Odd Permutations

A permutation is *even* (*odd*) if it can be expressed as a product of an even (odd, respectively) number of 2-cycles. This is not a definition as it stands because there are many ways of expressing a permutation as a product of 2-cycles (Lemma 3.5). For example, the permutation $(5, 6)(3, 4)$ equals, amongst others,

$$(1, 2)(4, 3)(5, 6)(2, 1), \quad (2, 3)(2, 4)(2, 3)(6, 5) \quad \text{and}$$

$$(5, 1)(2, 3)(2, 4)(1, 6)(1, 5)(3, 2),$$

and so they all represent the same permutation. But in each case the number of 2-cycles is even; this is typical of the general situation as we show now.

Suppose $o(X) = n$ and we are considering permutations on X. We introduce the following polynomial f which we use to 'codify' all possible 2-cycles. It is given by

$$f(x_1, \ldots, x_n) = \prod_{1 \le i < j \le n} (x_j - x_i). \tag{3.4}$$

Each 2-cycle (i, j) is associated with exactly one linear factor of f. Further, if σ is a permutation on X, we define the polynomial $f\sigma$ by

$$f\sigma(x_1, \ldots, x_n) = f(x_{1\sigma}, \ldots, x_{n\sigma}) = \prod_{1 \le i < j \le n} (x_{j\sigma} - x_{i\sigma}).$$

Clearly, as σ is a permutation on X, each factor $x_j - x_i$ of f in (3.4) occurs as a factor of $f\sigma$ and vice versa, but the sign of this factor may be altered by σ. Hence, for all σ we have $f\sigma = \pm f$, and we use this fact to define the terms *even* and *odd*.

Definition 3.7 (i) Using the notation set out above, we say that the permutation σ is *even* if $f\sigma = f$, and *odd* if $f\sigma = -f$.
  (ii) The *sign*, $\mathrm{sgn}(\sigma)$, of the permutation σ is given by

$$\mathrm{sgn}(\sigma) = \begin{cases} 1 & \text{if } \sigma \text{ is even,} \\ -1 & \text{if } \sigma \text{ is odd.} \end{cases}$$

We show now that our two 'definitions' agree, and begin by proving

Lemma 3.8 *If the permutation σ is a single 2-cycle, then $\mathrm{sgn}(\sigma) = -1$, that is $f\sigma = -f$ where f is given by (3.4).*

Proof Suppose σ is the 2-cycle (i, j) where $1 \leq i < j \leq n$. When σ is applied to f, the factor $x_j - x_i$ is replaced by $x_i - x_j$ introducing a minus sign. This gives the result as the remaining changes have no overall effect. For if $i \neq k \neq j$, the term $x_k - x_i$ is replaced by $x_k - x_j$, and $x_j - x_k$ is replaced by $x_i - x_k$. Some of these replacements introduce minus signs. If $k < i$, or if $j < k$, no sign changes occur, but if $i < k < j$, then

$$(x_k - x_i)\sigma = -(x_j - x_k) \quad \text{and} \quad (x_j - x_k)\sigma = -(x_k - x_i).$$

These sign changes occur in pairs and so when combined they have no overall effect, the result follows. $\qquad\Box$

Theorem 3.9 (i) *If σ, τ are permutations on X, then*

$$\text{sgn}(\sigma\tau) = \text{sgn}(\sigma)\,\text{sgn}(\tau).$$

(ii) *If $\sigma = \tau_1 \cdots \tau_k$ and each τ_i is a 2-cycle, then $\text{sgn}(\sigma) = (-1)^k$.*

Part (ii) shows that our two definitions of even and odd agree.

Proof (i) For $i \in X$, we have $(i\sigma)\tau = i(\sigma\tau)$ by associativity of composition (Appendix A), hence

$$f(\sigma\tau)(x_1, \ldots, x_n) = \prod_{1 \leq i < j \leq n} \left(x_{j(\sigma\tau)} - x_{i(\sigma\tau)} \right)$$

$$= \prod_{1 \leq i < j \leq n} \left(x_{(j\sigma)\tau} - x_{(i\sigma)\tau} \right) = (f\sigma)\tau(x_1, \ldots, x_n).$$

Now if $\text{sgn}(\sigma) = \text{sgn}(\tau) = 1$, then $f\sigma = f$, so $(f\sigma)\tau = f\tau = f$ which gives $f(\sigma\tau) = f$, and $\text{sgn}(\sigma\tau) = 1$ follows. The other cases can be treated similarly, and so (i) is proved.

(ii) This follows from (i) and Lemma 3.8. $\qquad\Box$

3.2 Permutation Groups

In the previous section, we studied properties of permutations, here we use these properties to construct a number of groups. Historically, these developments had a considerable influence on the progress of the theory as a whole; in the next chapter (Theorem 4.7), we show that every group can be treated as a group of permutations—a result due to Cayley. Note that there is nothing unique about permutations, in Problem 4.17 we show that every finite group can also be treated as a matrix group in many different ways; see also Web Section 4.7. First, we prove that the collection of all permutations on a set forms a group.

Theorem 3.10 *Suppose X is a set. The collection of all permutations on X, with composition as the operation, forms a group.*

Proof A permutation is a bijection on X. The facts that composition of two bijections is a bijection, and composition of maps is associative, are proved in Appendix A. The neutral element is the permutation which moves no elements of X (the identity map ι, page 281), and the inverse of a permutation σ is its reverse: If $a, b \in X$ and $a\sigma = b$, then $b\sigma^{-1} = a$; see page 43. This proves the theorem. $\qquad\square$

The group of all permutations on X is denoted by S_X. If $o(X) = n$ (where we usually take $X = \{1, 2, \ldots, n\}$), the group is denoted by S_n and called the *symmetric group* on n symbols. These groups have the following basic properties:

(a) $o(S_n) = n!$. There are $n!$ distinct bijections of an n-element set to itself.
(b) If $m \leq n$, then $S_m \leq S_n$. In S_n, consider all permutations which keep the same $n - m$ elements fixed. For instance, $S_4 \leq S_5$; in fact, S_5 contains five copies of S_4; see Problem 4.20(ii).
(c) $S_1 = \langle e \rangle$, S_2 is cyclic, and S_n is not Abelian if $n > 2$; for instance, the cycles $(1, 2)$ and $(1, 2, 3)$ do not commute.
(d) By Lemma 3.5, S_n is generated by its 2-cycles, it can also be generated by a two element set; see Problem 3.1.
(e) By Theorem 3.6, two elements $\sigma, \tau \in S_n$ are conjugate in S_n (Definition 2.28) if and only if they have the same cyclic structure—a result we use several times.

The subset (subgroup, see Theorem 3.11 below) of even permutations in S_n given by Definition 3.7 forms an important subgroup of S_n; for $n > 4$, this provides our first example of a class of non-Abelian simple groups. We begin with

Theorem 3.11 *For $n > 1$, the set of even permutations in S_n forms a normal subgroup of S_n with order $n!/2$.*

Proof A permutation is even if it can be expressed as an even number of 2-cycles (Theorem 3.9). Clearly, the identity permutation is even. Also the product of two permutations each a product of an even number of 2-cycles is itself a product of an even number of 2-cycles, and the inverse of a product of 2-cycles is the product written in reverse order. Hence, by Theorem 2.13, the set of even permutations forms a subgroup of S_n. It is normal by Theorems 2.29 and 3.6, as this last result shows that conjugation does not alter the cyclic structure of a permutation.

Exactly half of all permutations in S_n are even, hence the order of the subgroup is $n!/2$. We prove this as follows. Define a map θ from the set of even permutations to the set of odd permutations by

$$\sigma\theta = (1, 2)\sigma \quad \text{where } \sigma \text{ is even.}$$

This is a bijection, for if τ is odd, then $(1, 2)\tau$ is even and $(1, 2)\tau\theta = (1, 2)(1, 2)\tau = \tau$. Hence θ is surjective. A surjective map on a finite set onto itself is injective (Appendix A). The result follows. $\qquad\square$

This last property extends to all subgroups of S_n, see the example on page 77 and Problem 3.7(ii).

Alternating Groups

The subgroup of even permutations of S_n given by Theorem 3.11 is called the *alternating group* on n symbols, it is denoted by A_n. Note that $o(A_1) = o(A_2) = 1$, and $o(A_3) = 3$ as A_3 contains just two 3-cycles [$(1, 2, 3)$ and $(1, 3, 2)$] and the identity permutation. A_4 is the only non-Abelian alternating group that contains a (proper non-neutral) normal subgroup; see Problem 3.10 and page 158.

For $n > 4$, the groups A_n are simple,[3] and they form our second infinite collection of simple groups. We give an elementary proof of this simplicity result now; see comments at the end of the proof. The first step is to show that A_n is generated by its 3-cycles in the same way that S_n is generated by its 2-cycles (Lemma 3.5).

Theorem 3.12 *The group A_n is generated by its 3-cycles, provided $n > 2$.*

Proof By Theorem 3.9, every element in A_n is equal to a product of an even number of 2-cycles. We prove the result by showing that all products of two 2-cycles are products of 3-cycles. Suppose $1 \leq i, j, k, l \leq n$, and they are distinct. Then we have $(i, j)(i, j) = e$, $(i, j)(i, k) = (i, j, k)$ and $(i, j)(k, l) = (i, l, j)(j, k, l)$; the result follows.   □

We come now to the main simplicity proof. We begin by assuming the contrary, that is, a proper subgroup K exists satisfying $\langle e \rangle \neq K \triangleleft A_n$, then we show

(a) if K contains a 3-cycle, then K contains all 3-cycles, and so by Theorem 3.12, $K = A_n$; and
(b) K contains a 3-cycle if $n > 4$.

These two propositions prove the theorem because they show that A_n does not contain a proper non-neutral normal subgroup. First, we prove (a).

Lemma 3.13 *If $K \triangleleft A_n$ and K contains a 3-cycle, then $K = A_n$.*

Proof As $K \triangleleft A_n$, K contains conjugates of all of its elements by Theorem 2.29. Hence, by Theorem 3.12, we need to show that 3-cycles are conjugate in A_n. By Theorem 3.6, all 3-cycles are conjugate in S_n, but we cannot apply this directly because the conjugating element relating two 3-cycles may not belong to A_n; we must prove that all 3-cycles are conjugate by even elements. We show this directly using elementary permutation arguments, later

---

[3]For $n = 5$ this result is effectively due to Abel, it was a corollary of his work on the non-solubility of the quintic.

(Problem 5.25) we give another proof of this result which 'sheds more light' on the underlying structure.

We consider A_5 first. Two distinct 3-cycles will overlap by one or two elements. Suppose there is a single overlap then, relabelling if necessary, we can take the 3-cycles in the form

$$(1, 2, 3) \quad \text{and} \quad (1, 4, 5).$$

We have $(2, 4)(3, 5) \in A_5$ and

$$(2, 4)(3, 5)(1, 2, 3)(3, 5)(2, 4) = (1, 4, 5).$$

Similarly, if there is a double overlap, that is, the cycles are of the type $(1, 2, 3)$ and $(1, 2, 4)$, or $(1, 2, 3)$ and $(2, 1, 4)$, then $(1, 4, 3)$, $(3, 4, 5) \in A_5$ and

$$(5, 4, 3)(1, 2, 3)(3, 4, 5) = (1, 2, 4)$$

$$(3, 4, 1)(1, 2, 3)(1, 4, 3) = (2, 1, 4).$$

The result follows for A_5 by permuting the symbols $1, \ldots, 5$.

Secondly, we use this to prove the general result. If $n > 5$, then A_n contains copies of A_5 obtained by considering only those permutations that move the symbols of a fixed 5-element subset of $\{1, \ldots, n\}$. In one of these copies of A_5, moving the symbols $1, 2, 3, l, *$ only, we have $(1, 2, 3)$ is conjugate to $(1, 2, l)$, and so they are also conjugate in A_n. In a second copy, moving the symbols $1, 2, j, k, l$ only, we have $(1, 2, l)$ is conjugate to (j, k, l) in A_n. Hence $(1, 2, 3)$ is conjugate to (j, k, l) in A_n, and the lemma is proved when $n > 4$. (It is also true when $n = 3$ or 4, the reader should check this.)   □

Theorem 3.14 *If $n > 4$, then the group A_n is simple.*

Proof We use a similar method to that given in the previous proof. Suppose $K \lhd A_n$ and $\langle e \rangle \neq K$. As noted above ((b) on page 50), we need to show that K contains a 3-cycle, the theorem then follows by Lemma 3.13. Let τ ($\neq e$) be an element of K which moves the least number of symbols in the underlying set $\{1, \ldots, n\}$. We show that τ is a 3-cycle. It cannot be a 2-cycle because 2-cycles are odd, hence if it is not a 3-cycle, it must move at least four symbols. It cannot be a 4-cycle because 4-cycles are odd. Hence it satisfies one of:

(i) τ is a product of an even number of disjoint 2-cycles,
(ii) τ is a product of a cycle of length at least 3 and further disjoint cycle(s).

Therefore, τ is one of the following types τ_1 or τ_2, where σ_i is a cycle or a product of cycles whose entries are disjoint from those of its predecessors:

(i) $\tau_1 = (1, 2)(3, 4)\sigma_1$
(ii) $\tau_2 = (1, 2, 3, \ldots)(4, 5, \ldots)\sigma_2$.

In each case σ_i may be absent, and in (ii) the first cycle in τ_2 has length at least 3, also the second cycle may have length 2 or may be longer, but we write (one of) the longest cycles first. We use τ_i to construct a new permutation which is not e and which moves fewer symbols than τ_i, this contradicts our supposition and so proves the theorem. In both cases, this new permutation is

$$[\tau_i, \alpha] = \tau_i^{-1} \alpha^{-1} \tau_i \alpha \quad \text{where } \alpha = (3, 4, 5).$$

We show $[\tau_i, \alpha]$ moves fewer symbols than τ_i. Note that as $\tau_i \in K \lhd A_n$, all conjugates of τ_i by elements of A_n belong to K, and so $[\tau_i, \alpha] \in K$.

(i) Applying the permutations that make up $[\tau_1, \alpha]$ in turn, we have:

beginning with the base set	$1\,2\,3\,4\,5\,\ldots,$
applying τ_1^{-1} we obtain	$2\,1\,4\,3\,*\,\ldots,$
applying α^{-1} we obtain	$2\,1\,3\,5\,*\,\ldots,$
applying τ_1 we obtain	$1\,2\,4\,*\,*\,\ldots,$
applying α we obtain	$1\,2\,5\,*\,*\,\ldots.$

So if we apply $[\tau_1, \alpha]$ to our base set we obtain a permutation which fixes 1 and 2, and maps 3 to 5, that is, we have in K a non-neutral permutation which moves fewer symbols than τ_1, contradicting our assumption.

(ii) If we apply the permutation $[\tau_2, \alpha]$ to our base set $\{1, 2, 3, 4, 5, \ldots\}$ we obtain a permutation which fixes 2 and maps 3 to 4, and so again we have a contradiction; the reader should check this.

In both cases, we have constructed in K a non-neutral permutation moving fewer symbols than those moved by τ. Therefore, our assumption is false, τ is a 3-cycle, and the theorem follows. $\qquad\qquad\qquad\qquad\qquad\qquad\qquad$ $\square$

There are many proofs of this result in the literature. The proof given above has the advantage that it uses a minimal amount of 'apparatus', and so it can be presented here. Further proofs of this result are given in Problems 3.16 (using conjugation), 5.25 (using centralisers), and 6.16 (using the Sylow theory) and in Web Sections 3.6 and 14.1. Suzuki (1982, page 295) gives a proof that uses Bertrand's postulate! He shows that $o(A_n) = o(K)^2$ if $K \lhd A_n$, and then he applies the postulate (which states that for all positive integers m a prime p can be found lying between m and $2m$). Up to isomorphism, A_5 is the only non-Abelian simple group with order less than 168, see Problems 6.15 and 6.17, and Chapter 12.

3.3 Matrix Groups

Matrix algebra provides a wide range of group examples. Given a field F and a positive integer n, we consider the set of all $n \times n$ non-singular matrices with entries in F. (An $n \times n$ matrix A is non-singular if and only if another $n \times n$ matrix B

exists satisfying $AB = BA = I_n$. A standard theorem of linear algebra states that A is non-singular if and only if $\det A \neq 0$ where $\det A$ denotes the determinant of A.) This set of matrices forms a group called the *general linear group* over F, and it is denoted by $GL_n(F)$. The neutral element is the $n \times n$ identity matrix I_n, and inverses exist by definition. The normal subgroup (Theorem 3.15 below) of matrices with determinant 1 is denoted by $SL_n(F)$ and called the $n \times n$ *special linear group* over F.

We shall mainly be concerned with the case when F is finite, sec Section 12.2. Up to isomorphism, unique finite fields exist of order p^m for each prime number p and positive integer m; they are usually called *Galois fields*, and denoted by $\mathbb{F}_{p^m}$, after their discoverer É. Galois;[4] see page 229. The general linear group defined over a field with p^m elements, that is, $GL_n(\mathbb{F}_{p^m})$, is denoted by $GL_n(q)$ where $q = p^m$. Similarly, we write $SL_n(q)$ for $SL_n(\mathbb{F}_{p^m})$.

The basic properties of these groups are given by

Theorem 3.15 *Suppose F is a field and $n \geq 1$.*

(i)   *The set of matrices $GL_n(F)$ with matrix multiplication forms a group.*
(ii) *$SL_n(F) \lhd GL_n(F)$.*
(iiia) *If $q = p^m$, where p is prime and $m > 0$, then*

$$o(GL_n(q)) = q^{n(n-1)/2}(q^n - 1)(q^{n-1} - 1) \cdots (q - 1).$$

(iiib) *If $n > 1$, $o(SL_n(q)) = o(GL_n(q))/(q - 1)$, and $o(SL_1(q)) = 1$.*

Proof (i) and (ii) Both of these follow from the fact that $\det AB = \det A \det B$ for all $A, B \in GL_n(F)$, and its corollaries $\det A^{-1} = 1/\det A$, and $\det I_n = 1$.

(iiia) Let $A \in GL_n(q)$. As A is non-singular, every sequence of n elements of $\mathbb{F}_q$ is a possible top row of A except $\{0, 0, \ldots, 0\}$, that is, there are $q^n - 1$ possible top rows for A. Again as A is non-singular, a possible second row of A is a sequence of n elements of $\mathbb{F}_q$ which is not a linear multiple of the first row. There are q multiples of the first row (this includes the row $\{0, 0, \ldots, 0\}$), so $q^n - q = q(q^{n-1} - 1)$ second rows of A are possible. We can continue this process, the third row must not be a linear combination of the first two rows, and so on. The result follows by collecting terms.

(iiib) This is a consequence of (ii) and (iiia) as the number of non-zero elements of $\mathbb{F}_q$ is $q - 1$, see page 76, or we can use the following argument. In the last stage of the procedure given in (iiia), there are $q^n - q^{n-1} = q^{n-1}(q - 1)$ choices for the last row of the matrix. So there are q^{n-1} choices if we also stipulate that the determinant of the matrix in question has a particular value. This gives (iiib) if we choose this value to be 1; the reader should check this. $\square$

For example, we have $o(GL_3(2)) = 168$ and $o(SL_2(3)) = 24$.

---

[4] The fact that all finite fields have this form was first proved by E.H. Moore in 1893.

The groups $SL_n(q)$ give rise to our third class of simple groups. If $SL_n(q)$ is 'factored' by its centre then the resulting *linear group* which is denoted by $L_n(q)$ is simple, except when $n = 2$ and $q = 2$ or 3. The operation of forming a factor group will be discussed in the next chapter, and a simplicity proof will be given in Chapter 12. This result implies that $SL_n(q)$ is 'nearly' simple, its only (proper non-neutral) normal subgroup being its centre which has order $(q - 1, n)$. In some cases, $SL_n(q)$ is itself simple (when its centre has order 1), examples are $SL_2(4)$ (which is isomorphic to A_5, see Problem 6.16) and $SL_3(3)$ (Section 12.2).

Subgroups of $GL_n(q)$

The group $GL_n(q)$ has a number of important subgroups, one is $SL_n(q)$ as discussed above. Another type is the subgroup of permutation matrices discussed in Problem 3.12. A third type is the subgroup of 'upper triangular' matrices. (For a full list, see Dickson 2003, or Kleidman and Liebeck 1990.) A matrix $A \in GL_n(q)$ is called *upper triangular* if every entry in A below the main diagonal is zero, this subgroup is denoted by $UT_n(q)$. Lower triangular matrices can also be introduced. Note that each diagonal entry of a non-singular upper triangular matrix is non-zero because the determinant of this matrix equals the product of its diagonal elements. We have

Lemma 3.16 $UT_n(q) \leq GL_n(q)$.

Proof Clearly, $I_n \in UT_n(q)$, and the product of two upper triangular matrices is itself upper triangular. Hence, by Theorem 2.13, we need to show that the inverses are also upper triangular. Let $A = (a_{ij}) \in UT_n(q)$ (where a_{ij} is the (i, j)th entry in the matrix A) and so $a_{ij} = 0$ if $j < i$, and let $B = (b_{ij}) \in GL_n(q)$ where $AB = I_n$. The (i, j)th entry of AB is

$$s_{ij} = a_{ii}b_{ij} + \cdots + a_{in}b_{nj},$$

with $n - (i - 1)$ summands as $a_{i1} = \cdots = a_{i(i-1)} = 0$. If $j < n$, $s_{nj} = a_{nn}b_{nj} = 0$, and so $b_{nj} = 0$ because $a_{nn} \neq 0$. Secondly, if $j < n - 1$, then

$$s_{(n-1)j} = a_{(n-1)(n-1)}b_{(n-1)j} + a_{(n-1)n}b_{nj} = 0.$$

This gives $b_{(n-1)j} = 0$ when $j < n - 1$ by the first part and as $a_{(n-1)(n-1)} \neq 0$. Continuing this process for $i = n - 2, \ldots, 2$ completes the proof. $\square$

Suppose $A \in GL_n(q)$. We can find $U_1, U_2 \in UT_n(q)$ and a permutation matrix P (Problem 3.12) to satisfy

$$A = U_1 P U_2.$$

This result is known as the *Bruhat Decomposition Theorem*, and it has a number of useful applications. We shall give a simple proof in the case $n = 2$, the proof of the general case follows similar lines. Let $A = \left(\begin{smallmatrix} a & b \\ c & d \end{smallmatrix}\right)$. If $c = 0$, then the result follows by putting $U_1 = A$ and $P = U_2 = I_2$, the 2×2 identity matrix. So we may assume that $c \neq 0$. In this case, we let $P = \left(\begin{smallmatrix} 0 & 1 \\ 1 & 0 \end{smallmatrix}\right)$, $U_1 = \left(\begin{smallmatrix} x_1 & y_1 \\ 0 & 1 \end{smallmatrix}\right)$ and $U_2 = \left(\begin{smallmatrix} x_2 & y_2 \\ 0 & 1 \end{smallmatrix}\right)$. Then, if $U_1 P U_2 = A$ we have

$$\begin{pmatrix} x_1 & y_1 \\ 0 & 1 \end{pmatrix} \begin{pmatrix} 0 & 1 \\ 1 & 0 \end{pmatrix} \begin{pmatrix} x_2 & y_2 \\ 0 & 1 \end{pmatrix} = \begin{pmatrix} y_1 x_2 & y_1 y_2 + x_1 \\ x_2 & y_2 \end{pmatrix} = \begin{pmatrix} a & b \\ c & d \end{pmatrix}.$$

This gives $x_2 = c \neq 0$, $y_2 = d$, $y_1 = ac^{-1}$ and $x_1 = b - dac^{-1} \neq 0$, as $\det A \neq 0$. Now this defines both U_1 and U_2.

A 2×2 *transvection* is a matrix of the form $\left(\begin{smallmatrix} 1 & r \\ 0 & 1 \end{smallmatrix}\right)$ or $\left(\begin{smallmatrix} 1 & 0 \\ r & 1 \end{smallmatrix}\right)$ where $r \neq 0$; we shall consider these matrices in more detail in Chapter 12. Here we use them to derive a consequence of Bruhat's result as follows.

Theorem 3.17 *The group $GL_2(q)$ is generated by its diagonal matrices and its transvections.*

Proof By Bruhat's Theorem we need to show that we can construct the upper triangular and permutation matrices as products of diagonal matrices and transvections. We have

$$\begin{pmatrix} a & 0 \\ 0 & c \end{pmatrix} \begin{pmatrix} 1 & ba^{-1} \\ 0 & 1 \end{pmatrix} = \begin{pmatrix} a & b \\ 0 & c \end{pmatrix},$$

$$\begin{pmatrix} 1 & 1 \\ 0 & 1 \end{pmatrix} \begin{pmatrix} 1 & 0 \\ -1 & 1 \end{pmatrix} = \begin{pmatrix} 0 & 1 \\ -1 & 1 \end{pmatrix}, \quad \text{and}$$

$$\begin{pmatrix} 0 & 1 \\ -1 & 1 \end{pmatrix} \begin{pmatrix} -1 & 1 \\ 0 & 1 \end{pmatrix} = \begin{pmatrix} 0 & 1 \\ 1 & 0 \end{pmatrix},$$

which proves the result as there are only two permutation matrices in the 2-dimensional case. □

3.4 Group Presentation

The American mathematician W. von Dyck (1856–1934) in about 1880 introduced the 'presentation' of a group, as an example he derived the presentation of S_4 given in Problem 3.18. This work has led to a branch of the theory called *Combinatorial Group Theory*, we shall only give the basic ideas and definitions, the interested reader should consult Lyndon and Schupp (1976) or Chandler and Magnus (1982). In the second section of Chapter 2, we introduced briefly a method for defining groups using so-called *generators* and *relations*, here and in Web Section 4.7 we give the formal definitions and proofs. The method is important in the theory because many groups are best defined in these terms, see, for example, the group E discussed in Section 8.3. Although it is a representation, it is the nearest 'approach' to an abstract definition of the group.

Suppose we are given an *alphabet* or *list* A of *letters* or *symbols* which we usually assume to be finite (but this is not essential):

$$a_1, a_2, \ldots, b_1, \ldots, c_1, \ldots.$$

A *word* is defined as a finite sequence of letters written using concatenation, for example,

$$b_2, \quad c_3c_3b_1a_2c_1, \quad \text{and} \quad a_1a_1a_1a_1$$

are words. In this section, we use lower case characters at the beginning of the alphabet $a, b, c, \ldots$ for *letters*, and lower case characters at the end of the alphabet $x, y, z, \ldots$ for *words*. If the word x has the form $x = y_1 z y_2$, then z is called a *subword* of x (as are both y_1 and y_2). The operation is as suggested above, that is, concatenation. For example, if $x = a_i a_j b_k$ and $y = b_k c_l c_1 a_m$, then

$$xy = a_i a_j b_k b_k c_l c_1 a_m.$$

It follows immediately that the system of all words on a fixed alphabet A with the operation of concatenation forms a semigroup.

To construct a group we need to bring the neutral element and inverses into the system, and we do this as follows. The word containing no letters from A will act as the neutral element. Mainly for typographical reasons, it is necessary to introduce a symbol for this element, and so as previously we let e stand for it where $e \notin A$. Note that the symbol e and the blank symbol are synonymous. For the inverse operation to apply, the alphabets have the form $A \cup A'$, where $A \cap A' = \emptyset$ and there exists a bijection $'$ between A and A'. So the alphabet has the structure:

$$a_1, a_1', a_2, a_2', \ldots, b_1, b_1', \ldots, c_1, c_1', \ldots,$$

and the given prime bijection $'$ between A and A' satisfies $(a')' = a$, for all letters $a \in A$ and $a' \in A'$. The basic idea here is that we want a' to act as the inverse of a, *et cetera*. We define the set of *reduced words* on A by

Definition 3.18 A word x on the alphabet A (it is assumed that the second alphabet A' is also present; see above) is called a *reduced word* if (a) it contains no pairs of *consecutive* letters (symbols) of the form $a_i a_i', a_i' a_i, b_j b_j', \ldots$, that is, it contains no letter, with or without a prime $'$, immediately followed by its image under the $'$-map, and (b) all blank symbols have been removed except that if no alphabet letters remain, then a single blank symbol should remain.

For example,

$$c_3, a_1 b_2, a_1 a_1 a_1, e$$

are reduced, and

$$a_3 a_3', b_2 e, b_1 c_1 c_1', a_2 a_2 a_2' a_2'$$

are not reduced.

We now adapt the semigroup concatenation operation to construct a group operation. Suppose x and y are reduced words. To form their 'product', which we call the *reduced concatenation* of x and y, we first construct the semigroup concatenation xy of x and y, and then we remove (that is, replace by the empty word) all pairs of consecutive symbols of the form $a_i a'_i$, or $a'_i a_i$, or $b_j b'_j$, *et cetera*, that are formed by the concatenation, *or* by previous removals. For example, if $x = a'_1 a'_1 b_2 c'_1$ and $y = c_1 b'_2 a_1$, then

$$xy = a'_1 a'_1 b_2 c'_1 c_1 b'_2 a_1 = a'_1 a'_1 b_2 b'_2 a_1 = a'_1 a'_1 a_1 = a'_1;$$

and $yx = c_1 b'_2 a_1 a'_1 a'_1 b_2 c'_1 = c_1 b'_2 a'_1 b_2 c'_1$. (This construction is a generalisation of one that will be familiar to the reader. In the group of integers $\mathbb{Z}$, we have, for instance,

$$3 = 4 - 1 = 5 - 2 = \cdots = (n + 3) - n.$$

Note $(n + 3) - n = 3 + (n - n) = 3$ for all $n \in \mathbb{Z}$, that is, each time we encounter $n - n$ we replace it by 0, and delete it if other symbols are present. This is exactly mirrored in the general case described above.)

This system forms a group. Two words are equal if each can be obtained from the other by the insertion and/or deletion of pairs of consecutive symbols of the form aa' or $a'a$. To be more precise this procedure defines an equivalence relation on the set of words, and the group elements are the corresponding equivalence classes, see Web Section 4.7 (In our example using the group $\mathbb{Z}$ given above, we associate 3 with the set of differences $\{(n + 3) - n : n \in \mathbb{Z}\}$.) The set of words has a neutral element and is closed under the inverse operation, hence we need to establish associativity. We have

Theorem 3.19 *The system of reduced words on a fixed alphabet A (that is, $A \cup A'$, see page 56), with the operation of reduced concatenation defined above, forms a group.*

We shall not give a proof of this result here. This is best done once we have proved the First Isomorphism Theorem which is given in Chapter 4; see Web Section 4.7. It is also possible to give an *ad hoc* proof which splits the result into a number of particular cases, but it is somewhat unsatisfactory in that it does not illustrate the underlying structure, and it is not easy to be sure that all cases have been considered. Reader, try it.

The group defined by the above theorem is called the *free group* on the alphabet A. It is called *free* because there are no constraints on the words in the group other than those needed to form the group. It is necessarily infinite, for if $a \in A$ then $a, aa, aaa, \dots$ all belong to the group, and they are distinct. A range of new groups can be defined by introducing more constraints. Consider the following example. Let $A = \{a\}$, that is, the alphabet consists of the two letters a and a' where $aa' = a'a = e$, the empty word, and let G be the free group on A. Then the (reduced) elements are

$$\dots, a'a', a', e, a, aa, aaa, \dots,$$

and G is called the *infinite cyclic group*, it is denoted by $\mathbb{Z}$. We can introduce a relation in the form

$$a^n = e$$

where n is a positive integer and a^n stands for $aa \cdots a$ with n copies of a. It is easily seen that the elements of the new structure are

$$e, a, a^2, \ldots, a^{n-1},$$

and that it forms a group with n elements. It is called the *cyclic group of order n* and is denoted by C_n. We can treat the elements of C_n in the same way as those of the infinite cyclic group except each time we encounter a^n we delete it (replace it by the empty word) as we do for aa' and $a'a$. Further cyclic group properties are given in Section 4.3.

The procedure illustrated in this example can be applied to all free groups. Suppose A is an alphabet and R is a set of words on A (that is, on $A \cup A'$), then we can form the structure (group) H called a *presentation*. It is denoted by

$$H = \langle A \mid R \rangle, \tag{3.5}$$

and consists of the free group on A with the constraints (relations) $x = e$, for all $x \in R$. If $R = \{x_1, \ldots, x_k\}$, we usually write $x_1 = \cdots = x_k = e$ for R in this presentation. The elements of A are called the *generators* of H, for $x \in R$ the equation $x = e$ is called a *relation* of H, and this method of defining H is called a *presentation* of H. (Some authors use the term *relator* for a word x in R.) To study H we work in the free group on A, and each time we encounter a word in R we replace it by the empty word just as we do for $a_i a_i', a_i' a_i, \ldots$ in the free group. In the example above, we have $C_n = \langle a \mid a^n = e \rangle$ and a^n is replaced by the empty word each time it occurs. For all A and R the structure H forms a group, we shall prove this in Web Section 4.7 using the Isomorphism Theorems. We show that H can be defined as a 'factor group' of the free group on A. We also show that all groups can be treated in a similar manner.

It can be difficult, and in some cases impossible, to determine the properties of H (see (3.5)), its order, or even if it is finite or infinite. For a general discussion of this and related considerations, including the *Word Problem for Groups*, see Rotman (1994). One difficulty is of the following type: Given a collection of relations R and a generator a, we might be able to deduce both $a^m = e$ and $a^n = e$ where $(m, n) = 1$, this would give $a = e$ (use the Euclidean algorithm) and so a would be redundant. In the 'worst' case, the whole construction could collapse to the neutral group. To avoid this, and similar problems, in most cases it is necessary to find another representation of the group in question, as a matrix, permutation, or similar group, and to construct a bijection between the corresponding elements. We illustrate this in the following two examples. Todd and Coxeter have devised a method called *coset enumeration* which can be used to determine the structure of a group given by a presentation; see Coxeter and Moser (1984), Chapter 2.

Example 1 (Dihedral group D_n) Let $A_1 = \{a, b\}$ and $R_1 = \{a^n, b^2, (ab)^2\}$. Then $\langle a, b \mid a^n = b^2 = abab = e \rangle$ gives a presentation of the dihedral group D_n; see

Section 2.2. The order is $2n$ because it can easily be shown that every element of the free group on $\{a, b\}$ can be reduced to a member of the set

$$e, a, a^2, \ldots, a^{n-1}, b, ab, \ldots, a^{n-1}b$$

using the relations in R_1. There is no collapse, see above, because our first representation of this group was as the symmetry group of the regular polygon with n sides and $2n$ symmetries. The group can also be generated by two involutions, see Problem 3.20.

Example 2 (Dicyclic Group Q_n) Again let $A_1 = \{a, b\}$, and let $R_2 = \{a^n b^{-2}, abab^{-1}\}$, and so $Q_n = \langle a, b \mid a^n = b^2 = (ab)^2 \rangle$. We show first that the relation

$$a^{2n} = e \tag{3.6}$$

is a consequence of the relations in R_2; note we are assuming that the structure $\langle a, b \mid R_2 \rangle$ is a group, this is proved in Web Section 4.7. Hence we can apply basic group properties. Assume first that n is even. The relations R_2 give $a^n = b^2$ and $b = aba$, and so we have, replacing b by aba several times:

$$a^n = b^2 = (aba)(aba) = a^2 ba^4 ba^2 = \cdots = a^{n/2} ba^n ba^{n/2} = a^{n/2} b^4 a^{n/2} = a^{3n},$$

and so in this case (3.6) follows by cancellation; we leave it as an exercise for the reader to derive the remaining case. Now as in the first example we can show that a member of the group is of the form $a^t b^u$ where $0 \leq t < 2n$ and $0 \leq u \leq 1$, and so the group has order $4n$. Matrix representations of these groups are given in Problems 3.13 and 3.22. We use the term *quaternion group* for the dicyclic group Q_2, see Section 6.1. Also the term *generalised quaternion group* is used for the group Q_{2^n} (with order $2^{(n+2)}$) when $n = 2, 3, \ldots$; see Problem 3.22.

3.5 Problems

Problem ◆ 3.1 Show that S_n can be generated by each of the following sets:

(i) $\{(1, 2), (1, 3), \ldots, (1, n)\}$;
(ii) $\{(1, 2), (2, 3), \ldots, (n - 1, n)\}$;
(iii) $\{(1, 2), (1, 2, \ldots, n)\}$.

As every finite group G can be treated as a subgroup of S_m for some suitably chosen m where $m \leq o(G)$ (Cayley's Theorem, page 72), (iii) shows that every finite group is a subgroup of a group with two generators, or to put this another way, increasing the number of generators (above two) does not necessarily increase the complexity of the group. Also all finite non-cyclic simple groups have two element generating sets; see Cameron (1999).

(iv) Show that A_n, for $n > 4$, is generated by its involutions. Note that this property holds for all non-Abelian simple groups (Problem 2.28).

Problem 3.2 If $\sigma = (1,2,3)(4,5,6)$ and $\tau = (1,5,6)(2,3,4)$, find all α such that $\alpha^{-1}\sigma\alpha = \tau$; see the example given after Theorem 3.6.

Problem ♦ 3.3 List the conjugacy classes of (i) S_3, (ii) S_4, (iii) S_5, (iv) A_3, (v) A_4, and (vi) A_5. Note that (v) and (vi) are not straightforward, some permutation calculations are needed; for a more detailed explanation, see the subsection on the Class Equations in Chapter 5. Secondly, find the normal subgroups of (vii) S_4 and (viii) A_4.

Problem 3.4 Let σ, τ and ν be cycles in S_n.

(i)   Show that if σ and τ are disjoint and $\sigma\tau = e$, then $\sigma = e = \tau$.
(ii) If σ commutes with τ and τ commutes with ν, does it follow that σ commutes with ν?
(iii) Suppose σ and τ belong to S_X where $X = \{1, \ldots, n\}$ and both σ and τ move $i \in X$. Prove that if $i\sigma^r = i\tau^r$ for all $r \geq 0$, then $\sigma = \tau$.
(iv) Let p be a prime. Show that if $\sigma^p = \iota$, the identity permutation, then $\sigma = \iota$ or σ is a product (with possibly only one factor) of p-cycles.

Problem 3.5 Give formulas for the orders of the elements of (i) S_n, and (ii) A_n, and list them when $n = 7$.

Problem 3.6 Let p and q be primes where $p \mid q - 1$, let $\sigma = (1, 2, \ldots, q)$, and let

$$\tau_r = \begin{pmatrix} 1 & 2 & \cdots & q \\ r & 2r & \cdots & rq \end{pmatrix}, \quad \text{where } 0 < r < q,$$

and the lower row is to be read modulo q. Show that r can be chosen so that τ_r is a product of $(q-1)/p$ disjoint p-cycles (and one 1-cycle). With this choice of r prove that σ and τ_r generate a subgroup of S_q with order pq, and give a presentation on a 2-symbol alphabet. See also Problems 6.14, 7.21 and 9.14, and Web Section 14.3. (Hint. Use primitive roots, see Appendix B.)

Problem ♦ 3.7 (i) Show that S_n isomorphic to a subgroup of A_{n+2}.

(ii) Using Theorem 3.11 and its extension given in the example on page 77, show that if J is a simple subgroup of the symmetric group S_n, then it is also a (simple) subgroup of the alternating group A_n.

Problem 3.8 (Semi-regularity) A permutation σ on $X = \{1, \ldots, n\}$ is called *semi-regular* if (a) every element of X is moved by σ, and (b) σ can be expressed as a product of disjoint cycles all of the same length. The identity permutation is also called semi-regular (it is a product of n cycles each of length 1). (A permutation is called *regular* if it is semi-regular and transitive.)

(i) Prove that σ is semi-regular if and only if it is a power of an n-cycle.

(Hint, if $\sigma = (a_1, \ldots, a_r)(b_1, \ldots, b_r) \cdots (d_1, \ldots, d_r)$, consider the cycle

$$\theta = (a_1, b_1, \ldots, d_1, a_2, b_2, \ldots, d_2, a_3, \ldots, d_r).)$$

(ii) If τ is an n-cycle, show that τ^s is a product of (n,s) disjoint cycles each of length $n/(n,s)$ (note that (n,s) denotes the GCD of n and s). Deduce that if p is a prime, then each positive power of a p-cycle is either a p-cycle or the identity permutation.

Problem 3.9 (Maximal Subgroups of S_n) (i) Show that $S_k \leq S_n$, for $1 \leq k \leq n$, and determine a lower bound on the number of copies of S_k that occur in S_n; see also Problem 4.20(ii).

(ii) For $1 \leq k \leq n$, let $S_k \times S_{n-k}$ denote the set of elements in S_n with the form of a permutation using the symbols in $\{1, 2, \ldots, k\} = Y$ only, multiplied by a permutation using the symbols in $\{k+1, k+2, \ldots, n\} = Z$ only. (This is the direct product of S_k and S_{n-k}, see Chapter 7—you need to check that $S_k \lhd S_k \times S_{n-k}$, et cetera) Show that

$$S_k \times S_{n-k} \text{ forms a subgroup of } S_n.$$

If $k \neq n - k$, this subgroup is maximal in S_n. This can be proved as follows: Show that the subgroup generated by the elements of S_k, S_{n-k} and a 2-cycle (u, v) where $u \in Y$ and $v \in Z$ is the whole of S_n; you are not asked to prove this here, but you could try it, say when $n = 5$ or 6. Note that the same general argument applies if we replace the first set Y by an arbitrary k-element subset of $\{1, 2, \ldots, n\} = N$ provided the second set Z is replaced by its complement in N.

(iii)* By (ii) we have $S_n \times S_n$ is a subgroup of S_{2n}; show that it is not maximal by constructing a new subgroup of S_{2n} of order $2(n!)^2$. (Hint. Begin with $S_n \times S_n$, add an element of order 2, and then show that the new set forms a proper subgroup of S_{2n}.) As in (ii) the new subgroup constructed here is maximal in S_{2n}.

The properties described in (ii) and (iii) form part of the O'Nan–Scott Theorem which provides a complete description of the maximal subgroups of the symmetric groups. A number of these subgroups involve so-called wreath products which are introduced on page 156. For further details, the reader should consult Cameron (1999), page 107.

Problem 3.10 (Alternating Group A_4) There are four main ways to represent A_4 as follows:

(a) The group of even permutations on the set $\{1, 2, 3, 4\}$,
(b) The group with presentation

$$\langle a, b \mid a^2 = b^3 = (ab)^3 = e \rangle,$$

(c) The symmetry group of a regular tetrahedron (with four equilateral triangular sides), and
(d) $SL_2(3)$ 'factored by its centre'; see Chapters 4, 8 and 12.

(i) Using direct calculation show that (a), (b) and (c) define isomorphic groups, see Problem 4.4(iv) for (d).
(ii) Find the subgroups of A_4, and indicate which are normal.

Problem 3.11 (i) In this problem you are asked to construct a subgroup J of S_5 with order 20. Let $\sigma = (1, 2, 3, 4, 5)$. Choose a 4-cycle τ so that the subgroup $J = \langle \sigma, \tau \rangle$ has this order. One method is as follows. Show that the group $H = \langle a, b \mid a^5 = b^4 = e, ba^2 = ab \rangle$ has order 20, then find an isomorphism between H and J. It can be shown that J is a maximal subgroup of S_5, it is usually called *metacyclic* and denoted by $F_{5,4}$, see Theorem 6.18 and Web Section 6.5. Can a similar construction be undertaken in S_7?

(ii) Using Lagrange's Theorem (Theorem 2.27) and Problem 2.20 find the subgroups of A_5 of order less than 10. It does also have proper subgroups of order 10 and 12, but none larger; see page 101.

Problem 3.12 (Permutation Matrices) Permutation theory can be developed as a part of matrix algebra as follows. Given a permutation $\sigma \in S_n$, the $n \times n$ *permutation matrix* P_n is formed from the $n \times n$ identity matrix I_n by permuting its rows by σ, that is, the first row $(1, 0, \ldots, 0)$ becomes the 1σth row, the second row $(0, 1, 0, \ldots, 0)$ becomes the 2σth row, and so on. Prove the following properties:

(i)   Each row and each column of P_n contains a single '1' and $n - 1$ zeros.
(ii) The determinant of P_n, det P_n, equals ± 1.
(iii) The inverse of a permutation matrix is a permutation matrix, and so the set of all $n \times n$ permutation matrices forms a subgroup of $GL_n(F)$ for all fields F. Is this subgroup normal?
(iv) A permutation σ is even if and only if the determinant of the corresponding permutation matrix is positive.

Note that if the definition of the determinant function Δ uses the notions of even and odd permutations to determine the signs in the basic sums, then (iv) cannot be applied to define these permutations, but it can be so applied if Δ is defined as a multilinear alternating function which takes the value 1 on the identity matrix; see, for example, Rose (2002), Chapter 4.

Problem 3.13 Let $\eta = e^{\pi i/3}$ and $C = \left(\begin{smallmatrix} \eta & 0 \\ 0 & \eta^{-1} \end{smallmatrix} \right)$.

(i)   Calculate C^3, C^6 and C^{-1}.
(ii) Choose a 2×2 matrix D to satisfy $CD = DC^{-1}$ and $D^2 = C^3$.
(iii) Show that the group generated by C and D subject to the conditions in (ii) has order 12, and gives a representation of the dicyclic group Q_3.

This construction can easily be extended to give representations of the groups Q_{2n+1} for all positive integers n.

Problem 3.14* Show that $GL_2(4)$ is isomorphic to a subgroup of $GL_4(2)$ as follows. First, note that the set of non-zero elements of a field of four elements forms a multiplicative cyclic group of order 3. Second, use this fact to show that $GL_1(4)$ is isomorphic to a subgroup of $GL_2(2)$. Now repeat this procedure in the case under consideration using block multiplication of matrices. A fair knowledge of linear algebra is needed to complete this problem.

Problem ◆ 3.15 Suppose F is a field. Let $IT_n(F)$ be the subset of $UT_n(F)$ (the group of $n \times n$ upper triangular matrices, see page 54) of those matrices with 1 at each main diagonal entry—they are sometimes called 'unipotent', and let $IZT_{n,r}(F)$ be the subset of $IT_n(F)$ of those matrices whose ith superdiagonals consist entirely of zeros for $i = 1, \ldots, r$. We shall return to this example in Chapter 10 when discussing nilpotent groups. Show that

(i) $IT_n(F) \triangleleft UT_n(F)$.
(ii) $IZT_{n,r}(F) \triangleleft IT_n(F)$, and $IZT_{n,r}(F) \triangleleft IZT_{n,r-1}(F)$ if $r = 1, \ldots, n-1$.
(iii) If $o(F) = p$ (and so we work modulo p) show that $IT_n(F)$ is a Sylow p-subgroup of $GL_n(F)$, that is $o(IT_n(F)) = p^r$ where p^r is the largest power of p dividing $o(GL_n(F))$; see Section 6.2.
(iv) If $A \in IZT_{n,r}(F)$ and $B \in IT_n(F)$, then $[A, B] \in IZT_{n,r+1}(F)$.

Problem 3.16 Use Problem 3.3(vi) and Theorem 2.29(ii) to give another proof of the simplicity of A_5. You should begin by noting that the neutral element e belongs to all subgroups.

Problem 3.17 Show that the group $\mathbb{Q}$ does not have a finite generating set.

Problem 3.18 Prove that S_4 has the presentation $\langle a, b \mid a^4 = b^3 = (ab)^2 = e \rangle$. Show also that if the powers 4, 3 and 2 are permuted, then further presentations of S_4 are given.

Problem 3.19 (i) Show that the group $SL_2(p)$ possesses only one involution when p is an odd prime. Also note Problem 12.5.

(ii) Using a suitable computer program (or working by hand) show that the conjugacy classes (Definition 2.11(iii)) of the group $SL_2(5)$ have orders (and number of classes) 1(2), 12(4), 20(2) and 30, with nine classes in all.

(iii) Use (ii) to show that the only proper non-neutral normal subgroup of $SL_2(5)$ is isomorphic to C_2.

(iv) The maximal subgroups of $SL_2(5)$ are isomorphic to $SL_2(3)$, Q_5 and Q_3 with orders 24, 20 and 12, respectively. Find subgroups in $SL_2(5)$ isomorphic to these groups.

The group $SL_2(5)$ has a special connection with so-called Frobenius complements, this will be discussed in Web Section 14.3.

Problem ◆ 3.20 Let $D = \langle c, d \mid c^2 = d^2 = e \rangle$, $A = \langle c \rangle$, and $K = \langle cd \rangle$.

(i) Show that $A \leq D$, $K \triangleleft D$, $A \cap K = \langle e \rangle$, and $AK = D$ if $o(D) > 2$.
(ii) Use this to give new presentations of the dihedral groups D_n.

Proposition (i) shows that D is isomorphic to a semi-direct product of K by A; see Section 7.3. As noted earlier (page 26) involutions (that is, elements of order 2) play an important role in the theory, especially in simple group theory. This also applies to groups of the form D generated by two involutions; see, for example, Aschbacher (1986), page 242. Note also that the jump from two to three generators can have dramatic effect, for very large and complex groups exist which have a generating set with just three involutions—for example, $L_3(3)$ with order 5616, see pages 264 and 265, and Problem 12.10. Problems 2.28, 3.21* and 3.23 should also be consulted.

Problem 3.21* (A Presentation of S_{n+1} for $n > 1$) Let $a_1, \ldots, a_n$ be symbols satisfying the following conditions, where $\mathcal{Q}_{j,k}$ does not apply if $n = 2$,

$$\mathcal{P}_i : a_i^2 = e \qquad \text{for } 1 \le i \le n,$$
$$\mathcal{Q}_{j,k} : (a_j a_k)^2 = e \qquad \text{for } 1 \le k < j - 1 < n, \qquad (*)$$
$$\mathcal{R}_l : (a_l a_{l+1})^3 = e \qquad \text{for } 1 \le l < n.$$

(i) Show that (a) a_j and a_k commute if $1 \le k < j - 1 < n$, and (b) $a_l a_{l+1} a_l = a_{l+1} a_l a_{l+1}$ if $1 \le l < n$.

Let G be the group generated by the set $\{a_1, \ldots, a_n\}$ with the relations $(*)$, let H be the subgroup generated by the subset $\{a_1, \ldots, a_{n-1}\}$, and let Z be the set of cosets

$$Z = \{H, Ha_n, Ha_n a_{n-1}, \ldots, Ha_n a_{n-1} \ldots a_1\}.$$

(ii) Using (i), show that if $Hy \in Z$ and $1 \le i \le n$, then $Hya_i \in Z$. There are a number of cases to consider.
(iii) Use (ii) to show that $[G : H] \le n + 1$, and so by induction deduce $o(G) \le (n+1)!$.
(iv) By considering maps from G to S_{n+1} of the form $a_i \mapsto (i, i+1)$, show that G is a presentation of S_{n+1} using Problem 3.1 and (iii).

When S_{n+1} is represented in this way, that is, when it is generated by a set of involutions, it is known as a *Coxeter Group*. The theory of these groups has developed considerably in the past 30 years; see, for example, Suzuki (1982) and Björner and Brenti (2005).

Problem 3.22 Suppose m is a positive even integer.

(i) Let R_m be the subgroup of $GL_2(\mathbb{C})$ generated by

$$A_m = \begin{pmatrix} 0 & \omega \\ \omega & 0 \end{pmatrix} \quad \text{and} \quad B = \begin{pmatrix} 0 & 1 \\ -1 & 0 \end{pmatrix},$$

where ω is a primitive $2m$th root of unity. Prove $R_m \simeq Q_m$; see page 59.
(ii) Prove that Q_m has a unique involution C, and $Z(Q_m) = \langle C \rangle$.
(iii) Further, show that $Q_m/Z(Q_m) \simeq D_m$. (Note that factor groups are discussed in Chapter 4.)

If G is a 2-group (one whose order is a power of 2, see Chapter 6) and it has a single subgroup of order 2, then it has been shown that G is either cyclic or dicyclic; see Kurzweil and Stellmacher (2004), page 114.

Problem 3.23 Working in S_8, investigate the subgroup F generated by the permutations $a = (3, 4)(5, 6)$, $b = (1, 3)(2, 4)(5, 7)(6, 8)$ and $c = (1, 5)(2, 6)(3, 8)(4, 7)$. Construct a list of the elements with their orders, and so calculate $o(F)$. Secondly, list the subgroups of F, and determine their orders, their normality status, and which of them is the centre. Finally, find a presentation for F, that is, find a collection of properties of a, b and c from which all remaining properties can be derived. The calculations can be done by hand, or by using a computer algebra package which includes the basic permutation operations. See also Problem 8.12.

Problem 3.24 (Project—The Group $GL_2(3)$) A number of groups play a special role in the theory, A_5 being one of these. Another is the general linear group $GL_2(3)$, later (in Chapter 12) we shall describe some of its properties. Here you are asked to give representations in terms of the three main construction methods discussed in this chapter. In particular, (a) find three 2×2 matrices with orders 2, 3 and 8, respectively, which generate the group, (b) find three permutations of the set $\{1, 2, 3, 4, 5, 6, 7, 8\}$ which generate the group as a subgroup of S_8, and (c) show that the group has a presentation in the form

$$\langle a, b, c \mid a^8 = b^2 = c^3 = e, bab = a^3, bcb = c^2, c^2 a^2 c = ab, c^2 abc = aba^2 \rangle.$$

You will need to find inclusion maps between these systems, and then show that each contains 48 elements (Theorem 3.15). Note that there is no unique answer to this problem, and that it can be done by hand, but a suitable computer algebra package would help with the calculations. This project will be continued in Problem 6.23.

Chapter 4
Homomorphisms

During the past half century and more, one of the underlying 'themes' in mathematics has been the realisation that maps are equally as important as objects or sets, particularly those that preserve or transfer properties from one object or system to another. They have been called *natural maps*, *morphisms*, or sometimes *structure-preserving maps*. We use the first of these names as a general term for these maps. Therefore, as we have introduced groups, our basic objects of study, we must now discuss the natural maps between them. In group theory, they are called *homomorphisms*, and *isomorphisms* when the correspondence is 'exact'; isomorphisms were introduced informally on page 17 in Section 2.1.

Two groups can appear to be distinct, but are, in fact, identical from the group-theoretical point of view. For example, consider D_3, the dihedral group of the triangle defined on page 3, and S_3, the group of all permutations on the set $\{1, 2, 3\}$. They have identical group-theoretic properties, and a bijection θ between them that satisfies

$$gh\theta = g\theta h\theta \tag{4.1}$$

articulates this fact; equation (4.1) is called the *homomorphism equation*. The map $\theta : D_3 \to S_3$ can be defined as follows (it is not the only possible map; see Section 4.4). Referring to the definitions given on page 3, the rotation $\alpha \in D_3$ about the centre of the triangle by angle $2\pi/3$ clockwise is mapped to $(1, 2, 3)$, that is, we set $\alpha\theta = (1, 2, 3)$. Using (4.1) we have

$$\alpha^2\theta = \alpha\theta\alpha\theta = (1, 2, 3)^2 = (1, 3, 2),$$

and

$$\iota\theta = \alpha^3\theta = (\alpha\phi)^3 = (1, 2, 3)^3 = e$$

where ι denotes the 'identity map', see page 281. Also, the reflection $\beta \in D_3$ about a vertical axis is mapped to $(2, 3)$, that is, we set $\beta\theta = (2, 3)$. Then by (4.1) again

H.E. Rose, *A Course on Finite Groups*,
Universitext,
DOI 10.1007/978-1-84882-889-6_4, © Springer-Verlag London Limited 2009

we have

$$\alpha\beta\theta = \alpha\theta\beta\theta = (1,2,3)(2,3) = (1,3),$$

and

$$\alpha^2\beta\theta = \alpha^2\theta\beta\theta = (1,3,2)(2,3) = (1,2).$$

It is now easily seen that θ is a bijection and (4.1) holds for all elements of D_3; we say that D_3 and S_3 are *isomorphic*.

Isomorphisms are bijections; on the other hand, there is much to be gained by considering maps that preserve the group operation (they satisfy (4.1)) but are not necessarily bijective, these maps are called *homomorphisms*. For example, suppose $G_1 = GL_2(\mathbb{Q})$, $G_2 = \mathbb{Q}^*$ and $\phi : G_1 \to G_2$ is defined by

$$A\phi = \det A \quad \text{for } A \in G_1.$$

As $\det A \neq 0$ and $\det AB = \det A \det B$, for all $A, B \in G_1$, the map ϕ is a homomorphism, that is, it satisfies (4.1). We shall refer to this example repeatedly in the next few sections, and so we call it the *standard example*.

In this chapter, we define homomorphisms, isomorphisms and factor groups, derive their basic properties—the Isomorphism and Correspondence Theorems, discuss the cyclic groups, and introduce automorphisms (that is, isomorphisms of a group to itself). There are two Web Sections, 4.6 and 4.7, the first discusses a special kind of homomorphism called the *transfer* used in Web Section 6.5, and the second extends our work on presentations introduced in Chapter 3.

Maps—Left or Right

In algebraic contexts, we write maps and functions *on the right*, and in most cases we do not use brackets; that is, we write

$$a\phi \quad \text{and not} \quad \phi(a)$$

for the value of the map ϕ at the argument a. In the western world, we read and write from left to right, and so this is a more natural notation—when applying a map ϕ to an argument a, we *first* choose a in the domain, *then* we apply ϕ to obtain the value $a\phi$ in the range (co-domain); see Appendix A. This notational convention may seem strange at first, but it does make many constructions clearer, especially those involving composition or permutations. Also we use lower case Greek letters for maps or functions (including permutations) throughout. The argument of a map ϕ is given by the Roman letter or letters immediately to its left. For example, in the expression $abc\phi$ the argument is abc, but in the expression $a\phi b\psi$ the argument of ψ is b (and this expression is the product of $a\phi$ and $b\psi$). On some occasions we use brackets to aid clarity, so we might write $(abc)\phi$ for $abc\phi$, or $(a\phi)(b\psi)$ for $a\phi b\psi$, but $(a\phi b)\psi$ when the argument of ψ is $a\phi b$, that is when the argument is the product of $a\phi$ and b.

4.1 Homomorphisms and Isomorphisms

The basic notions are given by

Definition 4.1 Let G_1 and G_2 be groups, and let ϕ be a map from G_1 to G_2. The map ϕ is called a *homomorphism* from G_1 to G_2 if

$$gh\phi = g\phi h\phi \quad \text{for all } g, h \in G_1.$$

Note that the product gh on the left-hand side of this equation is in G_1, and the product $g\phi h\phi$ (or $(g\phi)(h\phi)$) on the right-hand side is in G_2.

Definition 4.2 Let ϕ be a homomorphism mapping G_1 to G_2.

(i) ϕ is called the *trivial homomorphism* if $a\phi = e$, for all $a \in G_1$;
(ii) ϕ is called an *isomorphism* if it is also a bijection from G_1 to G_2, the groups G_1 and G_2 are said to be *isomorphic*, and we write $G_1 \simeq G_2$ in this case;[1]
(iii) ϕ is called an *endomorphism* if it is a homomorphism of G_1 to itself;
(iv) ϕ is called an *automorphism* if it is an isomorphism of G_1 to itself.

Isomorphisms were first introduced by Jordan in 1865 whilst he was working on his proof of the Jordan–Hölder Theorem; see Chapter 9.

We use the word 'trivial' in (i) above to imply that the map trivialises, or destroys, all properties of the group except those associated with the neutral element. The identity map $\iota : G \to G$ which is given by $g\iota = g$ for all $g \in G$ is an example of an automorphism of G. The reader should also refer to the note on 'isomorphism classes' on page 17.

Examples One homomorphism was discussed on page 68 (the standard example), four more are given now; see also the examples on pages 17 and 84, and in Problem 4.1.

(a) If $G_1 = \mathbb{R}$, G_2 is the group of complex numbers having absolute value 1 with the operation complex multiplication, and $\phi_1 : G_1 \to G_2$ where

$$x\phi_1 = \cos x + i \sin x,$$

for $x \in \mathbb{R}$, then ϕ_1 is a homomorphism mapping G_1 to G_2. It is not an isomorphism because $x\phi_1 = (x + 2k\pi)\phi_1$ for each $k \in \mathbb{Z}$.
(b) If $G_1 = \mathbb{Z}/6\mathbb{Z}$ (the set $\{0, 1, 2, 3, 4, 5\}$ with operation addition modulo 6), $G_2 = (\mathbb{Z}/7\mathbb{Z})^*$ (the set $\{1, 2, 3, 4, 5, 6\}$ with operation multiplication modulo 7), and $\phi_2 : G_1 \to G_2$ is given by

$$a\phi_2 = 3^a \quad \text{modulo } 7,$$

for $a \in G_1$, then ϕ_2 is an isomorphism; the reader should check this.

---

[1] Some authors use the symbol $\cong$ in place of $\simeq$.

(c) If $G_1 = \mathbb{C}^*$, the multiplication group of the non-zero compex numbers, and $\phi_3 : G_1 \to G_1$ satisfies

$$z\phi_3 = z^2,$$

then ϕ_3 is an endomorphism. Reader, why is ϕ_3 not an automorphism?
(d) If $G_1 = \mathbb{Z}$ and $\phi_4 : G_1 \to G_1$ satisfies $a\phi_4 = -a$ for $a \in G_1$, then ϕ_4 is an automorphism of G_1.

The basic properties of these maps are given by the following lemmas.

Lemma 4.3 *Suppose ϕ is a homomorphism between the groups G_1 and G_2 with neutral elements e_1 and e_2, respectively.*

(i) $e_1\phi = e_2$.
(ii) *If $g \in G_1$ then $g^{-1}\phi = (g\phi)^{-1}$.*

Note that in (ii) the inverse on the left-hand side is in G_1, and the inverse on the right-hand side is in G_2.

> *Proof* (i) As $g = ge_1$ for all $g \in G_1$, we have $g\phi = ge_1\phi = g\phi e_1\phi$ and (i) follows by cancellation.
>   (ii) Using the group axioms and (i) we have
>
> $$e_2 = e_1\phi = (gg^{-1})\phi = g\phi g^{-1}\phi,$$
>
> the result follows as inverses are unique and two-sided (Theorem 2.5).   □

From now on we use the symbol e for the neutral element of every group.

Lemma 4.4 *Suppose $\phi : G_1 \to G_2$, $\psi : G_2 \to G_3$, and both ϕ and ψ are homomorphisms.*

(i)   *$\phi \circ \psi$ is a homomorphism mapping G_1 to G_3.*
(ii) *The image of ϕ is a subgroup of G_2.*
(iii) *If $H \leq G_1$ and ϕ' is the map ϕ with its domain restricted to H, then ϕ' is a homomorphism from H into G_2.*
(iv) *If ϕ is an isomorphism, so is ϕ^{-1}.*

The image of ϕ, see (ii), is denoted by $\text{im}\,\phi$. Also ϕ' in (iii) is sometimes written $\phi|_H$ and described as 'ϕ restricted to H'.

> *Proof* Straightforward, see Problem 4.3.   □

We now define the *kernel*, an important entity in the theory. It is a normal subgroup, the reader should review the material of these subgroups given in Chapter 2.

Definition 4.5 Let ϕ be a homomorphism mapping G_1 to G_2. The subset of G_1 defined by

$$\{a \in G_1 : a\phi = e\}$$

is called the *kernel* of ϕ, and it is denoted by $\ker \phi$.

In the standard example $\phi : GL_2(\mathbb{Q}) \to \mathbb{Q}^*$ where $A\phi = \det A$ (page 68), $\ker \phi$ is the set of 2×2 matrices A with determinant 1, the neutral element of $\mathbb{Q}^*$. Hence $\ker \phi$ equals $SL_2(\mathbb{Q})$ in this example. (Note $SL_2(\mathbb{Q}) \lhd GL_2(\mathbb{Q})$.)

The basic properties of the kernel are given by

Lemma 4.6 *Suppose* $\phi : G_1 \to G_2$ *is a homomorphism.*

(i) $\ker \phi \lhd G_1$, *see Definition* 4.5.
(ii) ϕ *is injective if and only if* $\ker \phi = \langle e \rangle$.

This lemma is useful in its own right, but it is also useful because it provides a second method (apart from Theorem 2.29) for showing that a subset of a group is a normal subgroup, that is, by showing the subset in question is the kernel of a homomorphism; see, for example, the proof of the N/C-theorem (Theorem 5.26).

Proof (i) The property $\ker \phi \leq G_1$ follows from Theorem 2.13, Definition 4.1, and Lemma 4.3. Now by Lemma 4.3 again, if $a \in G$ and $g \in \ker \phi$,

$$a^{-1}ga\phi = a^{-1}\phi g\phi a\phi = (a\phi)^{-1}ea\phi = e,$$

which shows that $a^{-1}ga \in \ker \phi$. Normality follows by Theorem 2.29.

(ii) Suppose first ϕ is injective. If $\ker \phi \neq \langle e \rangle$, we can find $c \in G_1$ such that $c \neq e$ and $c\phi = e = e\phi$; but as ϕ is injective, this implies $c = e$ which contradicts our assumption. Hence $\ker \phi = \langle e \rangle$. Conversely, if $\ker \phi = \langle e \rangle$ and $b\phi = c\phi$, for $b, c \in G_1$, then

$$e = (c\phi)^{-1}b\phi = c^{-1}b\phi, \quad \text{and so} \quad c^{-1}b \in \ker \phi.$$

But $\ker \phi = \langle e \rangle$, and so $b = c$. This holds for all $b, c \in G_1$, and so the result follows. $\square$

As an example we give a proof of Cayley's Theorem. It was first proved in 1850, and was a development of some work undertaken by Cauchy during the previous decade; see page 42. Note that there is nothing unique about the symmetric group here, for it can be shown that every group of order n is isomorphic to a group of $m \times m$ matrices, for some $m \leq n$, defined over an arbitrarily given field, see Problem 4.17. Also, in Web Section 4.7 we show that every group is a factor group of a free group.

Theorem 4.7 (Cayley's Theorem) *Every group is isomorphic to a subgroup of a symmetric group.*

Proof Let G be a group. For fixed $g \in G$, define the map $\sigma_g : G \rightarrow G$ by

$$a\sigma_g = ag \quad \text{for all } a \in G.$$

Using cancellation (Theorem 2.6), we see immediately that σ_g is a bijection; that is, σ_g is a permutation of the underlying set of G. Also, if $g, h \in G$,

$$a\sigma_{gh} = a(gh) = (ag)h = (a\sigma_g)\sigma_h = a(\sigma_g \circ \sigma_h) \quad \text{for all } a \in G. \qquad (4.2)$$

Hence we can define a map $\theta : G \rightarrow S_G$, where S_G is the group of permutations of the elements of the set G, by

$$g\theta = \sigma_g \quad \text{for all } g \in G,$$

and it is a homomorphism by (4.2). The result follows by Lemma 4.4(ii). $\square$

This result is usually not the best possible. For example, the group D_4 has order 8 and so, by Cayley's Theorem, it is isomorphic to a subgroup of S_8 (with order 40320). But, in fact, it is isomorphic to a subgroup of S_4 (with order 24), see Chapter 8. On the other hand, for a few groups the theorem is best possible, for example, Q_2 (which also has order 8) is isomorphic to no subgroup of S_n if $n < 8$.

Factor Groups

Cosets were defined in Chapter 2. Here we ask:

Is it possible to make the set of cosets of a subgroup H of G into a new group?

The answer is yes, but only when H is a *normal* subgroup of G (Definition 2.28). Consider the following simple example which is typical of the general situation. Let $G = \mathbb{Z}$ and $H = 2\mathbb{Z}$, the even integers under addition. Clearly $H \triangleleft G$, and there are just two cosets: the even integers $2\mathbb{Z}$, and the odd integers $2\mathbb{Z} + 1$. Now

> an even integer plus an even integer is even,
> an even integer plus an odd integer is odd, and $(*)$
> an odd integer plus an odd integer is even.

This suggests that we can treat the set of cosets in this example as a new two element group with the operation given by $(*)$, and this will characterise the terms 'even' and 'odd' when applied to the integers. Hence we make the following

Definition 4.8 Given $K \lhd G$, $g, h \in G$ and using the operation in G, we define the *coset product*, or *coset multiplication*, of gK and hK by

$$gKhK = ghK.$$

There is an important point concerning this definition which can cause some misunderstanding at first. By Lemma 2.22, if $j \in gK$ then $jK = gK$; that is, the *representative g* in the coset gK is *not unique*. Therefore, in the definition above we must check (in Theorem 4.9 below) that it is a product of cosets considered as entities in their own right, and it does not depend on the coset representatives g and h—we say the product is "well-defined".

Theorem 4.9 *Suppose $K \lhd G$.*

(i) *If $aK = a'K$ and $bK = b'K$, then $abK = a'b'K$.*
(ii) *The set of cosets of K in G with coset multiplication given by Definition 4.8 forms a group.*

Proof (i) The hypotheses and Lemma 2.22 give $k_1, k_2 \in K$ to satisfy

$$a' = ak_1 \quad \text{and} \quad b' = bk_2,$$

which shows that

$$a'b' = ak_1bk_2.$$

By Theorem 2.29 and as K is a normal subgroup, we can find $k_3 \in K$ to satisfy $k_1b = bk_3$, so

$$a'b' = ak_1bk_2 = abk_3k_2 \in abK,$$

and (i) follows by Lemma 2.22 again.

(ii) By (i), the set of cosets is closed under (well-defined) coset multiplication. Using the corresponding properties of G, coset multiplication is associative, the neutral element is K (as $eK = K$), and the inverse of the coset aK is $a^{-1}K$ (as $aKa^{-1}K = aa^{-1}K = K$); the result follows. $\square$

We give an example to show that normality is essential in this result. Let $G = D_3 = \langle c, d \mid c^3 = d^2 = e, dc = c^2d \rangle$ and let $J = \langle d \rangle$. We have $cJ = \{c, cd\}$ and $Jc = \{c, c^2d\}$, that is, J is not normal in G. Property (i) also fails, for if we let $a = c$, $a' = cd$ and $b = b' = c$, then $aJ = cJ = \{c, cd\} = a'J$ (as $d^2 = e$), and $bJ = cJ = b'J$. But $abJ = c^2J = \{c^2, c^2d\}$ whilst $a'b'J = cdcJ = \{d, e\} = J$ which shows that $abJ \neq a'b'J$.

Definition 4.10 The group of cosets given by Theorem 4.9(ii) is called the *factor group*, or sometimes the *quotient group*, of G by K, and it is denoted by G/K.

In some contexts, G/K is referred to as "G over K". We shall show below, especially in Theorem 4.17, that this 'fractional' notation is a reasonable choice but it needs to be treated with care. However, we can see already by Lagrange's Theorem (Theorem 2.27) that, if $o(G) < \infty$, then

$$o(G/K) = o(G)/o(K).$$

Example The notation $\mathbb{Z}/n\mathbb{Z}$ used in Chapter 2 can now be explained. As $n\mathbb{Z} \lhd \mathbb{Z}$ (the groups are Abelian) we can form the factor group $\mathbb{Z}/n\mathbb{Z}$ using Definitions 4.8 and 4.10. By the First Isomorphism Theorem to be proved below, this group is isomorphic to the group N of integers modulo n with addition modulo n as its operation, see page 18. The factor group $\mathbb{Z}/n\mathbb{Z}$ contains n cosets, they are $r + n\mathbb{Z}$, for $r = 0, \ldots, n - 1$, and the coset product mirrors the standard addition modulo n in N exactly. As $\mathbb{Z}/n\mathbb{Z}$ is isomorphic to N, we use this (distinctive) notation for both—a slight 'abuse of the notation', but as the groups are isomorphic, no problems should arise.

4.2 Isomorphism Theorems

We come now to what is probably the single most important collection of results in the theory—the Isomorphism Theorems. Note that the naming and numbering of the theorems given below is not universally accepted. Although many of the ideas, theorems and proofs had been 'known' for some time previously, they were first systematically formulated and proved in detail[2] by the German mathematician Emmy Noether (1882–1935) during the 1920s; and one of their first appearances in print was in *Moderne Algebra* by B.L. van der Waerden. This two-volume work was first published in the 1930s and it has had a considerable impact on the development of algebra in general during the twentieth century.

We begin by returning to the example concerning even and odd integers discussed on page 72. Define a map $\phi : \mathbb{Z} \to T_1$ (page 12) by

$$a\phi = 1 \quad \text{if } a \text{ is even}, \quad \text{and} \quad a\phi = -1 \quad \text{if } a \text{ is odd}.$$

Clearly, this is a surjective homomorphism mapping $\mathbb{Z}$ to T_1 with kernel $2\mathbb{Z}$, and

$$\mathbb{Z}/2\mathbb{Z} \simeq T_1,$$

see (∗) on page 72. This is typical of the general situation given by the following major result.

---

[2]A proof of the First Theorem appears in Burnside's classic text written a quarter of a century before; see also the comment below Definition 4.2.

Theorem 4.11 (First Isomorphism Theorem) *If G_1 and G_2 are groups, and ϕ :*
$G_1 \to G_2$ is a surjective homomorphism with kernel $\ker \phi = K$, then

$$G_1/K \simeq G_2.$$

Proof By Theorem 4.9 and Lemma 4.6, G_1/K is a group. We define a map
$\theta : G_1/K \to G_2$ as follows. If $a \in G_1$ then $aK \in G_1/K$, and we set

$$(aK)\theta = a\phi.$$

First, we need to show that θ is well-defined. Suppose $aK = a'K$, then by
Lemma 2.22 we have $a'a^{-1} \in K = \ker \phi$, and so $a'a^{-1}\phi = e$ which gives
$a'\phi = a\phi$. This shows that $aK\theta = a\phi = a'\phi = a'K\theta$ as required. The map
θ is surjective because ϕ is surjective. It is also injective, for if $aK\theta = bK\theta$,
then by definition $a\phi = b\phi$ which gives in turn $e = (a\phi)^{-1}b\phi = a^{-1}b\phi$ and
$a^{-1}b \in \ker \phi = K$. By Lemma 2.22 again, this shows that $aK = bK$, so θ is
injective, and therefore it is a bijection. For the homomorphism property we
have

$$
\begin{aligned}
\big((aK)\theta\big)\big((bK)\theta\big) &= a\phi b\phi && \text{by definition} \\
&= ab\phi && \text{as } \phi \text{ is a homomorphism} \\
&= (abK)\theta && \text{by definition} \\
&= \big((aK)(bK)\big)\theta && \text{by coset product,}
\end{aligned}
$$

and the theorem follows. $\qquad\square$

All is not lost if the homomorphism is not surjective, for we have the corollary
given below where the symbol $\preceq$ stands for 'is isomorphic to a subgroup of'.

Corollary 4.12 *If $\psi : G_1 \to G_2$ is a homomorphism with kernel $\ker \psi = K$, then*

$$G_1/K \preceq G_2.$$

Proof By Theorem 4.11, $G_1/K \simeq \operatorname{im} \psi$, and by Lemma 4.4(ii) we have
$\operatorname{im} \psi \leq G_2$, the corollary follows. $\qquad\square$

If in this corollary the group G_1 is simple, then it is isomorphic to a subgroup of G_2
provided ψ is not the trivial homomorphism.

Examples Let $G_1 = \mathbb{C}^*$, $G_2 = \mathbb{R}^*$, see page 18, and let ϕ be the absolute value
function given by

$$z\phi = |z| \quad \text{for } z \in \mathbb{C}^*.$$

This is a non-surjective homomorphism with image $\mathbb{R}^+$. Now $\ker \phi$ is the subgroup
of $\mathbb{C}^*$ of those complex numbers which have absolute value 1. Corollary 4.12 gives

$$\mathbb{C}^*/\ker\phi \preceq \mathbb{R}^*;$$

that is, associated with every non-zero complex number there is a unique non-zero real number, its absolute value. This value belongs to $\mathbb{R}^+$ a subgroup of $\mathbb{R}^*$.

Returning to the standard example we note that the det function is a surjective homomorphism mapping $GL_n(\mathbb{Q})$ onto $\mathbb{Q}^*$ with kernel $SL_n(\mathbb{Q})$, hence Theorem 4.11 gives

$$GL_n(\mathbb{Q})/SL_n(\mathbb{Q}) \simeq \mathbb{Q}^*. \tag{4.3}$$

Also if we replace the field $\mathbb{Q}$ by a finite field F with $o(F) = q$, then a similar isomorphism applies and we obtain $o(GL_n(F)) = o(SL_n(F))(q-1)$ by Lagrange's Theorem (Theorem 2.27) as F has $q - 1$ non-zero elements.

We single out for special mention the following homomorphism.

Definition 4.13 Let $K \lhd G$, and let ϕ be the surjective homomorphism $G \to G/K$ given by

$$g\phi = gK \quad \text{for all } g \in G,$$

where $K = \ker\phi$. The map ϕ is called the *natural homomorphism*, from G to the factor group $G/K = \operatorname{im}\phi$.

There are three further isomorphism theorems, see also Section 9.1. Some will not be required until later but as they are all consequences of the First Theorem (Theorem 4.11) we shall present them now. The Second and Third Theorems give conditions under which factor groups can be simplified, that is, parts cancelled out, whilst the remaining result, called the Correspondence Theorem, is not a single theorem but a collection of results which relate the properties of G 'above a normal subgroup K' to the properties of G/K, it or one of its extensions will be used many times in the sequel. We begin with a lemma about intersections and we restate the basic facts concerning products, see Theorem 2.30.

Lemma 4.14 *Suppose $H \leq G$ and $K \lhd G$.*

(i) $H \cap K \lhd H$.
(ii) $HK \leq G$ and $HK = KH$.
(iii) $HK \lhd G$ if we also have $H \lhd G$.
(iv) If $J \lhd H$ then $JK \lhd HK$.

Proof (i) By Theorem 2.15, we have $H \cap K \leq H$. For normality we argue as follows. Let $h \in H$. If $j \in H$ then $h^{-1}jh \in H$. Also as $K \lhd G$, $h^{-1}jh \in K$ if $j \in K$. Hence if $j \in H \cap K$, then $h^{-1}jh \in H \cap K$ and the result follows by Theorem 2.29.

For the remaining proofs, see Theorem 2.30 for (ii) and (iii), and Problem 4.6(iv) for (iv). $\square$

We come now to the Second Theorem, as it will be used mainly in our work on series in Chapters 9 to 11, it can be omitted on a first reading of the early chapters.

Theorem 4.15 (Second Isomorphism Theorem) *If $H \leq G$ and $K \lhd G$, then $K \lhd HK \leq G$ and*

$$H/H \cap K \simeq HK/K.$$

Proof As $K \lhd G$, by Lemma 4.14 we have $HK \leq G$. Also clearly $K \leq HK$, and so Problem 2.14(ii) gives $K \lhd HK$ and we can form the factor group HK/K. We define a map $\theta : H \to HK/K$ by

$$h\theta = hK \quad \text{for } h \in H.$$

As $hkK = hK$ for all $h \in H$ and $k \in K$, it follows that θ is a surjective map. It is also a homomorphism for if $h_1, h_2 \in H$ we have using coset multiplication (Definition 4.8)

$$h_1 h_2 \theta = h_1 h_2 K = h_1 K h_2 K = h_1 \theta h_2 \theta.$$

Hence the conditions for the First Isomorphism Theorem (Theorem 4.11) apply, and so to prove the theorem we need to show that $\ker \theta = H \cap K$. Now $h \in \ker \theta$ if, and only if, $hK = K$. If this equation holds then $h \in K$, and so $h \in H \cap K$. Conversely, if $h \in H \cap K$, then $h \in K$ and $hK = K$ clearly follows. Therefore, Theorem 4.11 now gives

$$H/\ker \theta = H/H \cap K \simeq HK/K. \qquad \square$$

Note that one consequence of this result is: If $H \leq G$, $K \lhd G$ and $H \cap K = \langle e \rangle$, then $HK/K \simeq H$.

Example We give a proof of a permutation result to illustrate the use of this theorem, see page 49. We show that if $G \leq S_n$ and G contains an odd permutation, then exactly half of the elements of G are even and half are odd. By Theorem 3.11, $A_n \lhd S_n$, and $G \leq S_n$ by hypothesis, so the Second Isomorphism Theorem gives

$$G/(G \cap A_n) \simeq GA_n/A_n = S_n/A_n.$$

This last equation follows because GA_n contains all even and at least one odd permutation, and so has an order larger than $o(S_n)/2$, which gives $GA_n = S_n$ by Problem 2.19. Now, as $o(S_n/A_n) = 2$ (Theorem 3.11), we have

$$o(G \cap A_n) = o(G)/2;$$

that is exactly half of the elements of G belong to A_n. So if G is simple and $G < S_n$, then we also have $G \leq A_n$.

Correspondence Theorem

The Correspondence Theorem which some authors call the 'Third Isomorphism Theorem' will be considered now. It is an important result with many applications, and it is a direct consequence of the First Isomorphism Theorem (Theorem 4.11). As noted above, it has a number of parts, some of which will be added later. Suppose θ is a surjective homomorphism mapping G_1 to G_2 with $\ker \theta = K$; that is, θ maps G_1 onto G_2, and K onto $\langle e \rangle$. Informally, the theorem says that the properties and structure of that part of G_1 which lies above K is exactly mirrored by the properties and structure of G_2, *and vice versa*. This is illustrated in the diagram below.

$$
\begin{array}{l}
G_1 \xrightarrow{\theta} G_2 \\[1.5em]
H \xrightarrow{\theta} H\theta \quad \text{and} \quad G_1/H \simeq G_2/H\theta \qquad \text{if } H \lhd G_1 \\[1.5em]
J\theta^{-1} \xrightarrow{\theta} J \quad \text{and} \quad G_2/J \simeq G_1/J\theta^{-1} \qquad \text{if } J \lhd G_2 \\[1.5em]
K \xrightarrow{\theta} \langle e \rangle
\end{array}
$$

There are also some upward inclusions. If H satisfies $K \leq H \leq G_1$, then $H\theta \leq G_2$, and if $H \lhd G_1$, then $H\theta \lhd G_1$; also if $J \leq G_2$, then $K \leq J\theta^{-1} \leq G_1$, *et cetera*. Hence we have

Theorem 4.16 (Correspondence Theorem) *Suppose G_1 and G_2 are groups, $H \leq G_1$, $J \leq G_2$, and θ is a surjective homomorphism mapping G_1 to G_2 with $\ker \theta = K$. This implies $G_1 \theta \simeq G_2$ and $K\theta = \langle e \rangle$.*

(i)   *If $K \leq H \leq G_1$, then $H\theta \leq G_2$.*
(ii) *If $K \leq H \lhd G_1$, then $H\theta \lhd G_2$ and $G_1/H \simeq G_2/H\theta$.*
(iii) *$K \leq J\theta^{-1} \leq G_1$.*
(iv) *If $J \lhd G_2$, then $J\theta^{-1} \lhd G_1$ and $G_1/J\theta^{-1} \simeq G_2/J$.*

Proof The proof is mainly a matter of checking that group axioms or subgroup conditions hold, we shall establish parts (i) and (iv) and leave the remaining parts for the reader to complete (Problem 4.13).

(i) By Lemma 4.3, $H\theta$ is a non-empty subset of G_2. Also, if $h_1, h_2 \in H$, then $(h_1\theta)^{-1}h_2\theta = h_1^{-1}h_2\theta \in H\theta$ as H is a subgroup of G_1 and θ is a homomorphism. This gives (i).

(iv) Suppose $g \in G_1$ and $k \in J\theta^{-1}$, and so $g\theta \in G_2$ and $k\theta \in J$. As $J \lhd G_2$, these properties show that

$$(g\theta)^{-1}k\theta g\theta \in J,$$

but $(g\theta)^{-1}k\theta g\theta = (g^{-1}kg)\theta$, hence $g^{-1}kg \in J\theta^{-1}$. By (iii), this gives $J\theta^{-1} \lhd G_1$. For the second part, define the map $\psi : G_1 \to G_2/J$ by

$$g\psi = (g\theta)J \quad \text{for } g \in G_1.$$

Now ψ is surjective as θ is surjective, and it is a homomorphism because

$$g_1g_2\psi = (g_1g_2\theta)J = (g_1\theta)(g_2\theta)J = (g_1\theta J)(g_2\theta J) = g_1\psi g_2\psi,$$

for $g_1, g_2 \in G_1$. The kernel of ψ is $\{g \in G_1 : (g\theta)J = J\}$. By Lemma 2.22,

$$(g\theta)J = J \quad \text{if and only if} \quad g\theta \in J \quad \text{if and only if} \quad g \in J\theta^{-1}.$$

As this holds for all $g \in G_1$, we obtain ker $\psi = J\theta^{-1}$. The result follows by applying the First Isomorphism Theorem (Theorem 4.11) to ψ. □

Example Suppose $G_1 = \mathbb{Z}$ (and so all subgroups are normal as this group is Abelian) and $G_2 = \mathbb{Z}/30\mathbb{Z}$. Secondly, suppose ϕ represents congruence reduction modulo 30, its kernel K equals $30\mathbb{Z}$. Now if, for example, $H = 10\mathbb{Z}$, then $K \lhd H \lhd G_1$, $H\theta \simeq \mathbb{Z}/10\mathbb{Z}$ and both G_1/H and $G_2/H\theta$ are isomorphic to the cyclic group of order 3. As an exercise the reader should take $J \simeq \mathbb{Z}/15\mathbb{Z}$ and apply parts (iii) and (iv) of Theorem 4.16.

The last isomorphism theorem follows easily from (iv) in the result above. It provides a further justification for the factor group notation.

Theorem 4.17 (Third Isomorphism Theorem) *If $K \lhd G$ and $K \lhd H \lhd G$, then $G/H \simeq (G/K)/(H/K)$.*

This is sometimes called the *Freshman's Theorem*; see Scott (1964).

Proof In (iv) of Theorem 4.16, put $G_2 = G/K$, $J = H/K$, and let θ be the natural homomorphism (Definition 4.13) from G to G/K. Note that $H/K \lhd G/K$, the reader should check this. We also have

$$(H/K)\theta^{-1} = H. \tag{4.4}$$

For if $g \in (H/K)\theta^{-1}$, then $g\theta = hK$ for some $h \in H$. But by definition $g\theta = gK$, and so $hK = gK$, and $g \in hK \subseteq H$ by Lemma 2.22. This gives $(H/K)\theta^{-1} \subseteq H$ and, as the reverse inclusion is given by definition, (4.4) follows. Finally, as the conditions in Theorem 4.16(iv) now apply, the result follows. □

Referring to the example above, if C_m denotes the cyclic group of order m, we have $G_1/K \simeq C_{30}$, $G_1/H \simeq C_3$ and $H/K \simeq C_{10}$, and Theorem 4.17 shows that $C_3 \simeq C_{30}/C_{10}$ (that is $3 = 30/10$!).

4.3 Cyclic Groups

In this section, we discuss the *cyclic groups* first introduced in Definition 2.19. The elements of a cyclic group are the powers (positive and negative) of a single generator a. If G is infinite, then all powers are distinct and G is the free group on the set $\{a\}$, but if $o(G) = n$ and $a \in G$, then $a^n = e$, see Theorem 4.20 below. The first theorem provides the basic facts.

Theorem 4.18 (i) *For each positive integer n there exists a cyclic group of order n.*
(ii) *All cyclic groups of order n are isomorphic.*
(iii) *All infinite cyclic groups are isomorphic to the group* $\mathbb{Z}$.
(iv) *Every homomorphic image of a cyclic group is cyclic.*

Proof (i) We give two examples. The group of integers modulo n, $\mathbb{Z}/n\mathbb{Z}$, see page 74, is cyclic of order n. Also the group of the complex numbers $e^{2k\pi i/n}$ for $0 \leq k < n$ with complex multiplication is another.
   (ii) If $G = \langle a \rangle$, $H = \langle b \rangle$, $o(G) = o(H) = n$, and $\phi : G \to H$ is defined by

$$a^t \phi = b^t \quad \text{for } t \in \mathbb{Z},$$

then ϕ is clearly an isomorphism between G and H.
   (iii) This is similar to (ii).
   (iv) If $G = \langle a \rangle$ and $\theta : G \to H$ is a surjective homomorphism, then $H = \langle a\theta \rangle$, that is H is cyclic with generator $a\theta$.   □

We write C_n for the abstract cyclic group of order n, that is, $C_n \simeq \langle a \mid a^n = e \rangle$, a group with a single generator a, say, and a single relation $a^n = e$; see the introduction to Chapter 3. Also we use $\mathbb{Z}$ as a notation for an infinite cyclic group, see (iii) above.

Theorem 4.19 (i) *The subgroups of* $\mathbb{Z}$ *are* $\langle e \rangle$ *and* $n\mathbb{Z}$, *one for each positive integer n.*
(ii) *All non-neutral subgroups of an infinite cyclic group are isomorphic to* $\mathbb{Z}$.

Proof Suppose H contains a non-zero integer, $a, b \in H$, and $H \leq \mathbb{Z}$. Then $ma + nb \in H$ for all $m, n \in \mathbb{Z}$. We can choose m and n so that the greatest common divisor c, say, of a and b satisfies $c = ma + nb$, $c \mid a$, $c \mid b$ and $c > 0$, hence $c \in H$; see Appendix B. (We usually use the notation (a, b) for c.) This shows that the least positive integer d, say, in H is a divisor of every element of H, and so $H = d\mathbb{Z}$, an infinite cyclic group. Both parts of the result follow by Theorem 4.18(iii).   □

Theorem 4.20 *Suppose G is a cyclic group of order n. It contains a cyclic subgroup H of order m if and only if m divides n, and when this happens H is unique.*

This and the previous result show that all subgroups of a cyclic group are cyclic. Also the proof below is not the shortest or most elegant for this result but it is given as an illustration of the Isomorphism Theorems 'in action'.

Proof We use the First Isomorphism and Correspondence Theorems. Define a map $\psi : \mathbb{Z} \to C_n$ as follows. Let j be a generator of the group C_n, and for let

$$a\psi = j^c \quad \text{where} \quad a \equiv c \pmod{n} \quad \text{and} \quad 0 \le c < n,$$

for $a \in \mathbb{Z}$. This is a surjective homomorphism with kernel $n\mathbb{Z}$, and so by the First Isomorphism Theorem we have $\mathbb{Z}/n\mathbb{Z} \simeq C_n$.

By Theorem 4.19, if $n \ne 0$ and J satisfies $n\mathbb{Z} \le J \le \mathbb{Z}$, then $J \simeq m\mathbb{Z}$ for some positive integer m. We have $m \mid n$. For if not, dividing m into n would give a positive remainder r satisfying $0 < r < m$ and $r \in m\mathbb{Z}$. This is impossible because, by Theorem 4.19, m is the smallest positive integer in $m\mathbb{Z}$. Hence by the Correspondence Theorem (with $G_1 = \mathbb{Z}$, $G_2 = C_n$ and $\theta = \psi$) we see that the subgroups of C_n are exactly the groups C_m where $m \mid n$. Now use Theorem 4.18(ii). □

We shall see later that the properties of cyclic groups given in Theorems 4.19 and 4.20 are special. In general, a group of order n can have many subgroups of order m, where $m \mid n$, or none at all. If $m = p^r$, $n = ms$ and $p \nmid s$, then the subgroup C_m is called the (unique) 'Sylow p-subgroup' of C_n; see Section 6.2. Also more generally, if $m \mid n$ and $(m, n/m) = 1$, then C_m is called a 'Hall' subgroup of C_n.

4.4 Automorphism Groups

Let G be a group and let $\operatorname{Aut} G$ denote the set of all automorphisms of G, see Definition 4.2. This set can be given a group structure using composition as the operation. The composition of two bijections is a bijection (Appendix A) and so the operation is well-defined and closed by Lemma 4.4(i). Composition is associative (Appendix A), the identity map ι on G acts as the neutral element, and the inverse of an isomorphism is also an isomorphism (Lemma 4.4(iv)). Thus $\operatorname{Aut} G$ forms a group under composition.

Definition 4.21 For a group G, the set of automorphisms of G with the operation of composition forms a new group called the *automorphism group* of G which is denoted by $\operatorname{Aut} G$.

We give an example now and another extended one at the end of the section, we shall also discuss the automorphism groups of the finite cyclic groups.

Example If $G \simeq \mathbb{Z}$, then $\operatorname{Aut} G \simeq C_2$ because there are only two automorphisms: For all $g \in G$, the first maps $g \mapsto g$, and the second maps $g \mapsto -g$; see Problem 4.19.

Some automorphisms are given by conjugation (Definition 2.28). Let $h \in G$ be fixed, then the map $\tau_h : G \to G$ given by

$$g\tau_h = h^{-1}gh, \quad \text{for } g \in G,$$

is an automorphism of G, the reader should check this. An automorphism of this type is called *inner*, and the set of all inner automorphisms of G is denoted by Inn G. It is a subgroup of Aut G as the following lemma shows.

Lemma 4.22 (i) *If G is a group, then* Aut G *is also a group.*
(ii) Inn $G \lhd$ Aut G.

Proof (i) This was proved above.
   (ii) The identity map is an inner automorphism (put $h = e$ in the definition), so Inn G is not empty. For $g, h \in G$, let τ_g and τ_h be the corresponding inner automorphisms, and let $a \in G$. Then

$$a(\tau_{gh}) = (gh)^{-1}a(gh) = h^{-1}(a\tau_g)h = (a\tau_g)\tau_h = a(\tau_g \circ \tau_h).$$

As this holds for all $a \in G$, we see that $\tau_g \circ \tau_h = \tau_{gh} \in \text{Inn}(G)$. Also $(\tau_g)^{-1} = \tau_{g^{-1}}$ (reader, check). Hence Inn $G \leq$ Aut G. Normality. Let $a, g \in G$, $\theta \in$ Aut G, and $a\theta^{-1} = b$, that is, $b\theta = a$. By Lemma 4.3(ii),

$$a(\theta^{-1}\tau_g\theta) = ((a\theta^{-1})\tau_g)\theta = (b\tau_g)\theta = (g^{-1}bg)\theta$$
$$= g^{-1}\theta b\theta g\theta = (g\theta)^{-1}a(g\theta) = a\tau_{g\theta}.$$

This holds for all $a \in G$, hence $\theta^{-1}\tau_g\theta = \tau_{g\theta} \in$ Inn G, the result follows. $\square$

Using Lemma 4.22, we can form the factor group Aut $G/$ Inn G; it is called the *outer automorphism group* denoted by Out G. Following normal practice, an automorphism which is not inner is called *outer*, so an outer automorphism is an element of one of the cosets of Aut $G/$ Inn G except Inn G itself. Note that if G is Abelian then Inn $G = \langle e \rangle$, and so all automorphisms except the identity automorphism are outer. Also Corollary 5.27 gives a formula for Inn G. It was conjectured by O. Schreier (1901–1929) that Out G is soluble (for a definition see Section 11.1) if G is simple. This has been shown to be true for finite groups but only by using CFSG (which is discussed in Chapter 12). The conjecture is false for non-simple groups, for example, Out$((C_2)^3) \simeq GL_3(2)$, a non-Abelian simple group; see page 264.
   For a particular group, it can be major task to find its automorphism group. Some examples are given in Chapter 8. For example, consider Aut S_n. It has been shown that, if $n \neq 2$ or 6, then Aut $S_n \simeq S_n$ and all automorphisms are inner (Problem 8.5). This is not true for S_6 which does possess outer automorphisms, and the order of Aut S_6 is twice the order of S_6; see Problem 4.20. An account of these results is given in Rotman (1994), pages 156 to 162.

Next in this section we show how to construct the automorphism group of a cyclic group. In some cases, it is also cyclic but not always.

Theorem 4.23 $\operatorname{Aut} C_n \simeq (\mathbb{Z}/n\mathbb{Z})^*$.

Proof Let g be a generator of C_n. A homomorphism mapping C_n to itself will map g to some power of g, so we define, for $m \in \mathbb{Z}$,

$$\theta_m : C_n \to C_n \quad \text{by} \quad g\theta_m = g^m.$$

Note that as θ_m is a homomorphism $g^r\theta_m = (g\theta_m)^r = g^{rm}$ for all integers r. As $g^n = e$ we only need to consider m in the range $0 \leq m < n$, and we can also exclude the case $m = 0$ because θ_0 is clearly not an automorphism unless $n = 1$. For $0 < m < n$, if θ_m is surjective, then it is an automorphism— a surjective map on a finite set to itself is also injective; Problem A.5. Also $(m, n) = 1$ if and only if m is congruent modulo n to an element of $(\mathbb{Z}/n\mathbb{Z})^*$.

Now if $(m, n) = 1$, then using the Euclidean Algorithm (Appendix B), integers r and s can be found to satisfy $rm + sn = 1$. Hence, as θ_m is a homomorphism,

$$g^r\theta_m = g^{rm} = g^{1-sn} = g,$$

as $g^n = e$. This shows that $g^{rt}\theta_m = g^t$ for $t = 1, \ldots, n$, hence θ_m is surjective. Conversely, if θ_m is surjective, then $g = g^u\theta_m = g^{um}$, for some $u \in \{1, \ldots, n\}$. This implies that $g^{1-um} = e$ and so $1 - um$ is a multiple of n. This can only happen if $(m, n) = 1$. Therefore, θ_m is an automorphism if and only if $(m, n) = 1$. Hence the map $\theta_m \mapsto m$ for $m \in (\mathbb{Z}/n\mathbb{Z})^*$ gives the required isomorphism. $\square$

The groups $(\mathbb{Z}/n\mathbb{Z})^*$ are Abelian, in Chapter 7 we shall show that they can be expressed as 'direct products' of cyclic groups. Also using the theory of *primitive roots*, a topic from number theory discussed briefly in Appendix B, we have

The group $(\mathbb{Z}/n\mathbb{Z})^*$ is cyclic if and only if $n = 2, 4, p^s$, or $2p^s$ where p is an odd prime and s is a positive integer, and their orders are as follows: $o((\mathbb{Z}/2\mathbb{Z})^*) = 1$, $o((\mathbb{Z}/4\mathbb{Z})^*) = 2$, and $o((\mathbb{Z}/p^s\mathbb{Z})^*) = o((\mathbb{Z}/2p^s\mathbb{Z})^*) = p^{s-1}(p - 1)$.

So in particular, the automorphism group of C_{p^s} where p is an odd prime and s is a positive integer is itself cyclic and has order $p^{s-1}(p - 1)$; a result due to Gauss. Further automorphism results are given in Problem 4.18, for elementary Abelian groups, and in Problems 6.18 and 7.10.

Overleaf we end this section with an extended example which illustrates some of the methods used to construct automorphism groups, it also provides some more examples of isomorphisms.

Example Show that Aut $D_4 \simeq D_4$ (also see Problem 4.19).

Let the dihedral group D_4 be given by $\langle a, b \mid a^4 = b^2 = e, bab = a^3 \rangle$. Note first that an automorphism maps an element of order k to an element of order k (this follows from Problem 4.5(i) and Lemma 4.4(iv)). Also D_4 has exactly two elements of order 4, that is a and a^3, and so automorphisms either map $a \mapsto a$ and $a^3 \mapsto a^3$, or map $a \mapsto a^3$ and $a^3 \mapsto a$. In either case, $a^2 \mapsto a^2$, and so a^2 is fixed by all automorphisms. Secondly, note that b, ab, a^2b and a^3b all have order two and so an automorphism could map b to b, or to ab, or to a^2b, or to a^3b. This suggests defining ϕ and ψ by

$$\phi : a \mapsto a \quad \text{and} \quad b \mapsto ab,$$

$$\psi : a \mapsto a^3 \quad \text{and} \quad b \mapsto b.$$

These give

$$a^k\phi = a^k, \quad a^k b\phi = a^{k+1}b, \quad \text{and} \quad a^k b^l \psi = a^{3k}b^l,$$

where k is to be read modulo 4, and l modulo 2. It is now an easy exercise to check that both ϕ and ψ are automorphisms, the reader should do this.

The remaining automorphisms can be generated using ϕ and ψ. Clearly, $\psi^2 = \iota$, the identity map (automorphism). Also $b\phi^2 = (b\phi)\phi = ab\phi = a\phi b\phi = a^2b$, and similarly $b\phi^3 = a^3b$ and $b\phi^4 = b$, and so $\phi^4 = \psi^2 = \iota$. For the remaining condition, we have

$$a\psi\phi\psi = (a^3\phi)\psi = a^3\psi = (a\psi)^3 = a^9 = a = a\phi^3,$$

$$b\psi\phi\psi = (b\phi)\psi = ab\psi = a\psi b\psi = a^3b = b\phi^3.$$

This shows that $\psi\phi\psi = \phi^3$. No other automorphisms are possible, see the comments in the paragraph above, and so the result follows.

Unlike both S_3 and S_4 (Problems 4.19(i) and 8.5), D_4 has both inner and outer automorphisms, see Corollary 5.27. As $a^{-1}aa = a$ and $a^{-1}ba = a^2b$, we see that the automorphism ϕ^2 corresponds to conjugation by a, and as $bab = a^3$ and $b^3 = b$, the automorphism ψ corresponds to conjugation by b. But ϕ has no such correspondence, and so forms an outer automorphism. By Lemma 4.22(ii), the set of inner automorphisms forms a normal subgroup of Aut D_4 which is isomorphic to the 4-group T_2 (page 19) in this case because $(\phi^2)^2 = \psi^2 = \iota$ and $\phi^2\psi = \psi\phi^2$. The four outer automorphisms are $\phi, \phi^3, \phi\psi$ and $\phi^3\psi$.

4.5 Problems

Problem 4.1 Show that the following maps are homomorphisms.

(i)   The maps (a), (b), (c), and (d) given on pages 69 and 70.
(ii) The trivial map (Definition 4.2).
(iii) The sgn map from S_n to C_2 (Definition 3.7 on page 47).

(iv) The determinant map from $GL_2(F)$ to F^* where F is a field (page 68).
(v) The projection map from $\mathbb{R}^2$ to $\mathbb{R}$ given by $(x, y) \mapsto x$.

Problem 4.2 Show that the following pairs of groups are isomorphic.

(i) $GL_2(2)$ and S_3, see Problem 2.20.
(ii) Let F be a field. The first group is F^*, the multiplicative group of F, and
      the second group has underlying set $F_1 = F \backslash \{1\}$ and the operation $*$ where
      $a * b = a + b - ab$ for $a, b \in F_1$. First, you will need to show that $(F_1, *)$ is a
      group.
(iii) $\mathbb{Q}^+$ and the additive group of all polynomials in the variable x with integer
      coefficients. (Hint. Use Unique Factorisation (Theorem B.6).)

Problem 4.3 (i) to (iv) Give proofs of the four parts of Lemma 4.4.
    (v) Suppose G is a group and X is a set. Given a bijection $\theta : G \to X$, construct
an operation on X so that θ becomes an isomorphism of G to the group formed by
X with this operation. Show also that this operation is unique.

Problem 4.4 An exercise working with cosets. Let $G = SL_2(3)$.

(i)   Find $Z(G)$ and show that it has order 2.
(ii) Write out a list of representatives of the cosets of $Z(G)$ in G.
(iii) Show that if E is a coset of $Z(G)$ in G and $E \neq Z(G)$, then either $E^2 = Z(G)$
      or $E^3 = Z(G)$, where E^k is defined using the coset product. Count the number
      of solutions in each case.
(iv) Use (iii) and Problem 3.10 to show that $G/Z(G) \simeq A_4$, that is, $SL_2(3)$ can be
      treated as an extension (Definition 9.9) of C_2 by A_4.

A similar argument can be used to show that the general linear group $GL_2(3)$ is an
extension of C_2 by S_4.

Problem ♦ 4.5 (i) Suppose $\phi : G \to G$ satisfies $g\phi = g^{-1}$ for all $g \in G$. Show that
ϕ is a homomorphism if and only if G is Abelian.
    (ii) Let G be a finite group and let $\theta : G \to G$ be an automorphism. Further
suppose (a) if $g \in G$ and $g\theta = g$, then $g = e$, and (b) θ^2 is the identity map ι on G.
Use these to show that $g\theta = g^{-1}$ for all $g \in G$, and so deduce that G is Abelian.
(Hint. First show that $\{a^{-1} \cdot a\theta : a \in G\} = G$.)

Problem ♦ 4.6 (Abelian Factor Groups and the Derived Subgroup) Before tackling
this problem the reader should revisit Problem 2.16 which gives the basic properties
of the derived subgroup.

(i)   Show that every factor group of an Abelian group is Abelian.
(ii) Prove that if $K \triangleleft G$, then

$$G/K \text{ is Abelian} \quad \text{if and only if} \quad G' \leq K,$$

where G' denotes the derived (or commutator) subgroup of G. This is an important fact we use many times.

(iii) Use (ii) to show that $S_n' = A_n$.
(iv)* Suppose $H_1, H_2 \leq G$, $H_2 \triangleleft H_1$ and $J \triangleleft G$. Show that $JH_2 \triangleleft JH_1$, and JH_1/JH_2 is Abelian if H_1/H_2 is Abelian.

Problem 4.7 Suppose $K \triangleleft G$ and $o(G/K) = n < \infty$.

(i) Show that if $g \in G$, then $g^n \in K$.
(ii) Suppose $(m, n) = 1$, $g \in G$ and $g^m \in K$, prove that $g \in K$.

Problem 4.8 (Perfect Groups) A group G is called *perfect* if it equals its derived subgroup, that is, if $G = G'$. Show that

(i) An equivalent definition is: No non-neutral factor group is Abelian.
(ii) If $H \leq G$ and H is perfect, then $H \leq G'$.

Problem ♦ 4.9 Let $\theta : G \to H$ be a homomorphism. Prove the following:

(i) If $a \in G$, then $(a^n)\theta = (a\theta)^n$, for all $n \in \mathbb{Z}$.
(ii) If $K \triangleleft G$, $K \subseteq \ker \theta$ and, for $a \in G$, $\theta' : G/K \to H$ is defined by

$$(aK)\theta' = a\theta,$$

then θ' is a homomorphism. You need to show that θ' is well-defined.
(iii) If $j_1, j_2 \in G$, then $[j_1, j_2]\theta = [j_1\theta, j_2\theta]$.

Problem 4.10 Suppose G is a finite group with the property

$$(ab)^n = a^n b^n,$$

for all $a, b \in G$ where n is some fixed integer larger than 1.

(i) Let

$$G_n = \{a \in G : a^n = e\} \quad \text{and} \quad G^n = \{c^n : c \in G\}.$$

Using a suitably chosen homomorphism show that both G_n and G^n are normal subgroups of G, and deduce $o(G^n) = [G : G_n]$.
(ii) Show that (a) if $n = 2$ then G is Abelian, and (b)* if $n = 3$ and $3 \nmid o(G)$, then G is again Abelian.

More is known, see Alperin (1969).

Problem ♦ 4.11 Let G be a finite group with a normal subgroup K satisfying

$$\bigl(o(K), [G : K]\bigr) = 1.$$

Using the Isomorphism Theorems show that K is the unique subgroup of G with order $o(K)$ by considering what happens to another such subgroup K_1 in G/K; see

Theorem 6.10. If π denotes the set of prime factors of $o(K)$, then K is called the π-*radical* of G, see Section 10.2, it is usually denoted by $O_\pi(G)$.

Problem 4.12 Suppose $K_1, \ldots, K_n$ are normal subgroups of G. Let $L = G/K_1 \times \cdots \times G/K_n$, the direct product of $G/K_1, \ldots, G/K_n$; see Section 7.1. (We can treat L as the group of ordered n-tuples $\{g_1 K_1, \ldots, g_n K_n\}$ for $g_i \in G$ with component-wise multiplication.) Define a map $\theta : G \to L$ by

$$g\theta = (gK_1, \ldots, gK_n) \quad \text{for } g \in G.$$

Show that θ is a homomorphism, and deduce $G/\bigcap_{i=1}^n K_i$ is isomorphic to a subgroup of L.

Problem ◆ 4.13 (i) Give proofs of the second and third parts of the Correspondence Theorem 4.16.

(ii) Suppose $K \lhd G$. Prove the following extension of the Correspondence Theorem: G/K is simple if and only if K is a maximal normal subgroup of G; that is, K is a proper normal subgroup of G, and no other normal subgroup J of G exists which satisfies $K < J < G$.

(iii) Suppose G is finite and not cyclic of prime order, and H is a maximal subgroup. Show that if H is also normal, then $[G : H]$ is prime, and give an example to show that normality is needed.

Problem 4.14 (Coset Enlargement) (i) Suppose J and K are normal subgroups of G, and $J \leq K$. Let $\xi : G/J \to G/K$ be defined by $(aJ)\xi = aK$ for $a \in G$. Show that ξ is a well-defined surjective homomorphism with kernel K/J. The map ξ is called the *enlargement of cosets map*.

(ii) Use (i) to reprove the Third Isomorphism Theorem (Theorem 4.17).

Problem 4.15 Let G be a finite Abelian group.

(i)   If $p \mid o(G)$, show how to find an element $g \in G$ of order p. (Hint. Write $o(G) = pn$ and use induction on n.) A second proof of this result is given in Theorem 6.2.

(ii) If $o(G) = mn$, $(m, n) = 1$, $H, J \leq G$, $o(H) = m$, $o(J) = n$ and both H and J are cyclic, show that G is also cyclic.

(iii) Give an example to show that the statement in (ii) is false if G is not Abelian.

Problem ◆ 4.16 (Properties of the Centre) Derive the following properties of the centre $Z(G)$ of a group G. Note that by Problem 2.14(ii) a subgroup of $Z(G)$ is normal in G. Suppose $K \lhd G$ throughout.

(i)   Construct an example to illustrate the following fact: There exists at least one group G with proper subgroups H and J which have the properties: $\langle e \rangle < H < J < G$, $Z(H) = Z(G) = \langle e \rangle$ and $Z(J) \neq \langle e \rangle$. One example uses symmetric groups.

(ii) If G is not Abelian, then $G/Z(G)$ is not cyclic—a useful fact.

(iii)   If $o(K) = 2$ then $K \leq Z(G)$.

(iv)   If $\theta : G \to J$ is a surjective homomorphism and $H \leq Z(G)$, then $H\theta \leq Z(J)$.

(v)   Show that if H is an Abelian subgroup of G, then $H Z(G)$ is also an Abelian subgroup of G.

(vi) $Z(K) \lhd G$; does it follow that $Z(K) \leq Z(G)$ (see Problem 4.22)?

(vii) If $K \leq J \leq G$, then $[J, G] \leq K$ if and only if $J/K \leq Z(G/K)$.

(viii) Finally, show that

$$\frac{Z(G)}{Z(G) \cap K} \leq Z\left(\frac{G}{K}\right).$$

Problem 4.17 Suppose F is a field and G is a finite group. Show that G is isomorphic to a subgroup of the general linear group $GL_n(F)$, for some n not larger than $o(G)$, using the following method. Let U denote the collection of all expressions of the form $z = \sum_{g \in G} m_g g$ where $m_g \in F$. Apply component-wise addition and scalar multiplication to show that U forms a vector space over F. Secondly, for $h \in G$ define a map $\theta_h : U \to U$ by

$$z\theta_h = \sum_{g \in G} m_g gh.$$

Show that θ_h defines an invertible linear map on U, and the collection of these maps forms a group isomorphic to $GL_n(F)$ where n is the dimension of the vector space U. This result can of treated as a 'matrix version' of Cayley's Theorem (Theorem 4.7), note that there is no restriction on the choice of field F.

Problem ♦ 4.18 (Elementary Abelian Groups) An Abelian group all of whose non-neutral elements have order p, for some fixed prime p, is called an *elementary Abelian group* (or sometimes an *elementary Abelian p-group*; see Chapter 6).

(i)   Let F denote the field $\mathbb{Z}/p\mathbb{Z}$ (page 18), and let G be an elementary Abelian p-group. On G define an 'addition' $\oplus$ by $x \oplus y = xy$, for $x, y \in G$, and a 'scalar multiplication' expressed using concatenation by $cx = x^c$, for $c \in F$ and $x \in G$. Prove that G with these new operations forms a vector space over F. We denote it by $\mathcal{G}$.

(ii)   Show that the subgroups of G correspond to the subspaces of $\mathcal{G}$.

(iii) If G is finite, then $\mathcal{G}$ will have a finite basis and finite dimension m, say. (This can be proved using some basic linear algebra, or it follows from Theorem 6.3.) Show that the automorphisms of G correspond to the linear maps on $\mathcal{G}$; and so deduce that

$$\text{Aut } G \simeq GL_m(p).$$

In Problem 7.14, we shall show that G is a direct product of m copies of $\mathbb{Z}/p\mathbb{Z}$, this will provide another proof of the result given in (i).

Problem 4.19 (i) Find the automorphism groups of the following groups: (a) $\mathbb{Z}$, (b) the 4-group T_2, (c) S_3, (d) C_4, and (e) C_{p^n} where p is an odd prime and n is a positive integer.

(ii) Is $\text{Aut}(D_8) \simeq D_8$?

(iii) Suppose $o(G) < \infty$. Show that $\text{Aut}(G) = \langle e \rangle$ if and only if $o(G) \leq 2$. (Hint. Show that the map defined by $a \mapsto a^{-1}$ is an automorphism of an Abelian group.)

Problem 4.20 (i) Let the map $\psi : S_6 \to S_6$ satisfy:

$$(1,2)\psi = (1,5)(2,3)(4,6),$$

$$(1,3)\psi = (1,4)(2,6)(3,5),$$

$$(1,4)\psi = (1,3)(2,4)(5,6),$$

$$(1,5)\psi = (1,2)(3,6)(4,5),$$

$$(1,6)\psi = (1,6)(2,5)(3,4).$$

Beginning with Problem 3.1(i), this can be extended to an automorphism of S_6; you are not asked to prove this but you could consider what is needed. Using this fact show that ψ^2 equals the identity map on S_6, and so provides an example of an outer automorphism of this group. For more details, see Rotman (1994), pages 156 to 167.

It can be shown that $\text{Aut } S_n \simeq S_n$ provided $n \neq 2$ or 6. When $n = 2$ use (iii) in Problem 4.18. But it is a fact that S_6 is the only non-Abelian symmetric group which has an outer automorphism; see page 82.

(ii) Use (i) and Problem 3.21* to show that S_6 has 12 subgroups isomorphic to S_5, note that six of them possess no 2-cycles and act transitively on six points. One member of this second set of six subgroups can be constructed as follows: Let H be the group generated by $(1, i+1)\psi$ for $i = 1, 2, 3, 4$, then choose four (of ten) products of three 2-cycles in H which satisfy the conditions of Problem 3.21* with $n = 4$.

Problem 4.21 The outer automorphism group for the group A_5 is isomorphic to C_2. Find an outer automorphism for A_5, and consider what would be needed to establish the previous statement. (Hint. Use Theorem 3.6.)

Problem ♦ 4.22 (Project—Characteristic Subgroups) In this project, you are asked to develop the notion of *characteristic subgroup* which is similar to, but stronger than, normality. A subgroup H of a group G is said to be *characteristic* in G if $H\phi \leq H$ for all $\phi \in \text{Aut } G$; this is denoted H char G. (Note that $H\phi = \{h\phi : h \in H\}$.) Prove the following statements.

(i)   It is sufficient to require $H\phi = H$.

(ii)   Normality is equivalent to: $H\nu \leq H$ for all inner automorphisms $\nu : G \to G$.

(iii)   Unlike normality, the characteristic relation is transitive.
(iv)   If J char K and $K \lhd G$, then $J \lhd G$, but does this proposition also follow if
          $J \lhd K$ and K char G?
(v)   A subgroup of a cyclic group is characteristic.
(vi) $Z(G)$ char G; we say "the centre of a group is a *characteristic subgroup*
          of G".
(vii)   The derived subgroup G' of G is a characteristic subgroup of G.
(viii) Give an example of a normal subgroup which is not characteristic.

Chapter 5
Action and the Orbit–Stabiliser Theorem

There is only a small intersection (mainly involving examples) between the material in this chapter and the next, with that in Chapter 7. Hence Chapter 7, which contains work on direct products and Abelian groups, can be read first.

In the last chapter, we introduced homomorphisms, they are maps that transfer properties from one group to another, and they satisfy the homomorphism equation (4.1). Here, given a set X we introduce new collections of maps that transform X to itself and which are governed by a group G; that is, for each $g \in G$ we define a map $\backslash g : X \to X$, and map composition 'corresponds' to the group operation. The map $\backslash g$ is a permutation of X and it is called an *action* of G on X. In many cases, X is closely related to G, but not always. It is also possible to define an action using homomorphisms; see Theorem 5.12. In one sense, this important notion has been part of the theory since its inception, but only in particular instances. If you look for the word 'action' in Scott's group theory text published in 1964—the standard introduction to the theory at that time—you will not find it. But you will find many entities now associated with actions, for example, 'centraliser' and 'normaliser'. About this time, and due in part to the work of Wielandt (1964), it was realised that a number of constructions have a similar basis, and emphasising their similarity would give new insights into the theory. Also at around this time, elegant proofs of some major theorems based on new and easily defined actions appeared; for instance, McKay's proof of Cauchy's Theorem (Theorem 6.2) or the main proof of the First Sylow Theorem (Theorem 6.7) both given in the next chapter. Nowadays actions form an important part to any introduction to group theory.

We begin this chapter by defining actions and two major associated entities: *orbits* and *stabilisers*. We then prove the Orbit–Stabiliser Theorem which leads to a number of important applications. Three major examples follow that introduce *centralisers* and *normalisers*, vital entities especially for the finite theory. In Web Section 5.4, we extend our work on permutations begun in Chapter 3, discuss transitive and primitive permutations, and prove Iwasawa's simplicity lemma, this work has applications in Chapter 12.

H.E. Rose, *A Course on Finite Groups*,
Universitext,
DOI 10.1007/978-1-84882-889-6_5, © Springer-Verlag London Limited 2009

5.1 Actions

We begin by considering an example. Let G be the group $(\mathbb{Z}/7\mathbb{Z})^*$, and let $X = \{\underline{1}, \underline{2}, \underline{3}, \underline{4}, \underline{5}, \underline{6}\}$. In this particular example, the set X and the group G have the same elements, and so we have underlined these elements when they are being treated as members of X. Consider the (right) multiplication of an element $\underline{x}$ of X by an element g of G, that is, $\underline{x} \cdot g$. Later we shall write this as $\underline{x} \backslash g$. We have

$$\underline{x} \cdot e = \underline{x},$$

that is, the (right) multiplication of elements of X by e does not alter X. Also, by associativity we have

$$\underline{x} \cdot (gh) = (\underline{x} \cdot g) \cdot h,$$

where $g, h \in G$, that is, applying gh to x is the same as first applying g to x, and then applying h to the result. We call this procedure an *action* of G on X, see Definition 5.1 below. Further, as $4 \in G$ we have

$$\underline{1} \cdot 4 = \underline{4}, \quad \underline{2} \cdot 4 = \underline{1}, \quad \underline{3} \cdot 4 = \underline{5}, \quad \underline{4} \cdot 4 = \underline{2}, \quad \underline{5} \cdot 4 = \underline{6} \quad \text{and} \quad \underline{6} \cdot 4 = \underline{3},$$

that is, the set X has been permuted by this procedure. There is nothing special about '4', the calculation works for all elements of G; the reader should try one. This procedure gives a map from the group G to the set S_X of all permutations of X, where X is the underlying set of G; see Theorem 5.2 below. The term *G-set* is sometimes used for the set X (that is, when G is acting on X), we will not use this notation because in some cases X and G are unrelated.

With the example above in mind we begin by stating the basic

Definition 5.1 Given a non-empty set X and a group G, we say G *acts* on X if, for each $g \in G$, there exists a map $\backslash g : X \to X$, and these maps satisfy

$$\begin{aligned} &\text{(i)} \quad x \backslash e = x, \\ &\text{(ii)} \quad x \backslash (gh) = (x \backslash g) \backslash h, \end{aligned} \tag{5.1}$$

for all $x \in X$ and $g, h \in G$. We call the map $\backslash g$ an instance of the *action* of the group G on the set X.

Notes (a) By (5.1), the group operation 'corresponds' to composition of actions, that is, the map $\backslash gh$ is defined to equal $\backslash g \circ \backslash h$.

(b) We usually follow the convention that groups and elements of groups are denoted by letters at the beginning of the alphabet, and sets and elements of sets are denoted by letters at the end of the alphabet.

(c) More formally, we can rewrite Definition 5.1 as follows: The function $\backslash$ that maps the set of pairs $X \times G$ to X, and which satisfies the two parts of (5.1), is called an *action* of G on X, see Theorem 5.12.

(d) The action defined above is a '*right action*'; we could also define a *left action* but as we write functions on the right, we shall only consider the former.

(e) Many authors use either xg, or sometimes x^g, for $x\backslash g$. Both concatenation and the exponential notation are used widely in many contexts, so for the sake of clarity, it seems preferable to use a new symbol, those readers used to the old concatenation notation can simply ignore the backslash. Personally, the author finds the exponential notation particularly confusing. Also we use the concatenation notation xg for the particular action called the *natural action* (Example (a) below), and so we need a distinct notation for the general case. In a recently published book by Isaacs (2008), the author uses $x \bullet g$ for our $x\backslash g$.

We shall give some examples below, but first we prove the following basic result.

Theorem 5.2 *Using the notation set out above, the map* $\backslash g : X \to X$ *is a permutation (bijection) of the set* X.

Proof Suppose $x, y \in X$ and $x\backslash g = y\backslash g$. Then by (5.1) we have

$$x = x\backslash e = x\backslash (gg^{-1}) = (x\backslash g)\backslash g^{-1} = (y\backslash g)\backslash g^{-1} = y\backslash (gg^{-1}) = y,$$

that is, the map $\backslash g$ is injective. Secondly, suppose $z \in X$ then, for $g \in G$,

$$z = z\backslash e = z\backslash (g^{-1}g) = (z\backslash g^{-1})\backslash g,$$

that is, $z\backslash g^{-1}$ is a preimage of z under the map $\backslash g$. Hence this map is also surjective, and so it is a permutation of X.   □

Examples We give four here, three more extended examples will be discussed in the next section; see also the proofs of Cauchy's and the First Sylow Theorems given in Chapter 6.

(a) Let G be a group and let X be the underlying set of G. The group G acts on X by *right multiplication* if we define, for $g \in G$ and $x \in X$,

$$x\backslash g = xg,$$

see the example given at the beginning of this section. This clearly satisfies the conditions (5.1), and the corresponding action is called the *natural action* on G.
(b) Let V be a vector space defined over a field F. The multiplicative group F^* of F acts on V by *scalar multiplication*, for if $a, b \in F^*$ and $v \in V$, the standard vector space axioms give

$$v\backslash 1 = v1 = v \quad \text{and} \quad v\backslash (ab) = v(ab) = (va)b = (v\backslash a)\backslash b.$$

(c) Let $G = \langle e \rangle$ and X be an arbitrary set, then G acts on X if we define $x\backslash e$ to equal x for all $x \in X$.
(d) Let $G = S_n$ and let $X = \{1, \ldots, n\}$, then if we define $x\backslash \sigma = x\sigma$ for $\sigma \in G$ and $x \in X$, we obtain a new action called the *permutation action*; see Theorem 5.2.

There are two important entities associated with an action which govern its basic properties. They are *orbits* and *stabilisers*, we introduce orbits first. Let the group G act on the set X. Define a relation $\sim$ on X by: If $x, y \in X$, then

$$x \sim y \quad \text{if and only if} \quad x \backslash g = y \quad \text{for some } g \in G.$$

To put this informally, x is related to y if we can 'get from x to y' by using an element of G.

Lemma 5.3 *The relation $\sim$ defined above is an equivalence relation.*

Proof We have $x \sim x$ as $x \backslash e = x$. Secondly, suppose $x \sim y$, that is, $x \backslash g = y$ for some $g \in G$. Then

$$y \backslash g^{-1} = (x \backslash g) \backslash g^{-1} = x \backslash (g g^{-1}) = x \backslash e = x,$$

and so $y \sim x$. Finally, suppose we also have $y \sim z$ with $y \backslash h = z$ for some $h \in G$. Then $x \backslash (gh) = (x \backslash g) \backslash h = y \backslash h = z$, that is, $x \sim z$. $\qquad \square$

Definition 5.4 (i) An equivalence class of the equivalence relation $\sim$ given in Lemma 5.3 above is called an *orbit* of the action of G on X. The orbit containing the element $x \in X$ is called the *orbit of x*, and it is denoted by $\mathcal{O}_G\{x\}$, or $\mathcal{O}\{x\}$ when it is clear which group G is being used.

(ii) An action of G on X is called *transitive* if there is only one orbit, that is, X itself, otherwise it is called *intransitive*.

By Lemma 5.3, X is a *disjoint* union of its orbits (Appendix A). The orbit of $x \in X$, $\mathcal{O}\{x\}$, is the subset of X of those elements that we can 'get to' starting with x and applying elements of G (that is, by applying the maps associated with the elements of G), so an action is transitive if we can 'get from' every member of X to every other member of X by applying elements of G. For instance, the action in Example (d) opposite is transitive because S_n contains all 2-cycles—if $y, z \in X$, then the 2-cycle (y, z) belongs to S_n and this 2-cycle maps y to z. But if in this example we change the group to $\langle (1, 2) \rangle \simeq C_2$, then the action would not be transitive if $n > 2$ because, for instance, no element of this group maps 1 to 3. Also, if we replace X by $X_1 = \{1, \ldots, n, n + 1\}$, the action of S_n on X_1 would again be intransitive because no permutation in S_n maps 1 to $n + 1$. An extension of transitivity is as follows. An action of G on X is called *k-transitive* if for all pairs of k-element subsets Y and Z of X, there exists an element $g \in G$ such that $\backslash g$ is a bijection between Y and Z. For example, the action given in Example (d) is k-transitive if $k \leq n$. See Web Section 5.4 and Section 12.4.

We have defined both orbits and cycles (Definition 3.3), the notion of an orbit in a general group is related to the notion of a cycle in a symmetric group. If $n > 1$ and

$\sigma \in S_n$ then, by Theorem 3.4, σ can be expressed as a (disjoint) product of cycles

$$\sigma = (a_1, \ldots, a_j)(b_1, \ldots, b_k) \cdots$$

where $j + k + \cdots = n$. Let $H = \langle \sigma \rangle$, the cyclic subgroup of S_n generated by σ, then H acts on $X = \{1, \ldots, n\}$ by setting $x \backslash h = xh$ for $h \in H$ and $x \in X$. Here h has the form σ^t for some $t \in \mathbb{Z}$, and so the orbit of a_1, say, is $\{a_1 \sigma^t : t = 0, 1, 2, \ldots\}$ which is exactly the cycle (in σ) containing a_1. Hence in this particular example, orbits and cycles coincide.

Our second new entity is the *stabiliser* which we introduce by

Definition 5.5 Given a group G acting on a set X (Definition 5.1), and $x \in X$, the subset of G,

$$\{g \in G : x \backslash g = x\},$$

is called the *stabiliser* of x in G; it is denoted by $\mathrm{stab}_G(x)$.

The stabiliser of x is the subset (subgroup, see Lemma 5.6) of G of those elements whose associated maps 'do not move' x. For instance, if we let $x = 1$ in Example (d) on page 93, then the stabiliser of 1 is the set of all permutations in S_n which leave the element 1 fixed. This is clearly a subgroup of S_n isomorphic to S_{n-1}, and is an instance of

Lemma 5.6 *Using the notation set out in Definition 5.5, for $x \in X$,*

$$\mathrm{stab}_G(x) \leq G.$$

Proof Clearly, $e \in \mathrm{stab}_G(x)$ as $x \backslash e = x$ by (5.1). If $g, h \in \mathrm{stab}_G(x)$, then $x \backslash g = x = x \backslash h$ and, by (5.1) again, we have

$$x \backslash (gh^{-1}) = (x \backslash g) \backslash h^{-1} = (x \backslash h) \backslash h^{-1} = x.$$

This shows that $gh^{-1} \in \mathrm{stab}_G(x)$, now use Theorem 2.13. $\qquad \square$

We come now to the Orbit–Stabiliser Theorem, the main result in this chapter. It has a number of applications and, considering its importance, it is remarkably easy to prove.

Theorem 5.7 (Orbit–Stabiliser Theorem) *If G acts on a set X, $x \in X$, and $\mathcal{O}_G\{x\}$ is the orbit of x, then*

$$o(\mathcal{O}_G\{x\}) = [G : \mathrm{stab}_G(x)].$$

Proof Note first $\text{stab}_G(x) \leq G$ by Lemma 5.6. We define a map γ from $\mathcal{O}_G\{x\}$ to the set of right cosets of $\text{stab}_G(x)$ in G, and the theorem follows by showing that this map is a bijection. The map γ is given by

$$(x\backslash g)\gamma = \big(\text{stab}_G(x)\big)g,$$

where $g \in G$, and so $x\backslash g \in \mathcal{O}_G\{x\}$. First, we show that this map is well-defined. Suppose $x\backslash g = x\backslash h$, then as in the proof above we have

$$x\backslash\big(gh^{-1}\big) = (x\backslash g)\backslash h^{-1} = (x\backslash h)\backslash h^{-1} = x\backslash e = x$$

by (5.1), that is, $gh^{-1} \in \text{stab}_G(x)$. By Lemma 2.22, this gives $(\text{stab}_G(x))g = (\text{stab}_G(x))h$, as required. Second, note that γ is clearly surjective, for if $g \in G$, then one preimage of $(\text{stab}_G(x))g$ is $x\backslash g$. Last, we show that it is injective. Suppose

$$\big(\text{stab}_G(x)\big)g = \big(\text{stab}_G(x)\big)h,$$

then, as above, this shows $gh^{-1} \in \text{stab}_G(x)$, and so $x\backslash(gh^{-1}) = x$. Hence

$$x\backslash h = \big(x\backslash gh^{-1}\big)\backslash h = x\backslash\big(gh^{-1}h\big) = x\backslash g,$$

that is, γ is injective. The theorem now follows. $\square$

Referring back to Example (d) on page 93, let $x = n$. The orbit of n is $\{1, \ldots, n\}$ (as the action is transitive), and so $o(\mathcal{O}\{n\}) = n$. We noted above that $\text{stab}_{S_n}(n) \simeq S_{n-1}$. Hence we have by the Orbit–Stabiliser Theorem, and as $o(S_n) = n!$,

$$o\big(\mathcal{O}\{n\}\big) = n = [S_n : S_{n-1}] = \big[S_n : \text{stab}_{S_n}(1)\big].$$

This is a easy example but it does provide an illustration of the theorem 'at work'.

As a second application of this theorem we prove the following useful result. Another proof was given in Problem 2.27. Note that if H and J are both non-normal subgroups of G, then HJ is not a subgroup of G, see the example below. But if either H or J is normal, then HJ is a subgroup (Theorem 2.30) and the result follows using the Second Isomorphism and Lagrange's Theorems.

Theorem 5.8 *If G is a finite group and $H, J \leq G$, then*

$$o(HJ)o(H \cap J) = o(H)o(J).$$

Proof We define an action. Let $X = \{Hg : g \in G\}$, the set of right cosets of H in G. The subgroup J acts on X by right multiplication if we set

$$Hg\backslash j = Hgj \quad \text{for } j \in J.$$

This is an action since $Hg\backslash e = Hge = Hg$ and $Hg\backslash j_1 j_2 = Hgj_1 j_2 = (Hgj_1)\backslash j_2 = (Hg\backslash j_1)\backslash j_2$. The orbit $\mathcal{O}\{H\}$ of H is $\{H\backslash j : j \in J\}$, and this equals HJ, and so $o(HJ) = o(H) \times o(\mathcal{O}\{H\})$. (Note that HJ is a disjoint (by Lemma 2.23) union of right cosets of H, and the orbit of H under this action is the union of those cosets of H we can 'get to' starting with H itself and applying elements j in J.) Further, the stabiliser of H, $\mathrm{stab}_J(H)$, equals $\{j \in J : Hj = H\}$, and so $\mathrm{stab}_J(H) = H \cap J$ (as $Hj = H$ if and only if $j \in H$). Hence, using the equation for $o(HJ)$ above, the Orbit–Stabiliser Theorem gives

$$\frac{o(HJ)}{o(H)} = o(\mathcal{O}\{H\}) = \left[J : \mathrm{stab}_J(H) \right] = \frac{o(J)}{o(H \cap J)},$$

using the equation for $\mathrm{stab}_J(H)$ and Lagrange's Theorem (Theorem 2.27) for the last identity, the result follows. $\qquad\square$

Example Suppose $G = D_3 \simeq \langle a, b \mid a^3 = b^2 = e, a^2 b = ba \rangle$, $H = \langle b \rangle \leq G$ and $J = \langle ab \rangle \leq G$. Then $o(H) = o(J) = 2$, $H \cap J = \langle e \rangle$, and so $o(HJ) = 4$, by Theorem 5.8. Clearly, HJ is not a subgroup of G (as $4 \nmid 6$, also neither H nor J is normal), but it is a union of *two* cosets. For as $ab = ba^2$, we have $Hab = Hba^2 = Ha^2$ and so $HJ = H \cup Ha^2$. Note that we also have $HJ = J \cup bJ$.

We have shown (Theorem 5.2) that an action of a group G on a set X is a collection of permutations of X; that is, the action provides a map from G to S_X. We shall develop this further.

Definition 5.9 Let the group G act on the set X. The map $\nu : G \to S_X$ given by

$$g\nu = \backslash g, \quad \text{for all } g \in G,$$

is called the *permutation representation* of G for this action.

Lemma 5.10 *The map ν given by Definition 5.9 is a homomorphism.*

Proof For all $g, h \in G$ and $x \in X$, we have by (5.1),

$$x\backslash(gh) = (x\backslash g)\backslash h = x(\backslash g \circ \backslash h)$$

by the definition of composition ($\circ$) of functions. The lemma follows as this holds for all $x \in X$. $\qquad\square$

This leads to the following useful

Theorem 5.11 *Let G act on X with permutation representation ν as defined above, then*

$$\ker \nu = \bigcap\nolimits_{x \in X} \mathrm{stab}_G(x).$$

Proof This is an immediate consequence of the definitions. As ν is a homomorphism (Lemma 5.10), its kernel is the set of those $g \in G$ for which $\backslash g$ is the identity permutation in S_X, that is, $x\backslash g = x$ for all $x \in X$. But $\mathrm{stab}_G(x)$ is the set of those $g \in G$ for which $x\backslash g = x$, hence $\bigcap_{x \in X} \mathrm{stab}_G(x)$ is the set of those $g \in G$ for which $x\backslash g = x$ *for all* $x \in X$; that is, the kernel of ν. $\square$

Referring again to Example (d) on page 93, if $k \in \{1, \ldots, n\}$, then $\mathrm{stab}_{S_n}(k)$ is the set of all permutations which fix k. Hence the intersection of these stabilisers for $k = 1, \ldots, n$, is $\langle e \rangle$, and so the kernel of the permutation representation in this case is the neutral subgroup. This reflects the fact that S_n has very few normal subgroups.

The converse of the last result is also valid as we show now. It can be used as an alternative definition of the action of a group G on a set X.

Theorem 5.12 *Suppose* $\sigma : G \to S_X$ *is a homomorphism of G to the group of all permutations on the set X. The map defined by $\backslash g = g\sigma$, for all $g \in G$, is an action of G on X, and the permutation representation of this action is identical to σ.*

Proof For $x \in X$, we have $x(e\sigma) = x$ (as $e\sigma$ is the identity permutation ι on X) and, by composition of maps and as σ is a homomorphism,

$$\big(x(g\sigma)\big)(h\sigma) = x(g\sigma \circ h\sigma) = x\big((gh)\sigma\big).$$

Hence, if we define

$$x\backslash g = x(g\sigma),$$

for all $g \in G$ and $x \in X$, these equations show that $\backslash g$ is an action of G on X. Let ν be the permutation representation of this action, that is, for $g \in G$ and $x \in X$, $x\backslash g = x(g\nu)$. Combining these facts gives $x(g\nu) = x(g\sigma)$ for all $x \in X$, hence $g\nu = g\sigma$ for all $g \in G$, which shows that $\nu = \sigma$. $\square$

Restricted Actions

Here we ask: What is the relationship of $\mathrm{stab}_H(x)$ to $\mathrm{stab}_G(x)$ when $H \leq G$? To answer this question we first need to consider the subset of those elements which are fixed by an action.

Definition 5.13 Let G act on the set X. We set

$$\mathrm{fix}(G, X) = \big\{x \in X : x\backslash g = x \text{ for all } g \in G\big\};$$

it is called the *fixed set* of X under the action of G.

Note that fix(G, X) is a *subset* of X, and so it is *not* a group; for example, it is empty when the action is transitive, also we have the equivalent definitions

$$\text{fix}(G, X) = \{x \in X : \mathcal{O}\{x\} = \{x\}\} = \{x \in X : \text{stab}_G(x) = G\}. \qquad (5.2)$$

If G and X are given by the first example in this chapter (page 92), then fix$(G, X) = \emptyset$, but if we change X to $X' = \{\underline{1}, \dots, \underline{7}\}$, then fix$(G, X') = \{\underline{7}\}$.

Let G act on a set X and let $H \leq G$, we say H acts on X by *restriction of the action* of G on X if we ignore those maps $\backslash g$ in the action of G on X where $g \notin H$, and only consider those maps $\backslash h$ where $h \in H$. This is clearly an action because H is a (sub)group. For example, if $G = \mathbb{Z}$, $H = 2\mathbb{Z}$, X is the underlying set of G (the integers) and the action of G on X is the natural one given by $x \backslash g = xg$, then the orbit of x under the action of G is the set of all integer multiples of x, whilst the orbit of x under the restricted action (by H) is the set of all even integer multiples of x.

We have, for $x \in X$ and $H \leq G$,

$$\text{stab}_H(x) = \text{stab}_G(x) \cap H, \qquad (5.3)$$

by definition of the restricted action. We also have

Lemma 5.14 *If $H \leq G$, G acts on X, and H acts on X by restriction of the action of G, then*

$$x \in \text{fix}(H, X) \quad \text{if and only if} \quad H \leq \text{stab}_G(x).$$

Proof We have

$$
\begin{array}{llll}
x \in \text{fix}(H, X) & \text{if and only if} & \text{stab}_H(x) = H & \text{by (5.2)} \\
& \text{if and only if} & \text{stab}_G(x) \cap H = H & \text{by (5.3)} \\
& \text{if and only if} & H \leq \text{stab}_G(x),
\end{array}
$$

by Problem 2.5. $\qquad\qquad\qquad\qquad\qquad\qquad\qquad\qquad\qquad\qquad\qquad\qquad\quad\square$

For an example, see Problem 5.3(i). One consequence of this problem is: If G is a p-group (Section 6.1) and $p \nmid o(X)$, then there exists $x \in X$ which is moved by no $g \in G$, that is, fix(G, X) is not empty.

5.2 Three Important Examples

In this section we discuss three action examples, the first involves cosets, and the second and third use conjugation of elements and subgroups, respectively. All three introduce major new concepts and theorems which will be used widely in the following chapters, and as noted above all three have a long history in the theory.

Coset Action

For the first example, choose H, a subgroup of G, and let X be the set of right cosets of H in G. Given $g \in G$ and $Hx \in X$, we define

$$(Hx)\backslash g = Hxg. \tag{5.4}$$

This is an action because $(Hx)\backslash e = Hxe = Hx$ and

$$\big((Hx)\backslash g\big)\backslash h = (Hxg)\backslash h = Hxgh = (Hx)\backslash(gh).$$

Further, it is a transitive action. For if $Hx, Hy \in X$, then

$$(Hx)\backslash x^{-1}y = \big((Hx)x^{-1}\big)\backslash y = Hy,$$

and so there is only one orbit, that is, X itself. Also

$$\text{stab}_G(Hx) = x^{-1}Hx$$

because

$$\text{stab}_G(Hx) = \big\{g \in G : (Hx)\backslash g = Hx\big\} = \big\{g \in G : Hxg = Hx\big\}$$

$$= \big\{g \in G : xgx^{-1} \in H\big\} = x^{-1}Hx,$$

by Problem 2.23. Hence, using the Orbit–Stabiliser Theorem (Theorem 5.7), we obtain

$$\big[G : x^{-1}Hx\big] = \big[G : \text{stab}_G(Hx)\big] = o(X) = [G : H] \tag{5.5}$$

because there is only one orbit which in this case is the set of right cosets of H in G, this gives another proof of Problem 2.23(ii).

If ν_H is the permutation representation of this action, then by Theorem 5.11,

$$\ker \nu_H = \bigcap\nolimits_{x \in G} x^{-1}Hx,$$

see the comment below Definition 2.21. The entity on the right-hand side of this equation is called the *core* of H in G, core(H) which was defined in Problem 2.24; it is the largest normal subgroup of G contained in H. If $[G : H] = n < \infty$, then $S_X \simeq S_n$, and ν_H gives a homomorphism from G into S_n. Hence we have

Theorem 5.15 (i) *If $H < G$ and $[G : H] = n < \infty$, then there exists an injective homomorphism from $G/\text{core}(H) = G/(\bigcap_{x \in G} x^{-1}Hx)$ into S_n.*
(ii) *If $o(G/\text{core}(H)) = m$, then $n \mid m$ and $m \mid n!$.*

Proof (i) This follows immediately from Theorem 5.11 and Corollary 4.12.
(ii) Both of these properties follow from (i) and Problem 2.15(i) with $J = \text{core}(H)$.   □

This is a useful result, especially when n is small, for it shows that if a group G has only a few normal subgroups, then its total number of subgroups is also restricted.

Example Subgroups of A_5. If G is simple and $H < G$, then

$$\bigcap_{x \in G} x^{-1} H x = \langle e \rangle,$$

as this intersection forms a normal subgroup of G contained in H. Hence by Theorem 5.15, there is an injective homomorphism from G to S_n, and so $o(G) \leq o(S_n) = n!$. Therefore, if $o(G) > n!$, G does not contain a subgroup (normal or not) of index n. For instance, consider $G = A_5$ with order 60. Suppose $H < A_5$, and $[A_5 : H] = n$. As $n! \geq 60$ only if $n > 4$, the theorem shows that A_5 cannot have a subgroup of index 2, 3 or 4, so it cannot contain a subgroup of order 30, 20 or 15. In this case, we say that A_5 is not *reverse Lagrange*. It does contain a number of subgroups of order 12 (with index 5). The group A_4 is also not reverse Lagrange, reader why?

We give an application of Theorem 5.15 here, more will follow later.

Theorem 5.16 *If G is an infinite group, $H \leq G$, and $[G : H] < \infty$, then G contains a normal subgroup K which satisfies $K \leq H$ and G/K is finite.*

Proof Take $K = \bigcap_{g \in G} g^{-1} H g$ in Theorem 5.15 and use Problem 2.23. The factor group G/K is finite because this theorem gives an injective homomorphism into a finite symmetric group. $\square$

This shows that if an infinite group has a subgroup of finite index (and so is infinite), then it also has a normal infinite subgroup of finite index—an important fact concerning infinite groups. This also shows that an infinite simple group has no subgroups of finite index.

Centralisers and Class Equations

Our second extended example involves conjugation, and introduces a number of new concepts and constructions. Let G be a group and let X be the underlying set of G. If $g, x \in G$, then $g^{-1} x g$ is called the *conjugate* of x by g (Definition 2.28). The group G acts on its underlying set G by conjugation if we define

$$x \backslash g = g^{-1} x g, \tag{5.6}$$

for all $g, x \in G$. The operation defined by (5.6) is an action; for clearly $x \backslash e = e^{-1} x e = x$, and we have

$$x \backslash (gh) = (gh)^{-1} x gh = h^{-1} (g^{-1} x g) h = h^{-1} (x \backslash g) h = (x \backslash g) \backslash h.$$

The action (5.6) is called the *conjugacy action* on G, three important entities are associated with it as follows.

Definition 5.17 Using the conjugacy action defined above, the orbit of x under this action is called the *conjugacy class* of x and it is denoted by $C\ell_G\{x\}$, that is,

$$C\ell_G\{x\} = \left\{ g^{-1}xg : g \in G \right\}.$$

This is, of course, the set of conjugates of x in G. When it is clear which group is involved, we write $C\ell\{x\}$ for $C\ell_G\{x\}$.

Definition 5.18 The stabiliser of x in G under the conjugacy action (5.6) is called the *centraliser* of x in G, it is denoted by $C_G(x)$; that is,

$$C_G(x) = \text{stab}_G(x) = \left\{ g \in G : g^{-1}xg = x \right\}.$$

Note this is equivalent to $C_G(x) = \{ g \in G : xg = gx \}$; therefore, the centraliser of x in G is the subgroup (Lemma 5.6) of those elements $g \in G$ which commute with x.

Applying the Orbit–Stabiliser Theorem (Theorem 5.7) we have directly

Theorem 5.19 *Using the conjugacy action (5.6) defined above, if $x \in G$, then*

$$o\bigl(C\ell_G\{x\}\bigr) = \bigl[G : C_G(x) \bigr].$$

By Lagrange's Theorem (Theorem 2.27), this shows that the order of a conjugacy class of a finite group G divides the order of G. It also shows that the set of elements that commute with a fixed element a, say, forms a subgroup. Both of these facts have important ramifications; for the first, see Lemma 5.21 below.

The third entity associated with the conjugacy action is the centre, see Definition 2.32. If τ denotes the permutation representation of the conjugacy action defined above, then the kernel, $\ker \tau$, is just the *centre* of the group $Z(G)$; that is,

$$Z(G) = \ker \tau = \bigcap_{x \in G} C_G(x).$$

Note also that, using Definitions 2.32 and 5.17, we have

$$x \in Z(G) \quad \text{if and only if} \quad C\ell_G\{x\} = \{x\}. \tag{5.7}$$

Example We construct the conjugacy classes, centralisers and centre of the dihedral group D_3. Let D_3 be given by (Section 3.4)

$$\langle a, b \mid a^3 = b^2 = e, ba = a^2b \rangle.$$

We have

$$Cl\{a\} = \{g^{-1}ag : g \in D_3\} = \{a, a^2\}, \quad o(Cl\{a\}) = 2,$$

$$C_{D_3}(a) = \{g \in D_3 : ga = ag\} = \{e, a, a^2\} = \langle a \rangle, \quad o(C_{D_3}(a)) = 3,$$

$$Cl\{b\} = \{b, ab, a^2b\}, \quad o(Cl\{b\}) = 3,$$

$$C_{D_3}(b) = \{e, b\} = \langle b \rangle, \quad o(C_{D_3}(b)) = 2.$$

The reader should check these statements and complete the remaining cases, they provide applications of the Orbit–Stabiliser Theorem. Note finally that $Z(D_3) = \langle e \rangle$ as $C_{D_3}(a) \cap C_{D_3}(b) = \langle e \rangle$.

Putting these ideas together we can introduce the *Class Equations* by

Theorem 5.20 (Class Equations) *Suppose G is a finite group, and $Cl_G\{y_1\}, \ldots,$ $Cl_G\{y_k\}$ is a complete list of the conjugacy classes of G whose orders are larger than 1.*

(i) $G = Z(G) \,\dot{\cup}\, \bigcup_{i=1}^{k} Cl_G\{y_i\}$ *(disjoint unions).*

(ii) $o(G) = o(Z(G)) + \sum_{i=1}^{k} o(Cl_G\{y_i\}) = o(Z(G)) + \sum_{i=1}^{k} [G : C_G(y_i)].$

Proof (i) This follows immediately from Lemma 5.3 and (5.7) using Definition 5.17—G is a disjoint union of its conjugacy classes.

(ii) The first equation is given by (i) as the unions are disjoint, and the second follows from Theorem 5.19. □

The equations in (i) and (ii) above are called the *Class Equations* for G. Also the positive integer $o(Z(G)) + k$, that is, the number of conjugacy classes of G including the singleton classes counted in $o(Z(G))$, is called the *class number* of G and it is denoted by $h(G)$. See Appendix C for some examples.

Next we give some applications of the Class Equations, more will follow later.

Lemma 5.21 *If p is prime and $o(G) = p^n$, then $Z(G) \neq \langle e \rangle$.*

This lemma has important implications for p-group theory; see Section 6.1.

Proof If G is Abelian there is nothing to prove, and if not, then at least one y_i exists with $o(Cl_G\{y_i\}) > 1$. Referring to the notation given in Theorem 5.20, we have by Theorems 5.19 and 2.27

$$p \mid [G : C_G(y_i)],$$

for $i = 1, \ldots, k$, as these indices are larger than 1 by definition of y_i. Hence by the second Class Equation we have $p \mid o(Z(G))$ because $p \mid o(G)$ by hypothesis. Now $o(Z(G)) > 0$ (as $e \in Z(G)$), and so this shows that $o(Z(G)) \geq p$ and therefore $o(Z(G))$ cannot equal 1; the result follows. □

Theorem 5.22 (i) *If $o(G) = p^n$, then G is not simple, provided $n > 1$.*
(ii) *If p is prime and $o(G) = p^2$, then G is Abelian.*

If $o(G) = p$, then G is simple and cyclic (and so Abelian) by Theorem 2.34, and (ii) deals with the case $o(G) = p^2$. But there exist non-Abelian groups with order p^n for $n > 2$; see Section 6.1 and Problem 6.16.

Proof (i) This follows by Lemmas 2.31 and 5.21 as $\langle e \rangle < Z(G) \triangleleft G$.

(ii) Suppose G is not Abelian, then $Z(G) < G$. Hence, by Lagrange's Theorem (Theorem 2.27) $o(Z(G)) = 1$ or p. But, by Lemma 5.21, it cannot equal 1, hence $o(Z(G)) = p$. Therefore $o(G/Z(G)) = p$, and so $G/Z(G)$ is cyclic by Theorem 2.34. As $Z(G)$ is also cyclic in this case, Problem 4.16(ii) now shows that G is Abelian.   □

In Chapter 7, we show that if $o(G) = p^2$, then G is cyclic or a 'product' of two cyclic groups of order p, that is, elementary Abelian, see Problem 4.18.

Centralisers played a vital role in the solution of CFSG, see Chapter 12. Brauer and Fowler (1955) showed that for a given finite group G there can only be finitely many simple groups H which contain an involution (element of order 2) a and have the property:

$$C_H(a) \simeq G.$$

In many cases, G can be taken to be quite small and only a few simple groups H are involved, several simple groups can be characterised using this result; see the discussion on page 265.

We can extend the notion of centraliser of subsets of a group as follows.

Definition 5.23 The *centraliser* $C_G(X)$ of a subset X of a group G is given by

$$C_G(X) = \big\{ g \in G : ag = ga \text{ for all } a \in X \big\}.$$

The basic properties are as follows; proofs are left as exercises for the reader, see Problem 5.8. Note that as the size of X increases, the size of $C_G(X)$ decreases.

(i)   If $X \subseteq G$ then $C_G(X) \leq G$.
(ii) $C_G(G) = Z(G)$.
(iii) If $H \leq G$ then $C_G(H) = \bigcap_{h \in H} C_G(h)$.
(iv) If $J \leq H \leq G$ and $H \leq C_G(J)$, then $J \leq Z(H)$.

If we refer back to the example on page 102 we see that, using (iii) above, $C_{D_3}(\langle a \rangle) = \langle a \rangle$, $C_{D_3}(\langle b \rangle) = \langle b \rangle$ and $C_{D_3}(\{a, b\}) = \langle e \rangle$ as the reader can easily check. See also Problem 5.26.

Normaliser

Our last extended action example also uses conjugation but now applied to subgroups. This will introduce the *normaliser*—one of the most important entities in group theory, and one with a long history in the theory.

If $H \leq G$ and $g \in G$, then

$$g^{-1}Hg = \{g^{-1}hg : h \in H\},$$

is a subgroup of G which is called a *conjugate subgroup* of H in G (Problem 2.23). Using this notion we can define an action of G on the set of all subgroups H of G by

$$H \backslash g = g^{-1}Hg \quad \text{for } g \in G. \tag{5.8}$$

This is an action because $H \backslash e = e^{-1}He = H$ and, for $g, h \in G$, $H \backslash gh = (gh)^{-1}Hgh = h^{-1}(g^{-1}Hg)h = (H \backslash g) \backslash h$. The orbit of H under this action is the set of subgroups of G which are conjugate to H in G (Problem 2.23 again), and the stabiliser is given by

Definition 5.24 For $H \leq G$, the stabiliser of H in G under the action (5.8) defined above, that is,

$$\text{stab}_G(H) = \{g \in G : g^{-1}Hg = H\},$$

is called the *normaliser* of H in G, and it is denoted by $N_G(H)$.

The normaliser $N_G(H)$ clearly contains H, and it is the largest subgroup of G in which H is normal; the reader should check this, see Problem 5.13. Note that $N_G(\langle e \rangle) = G = N_G(G)$, so in particular it is *not* true in general that if $H < J$ then $N_G(H) < N_G(J)$. The normaliser has many uses in the theory as we shall see in the sequel; note the similarities and differences with the centraliser. The basic properties are given by

Lemma 5.25 *Suppose $H \leq G$.*

(i) $H \triangleleft N_G(H) \leq G$.
(ii) $N_G(H) = G$ *if and only if* $H \triangleleft G$.
(iii) *The number of conjugates of H in G equals* $[G : N_G(H)]$.

   Proof (i) Clearly $H \leq N_G(H)$, as $h^{-1}Hh = H$ if $h \in H$; also $N_G(H) \leq G$ by definition and Lemma 5.6. Theorem 2.29 gives normality.
   (ii) If $N_G(H) = G$ use (i), and if $H \triangleleft G$, then $g^{-1}Hg = H$ for all $g \in G$, that is $N_G(H) = G$.
   (iii) This follows immediately from the Orbit–Stabiliser Theorem (Theorem 5.7).   □

Example Let $G = S_4$ and $H = \langle (1, 2, 3) \rangle \simeq C_3$. We have

$$N_G(H) = \langle (1, 2, 3), (1, 2) \rangle \simeq S_3,$$

for $(1, 2)(1, 2, 3)(1, 2) = (1, 3, 2)$, and so $(1, 2) \in N_G(H)$, *et cetera*. Hence $H < N_G(H) < G$. Sometimes H equals $N_G(H)$ (in this case, we say H is *self-normalising*), for example, when H is a maximal non-normal subgroup of G. In other cases, $H < N_G(H) = G$, this holds when H is a proper normal subgroup of G.

The final pair of results in this section are easily proved and have a number of useful applications; see Section 4.4 for the basic properties of automorphisms.

Theorem 5.26 (N/C-Theorem) *Suppose $H \leq G$.*

(i) $C_G(H) \lhd N_G(H)$.
(ii) $N_G(H)/C_G(H)$ *is isomorphic to a subgroup of* Aut H.

The factor group $N_G(H)/C_G(H)$ is called the *automiser* of H in G.

Proof For $g \in G$, let ϕ_g be the (inner) automorphism given by $a\phi_g = g^{-1}ag$ for $a \in G$, and define $\theta : N_G(H) \to$ Aut H by

$$j\theta = \phi_j|_H \quad \text{for } j \in N_G(H); \tag{5.9}$$

that is, $j\theta$ is ϕ_j with its domain restricted to the subgroup H. Note that $j\theta \in$ Aut H because $j \in N_G(H)$, as $j^{-1}hj \in H$ for $h \in H$, and $\phi|_H$ is defined by conjugation. Also θ is a homomorphism (as the reader can check). Now

$$a \in \ker \theta \quad \text{if and only if} \quad \phi_a|_H \text{ is the identity map on } H,$$
$$\text{if and only if} \quad h^{-1}ah = a \text{ for all } h \in H,$$
$$\text{if and only if} \quad a \in C_G(H),$$

using Definition 5.23. Hence $\ker \theta = C_G(H)$, (i) follows by Lemma 4.6, and (ii) follows by Corollary 4.12. $\square$

Example For this example, the reader will need to refer to Section 8.2 where the group $SL_2(3)$ is discussed; we show here that this group has a factor group isomorphic to A_4; see Problem 3.10. Let $G = SL_2(3)$ and $H = \langle \left(\begin{smallmatrix} 2 & 2 \\ 2 & 1 \end{smallmatrix}\right), \left(\begin{smallmatrix} 1 & 2 \\ 2 & 2 \end{smallmatrix}\right) \rangle \simeq Q_2$. We have $H \lhd G$, and so $N_G(H) = G$, $C_G(H) = Z(G) \simeq C_2$, and Aut $H \simeq S_4$ (Problem 6.18). In this case, the N/C-theorem gives $G/Z(G) \preceq$ Aut $H \simeq S_4$. But $o(G/Z(G)) = 12$, and we have $G/Z(G) \simeq A_4$ because the only subgroup of S_4 of order 12 is A_4 (Problem 3.3(vii)).

The last result in this chapter provides an essential first step in the construction of the automorphism group of a group.

Corollary 5.27 $G/Z(G) \simeq \mathrm{Inn}\, G$.

Proof Put $H = G$ in Theorem 5.26. We have $N_G(G) = G$ (as the normaliser of a subgroup always contains that subgroup), $C_G(G) = Z(G)$ (see (ii) on page 104), and $\theta \in \mathrm{Inn}\, G$ using (5.9) with $H = G$. Note also that the map θ given in (5.9) is surjective in this case.   □

For example, this corollary shows that $\mathrm{Inn}\, S_4 \simeq S_4$ as this group is centreless. In fact, we have $\mathrm{Aut}\, S_4 \simeq S_4$, but this is harder to prove, see Problem 8.5. Some further applications are given in Chapter 8, and in the example at the end of Section 4.4.

5.3 Problems

Problem 5.1 (i) Suppose $G \leq S_4$ and G acts naturally on the set $\{1, 2, 3, 4\}$, see (d) on page 94. Construct the orbits and stabilisers of this action when (i) $G = \langle(1, 2, 3)\rangle$, (ii) $G = \langle(1, 2, 3, 4)\rangle$, (iii) $G = \langle(1, 2)(3, 4), (1, 3)(2, 4)\rangle$, (iv) $G = \langle(1, 2), (3, 4)\rangle$, and (v) $G = A_4$.

(vi) Find the orbits and stabilisers of the action given in Example (b) on page 93 when $\dim V = n$ and $n > 1$.

(vii) Let $\mathbb{Q}[x_1, \ldots, x_n]$ denote the set of all polynomials with rational coefficients in the variables $x_1, \ldots, x_n$. For $f(x_1, \ldots, x_n) \in \mathbb{Q}[x_1, \ldots, x_n]$ and $\sigma \in S_n$, define

$$f(x_1, \ldots, x_n) \backslash \sigma = f(x_{1\sigma}, \ldots, x_{n\sigma}).$$

Show that this defines an action of S_n on the set $\mathbb{Q}[x_1, \ldots, x_n]$, and describe the orbits and stabilisers. Hence prove that the order of the set of n-variable polynomials of the form $f(x_1, \ldots, x_n) \backslash \sigma$, where f is fixed, is a divisor of $n!$.

Problem 5.2 Let G be a group with subgroups H and J, and suppose we have the identity $([G : H], [G : J]) = 1$. Show that (i) $G = HJ$, and (ii) $[G : H \cap J] = [G : H][G : J]$. (Hint. Use Theorem 5.8.)

Problem ♦ 5.3 Suppose $o(G) = p^n$ where p is a prime, and so G is a p-group, see Section 6.1.

(i) If G acts on a finite set X using Definition 5.13 show that

$$o\big(\mathrm{fix}(G, X)\big) \equiv o(X) \pmod{p}.$$

(ii) Prove that if $\langle e \rangle \neq J$ and $J \lhd G$, then $J \cap Z(G) \neq \langle e \rangle$—a useful result, note that it provides a generalisation of Lemma 5.21.

Problem 5.4 Suppose G acts on a set X.

(i) If $x, y \in X$ and $y = x \backslash g$ for some $g \in G$, show that $\text{stab}_G(y) = g^{-1} \text{stab}_G(x) g$, and so deduce the result: $o(\text{stab}_G(x)) = o(\text{stab}_G(y))$.

(ii) For $g \in G$, let $\text{fix}(g, X)$ denote the subset of X of those x which are fixed by g (so $x \backslash g = x$). Show that if m is the number of orbits of this action, then

$$m = (1/o(G)) \sum_{g \in G} o(\text{fix}(g, X)).$$

(iii) Using (ii), show that if $1 < o(X) < \infty$ and the action is transitive, then there exists $g \in G$ with no fixed points.

Problem 5.5 (i) Use Theorem 5.19 to show that $o(G) = 2$ when G has just two conjugacy classes.

(ii) Show that if G is a finite non-Abelian group, $1 < [G : H] < 5$ and $H \leq G$, then G is not simple.

(iii) Let G be a finite group and let $h(G)$ denote its class number, see page 103. Show that the total number of ordered pairs (a, b) which satisfy $ab = ba$, where $a, b \in G$, is $h(G) \cdot o(G)$.

Problem ♦ 5.6 If p_0 is the smallest prime dividing $o(G)$, $K \leq G$ and $[G : K] = p_0$, prove that $K \lhd G$. (Hint. Use Theorem 5.15.)

Problem 5.7 (i) If $\mathcal{C}$ is a conjugacy class of a group G, show that the set $\mathcal{C}^{-1} = \{a^{-1} : a \in \mathcal{C}\}$ is another conjugacy class of G, and vice versa.

(ii) Construct the conjugacy classes, and their inverses as given by (i), for the groups D_4, A_4 and $SL_2(3)$.

Problem 5.8 (Properties of the Centraliser) (i) to (iv) Prove the centraliser properties listed on page 104.

(v) Let $H \leq G$ and $g \in G$. Show that $C_H(g) = C_G(g) \cap H$, and

$$g^{-1} C_G(H) g = C_G(gHg^{-1}).$$

(vi) Prove $C_G(H) \leq N_G(H)$ without using Theorem 5.26.

(vii) Let $H \leq G$. Show that $C_G(H) = \langle e \rangle$ if, and only if, $Z(J) = \langle e \rangle$ for all J satisfying $H \leq J \leq G$.

Problem 5.9 (i) Find the centralisers of the elements of D_6 and S_4, and of the subgroups of S_4.

(ii) Find the normalisers of the subgroups of S_4; see Section 8.1 and Lemma 5.25(iii).

(iii) Write out the Class Equations (Theorem 5.20) for A_5 explicitly; your answer should include a description of all centralisers involved.

Problem 5.10 Let $\tau = (1, \ldots, m)$ be an m-cycle in S_n where $n \geq m > 1$. Show that $C_{S_n}(\tau) \simeq \langle \tau \rangle \times S_{n-m}$, where we assume that $S_0 = \langle e \rangle$. The direct product notation $\times$ is defined in Section 7.1.

Problem 5.11 (i) Suppose G is a non-Abelian group. Show that if $a \in G \backslash Z(G)$, then $\langle a \rangle Z(G)$ is an Abelian subgroup of G which properly contains $Z(G)$.

(ii) Suppose $H, J \leq G$ where H is Abelian. The subgroup H is called *maximal Abelian* if J is not Abelian whenever $H < J$. Show that H is maximal Abelian if, and only if, $C_G(H) = H$.

Problem 5.12 Let A and B be subsets of the group G. Show that

(i) if $A \subseteq B$ then $C_G(B) \leq C_G(A)$,
(ii) $A \subseteq C_G(C_G(A))$,
(iii) $C_G(C_G(C_G(A))) = C_G(A)$.

Problem ◆ 5.13 (Properties of the Normaliser) Let $H \leq G$. Prove that

(i) $N_G(H)$ is the largest subgroup of G in which H is normal.
(ii) $N_G(g^{-1}Hg) = g^{-1}N_G(H)g$, for all $g \in G$.
(iii) If $H \leq J \leq G$, then $N_J(H) = N_G(H) \cap J$.
(iv) If $K \leq G$, then $[H, K] \leq H$ if, and only if, $K \leq N_G(H)$.
(v)   Let p be a prime. If $J \leq G$, $o(J) = p^r$ for some $r > 0$ and $p \mid [G : J]$, so J is a non-Sylow p-subgroup of G (Chapter 6). Prove that $J < N_G(J)$, that is, J is not self-normalising. (Hint. Use Problem 5.3.)

Problem 5.14 Let D denote the subgroup of diagonal matrices in $G = GL_2(\mathbb{Q})$. Find $N_G(D)$. Do the same calculation for $GL_3(\mathbb{Q})$.

Problem 5.15 (i) Suppose $H < G$ and G is finite. Show that there exists $a \in G$ which does not belong to any conjugate of H in G. (Hint. One method uses Lemma 5.25.)

(ii) Show that if G is finite and all of the maximal subgroups of G are conjugate, then G is cyclic; see Problem 2.13.

Problem 5.16 Throughout this problem $K \lhd G$ and we say that K is *central* if $K \leq Z(G)$. Use the N/C-theorem (Theorem 5.26) to prove the following propositions.

(i)   If G is perfect (Problem 4.8) and K is cyclic, then K is central.
(ii) If p_0 is the smallest prime dividing $o(G)$ and $o(K) = p_0$, then K is central.
(iii) If G is infinite and K is finite, then $G/C_G(K)$ is finite. Deduce that if the only finite factor group of G has order 1, and $o(K) < \infty$, then again K is central.

Problem 5.17 (i) Suppose $K \lhd H \leq G$ and $J = C_G(K)$. Show that (a) $H \leq N_G(K)$, and (b) if J is self-normalising (that is, $J = N_G(J)$), then $K \leq Z(H)$.

(ii) Let $H \leq G$. Prove that $C_G(H) = N_G(H)$ if and only if $H \leq Z(N_G(H))$. These properties are used in the statement of Burnside's Normal Complement Theorem (Theorem 6.17).

Problem 5.18 In this problem, you are asked to show that a group G of order 15 is Abelian using the Class Equations; see also Problem 7.21. The method is as follows: Assume the contrary and use Problem 4.16(ii) to show that $Z(G) = \langle e \rangle$, then use the Class Equations (Theorem 5.20) to show that G has exactly one conjugacy class of order 5 consisting of elements of order 3, and then obtain a contradiction.

Problem 5.19 Suppose $G = GL_2(3)$ and so $o(G) = 48$, see Theorem 3.15 and Problem 6.23.

(i)   Find the centre $Z(G)$ and show that $o(Z(G)) = 2$.
(ii) Let H denote the set of upper triangular matrices in G (that is, $H = UT_2(3)$ which has elements of the form $\left(\begin{smallmatrix} a & b \\ 0 & c \end{smallmatrix} \right)$ where $a \neq 0 \neq c$ and we work modulo 3). Show that $Z(G) \leq H \leq G$, and $o(H) = 12$.
(iii) Prove that $\text{core}(H) = Z(G)$ (Problem 2.24).
(iv) Use Theorem 5.15 to prove that $G/Z(G) \simeq S_4$, see Problem 4.4.

Problem 5.20 Suppose $C_G(a) \lhd G$. Use Problem 2.25 to show that a belongs to a normal Abelian subgroup of G. Is the converse false?

Problem 5.21 Suppose $o(G)$ is finite, and $a, b \in G$.

(i)   Show that the number of elements g in G satisfying $g^{-1}ag = b$ equals $o(C_G(a))$.
(ii) Deduce $o(C_G(a)) \geq o(G/G')$ where G' denotes the derived subgroup of G.
(iii) Now suppose $K \lhd G$, $[G : K] = p$ (p a prime), and $c \in K$ with the property: $C_K(c) < C_G(c)$. Show that if b is conjugate to c in G, then b is also conjugate to c in K.

Problem ♦ 5.22 Let $r = h(G)$ be the class number of the finite group G, see page 103. Show that

(i)   If p_0 is the smallest prime dividing $o(G)$ and $rp_0 > o(G)$, then $Z(G) \neq \langle e \rangle$. (Hint. Use the Class Equations.)
(ii) If G is not Abelian, then $r > o(Z(G)) + 1$. (Hint. Use Problem 4.16(ii).)
(iii) If $o(G) = p^3$ and G is not Abelian, then $G' = Z(G)$, $o(Z(G)) = p$ and $r = p^2 + p - 1$. (Hint. Apply Problem 4.6 and Lemma 2.31, and show that no conjugacy class has order p^2 using Problem 5.21.)

Problem 5.23 (i) Show that elements of the same conjugacy class have conjugate centralisers.

(ii) If $n_1, \dots, n_r$ is a list of the orders of the centralisers of elements of distinct conjugacy classes of G, prove that $n_1^{-1} + \cdots + n_r^{-1} = 1$.

(iii) Deduce there are only finitely many groups with class number r using the fact that there are only finitely many ways of writing a positive integer as a sum of reciprocals, see Problem B.4.

(iv) Find all groups with class number 3.

Problem 5.24 Suppose G is finite and $H \leq G$. Show that

$$o\left(\bigcup_{g \in G} g^{-1} H g\right) \leq 1 + o(G) - [G : H].$$

Use this result to show that if $H \leq G$ and H contains at least one element of each conjugacy class of G, then $H = G$.

Problem 5.25 Use the following method ((a) to (d)) to show that if $K \triangleleft A_n$, $n > 4$, and K contains a 3-cycle, then $K = A_n$. With Theorem 3.14 this provides a new proof of the simplicity of A_n when $n > 4$. Throughout suppose σ is a 3-cycle in K.

(a) Show that $C_{S_n}(\sigma) > C_{A_n}(\sigma)$.
(b) Secondly, show that $C_{A_n}(\sigma) = C_{S_n}(\sigma) \cap A_n$.
(c) Using Theorem 5.8 and (b), deduce $[C_{S_n}(\sigma) : C_{A_n}(\sigma)] = 2$.
(d) Lastly, show that $C\ell_{S_n}\{\sigma\} = C\ell_{A_n}\{\sigma\}$, and use Theorem 3.6.

Problem 5.26 (Project—Centralisers and Normalisers of Groups of Order 24) Let $G_1 = S_4$ and $G_2 = SL_2(3)$, see Chapters 3 and 8. First, calculate the centralisers of each of the elements of these groups. Second, calculate the centralisers and normalisers of each of the subgroups. Also check that the properties (i) to (iv) given on page 104 apply. Two major theorems in the theory are connected to the relationship between subgroup centralisers and normalisers: the N/C-theorem (Theorem 5.26) and Burnside's Normal Complement Theorem (Theorem 6.17). Check that these results apply to the groups G_1 and G_2, and their centralisers and normalisers.

Chapter 6
p-Groups and Sylow Theory

There are important connections between number theory and finite group theory, results in one theory have vital applications in the other, and *both* theories benefit from this interaction. Lagrange's Theorem shows that a major invariant of a finite group is its order and the prime factorisation of the order. We shall develop this aspect of the theory here; the number-theoretic results we use are discussed in Appendix B. Elements of order two play a central role, they are called *involutions*. One reason is that an involution is its own inverse; but this is not the only special property. For example, all groups with exponent 2 ($a^2 = e$ for all a in the group) are Abelian (Corollary 2.20), and in Problem 2.28 we showed that a non-Abelian simple group is generated by its involutions. But perhaps the most remarkable fact concerning the number two in the theory is the theorem of Feit and Thompson (1963) which states that no non-cyclic group of odd order can be *simple*. In fact, more is true—a group of odd order must be 'soluble', see Chapter 11.

First in this chapter, we consider groups whose orders are powers of a particular prime number *p*—the *p-groups*. These groups have a number of useful properties, for example, they have normal subgroups for all orders dividing the group order. Also the number of groups of order p^n increases exponentially with n, some details are given on page 118.

We have noted previously that the converse of Lagrange's Theorem is not true in general (page 101). But it is true for prime powers, that is, for subgroups which are *p*-groups. This is the beginning of the Sylow theory which we develop in the second section of the chapter; it is central to finite group theory. We shall give a number of applications. In Section 6.3, we show that if $o(G)$ has up to three not necessarily distinct prime factors, then G is not simple and it can be determined completely; we also prove a remarkable result due to Frattini—the Frattini argument—and introduce 'nilpotent groups', which have many properties in common with *p*-groups including being 'reverse Lagrange'. We give some more applications in Web Section 6.5, these will include a proof of Burnside's Normal Complement Theorem (Theorem 6.17) and a description of those groups all of whose Sylow subgroups are cyclic, see page 130.

Throughout this chapter *p* denotes a prime number.

H.E. Rose, *A Course on Finite Groups*,
Universitext,
DOI 10.1007/978-1-84882-889-6_6, © Springer-Verlag London Limited 2009

6.1 Finite p-Groups

In this section, we develop the basic properties of finite p-groups, and begin with

Definition 6.1 Let p be a fixed prime. A group G is called a *p-group* if all of its elements have orders which are powers of p.

The neutral group $\langle e \rangle$ is a p-group for all primes p as $p^0 = 1$. We shall give some more examples at the end of this section.

In the finite case, Definition 6.1 can be replaced by

$$G \text{ is a finite } p\text{-group} \quad \text{if and only if} \quad o(G) \text{ is a power of } p.$$

This follows from Cauchy's Theorem (Theorem 6.2) for which we give three proofs. The first is due to J. McKay and uses a simple action argument, the second uses the Class Equations (Theorem 5.20) but it has the disadvantage that the Abelian case must be treated separately. The result also follows directly from the First Sylow Theorem (Theorem 6.7). Cauchy proved this result for permutation groups in 1845 as part of a series of papers on the properties of permutations; see Section 3.1. In all probability, the first proof for general groups was given by Jordan in the 1870s.

Theorem 6.2 (Cauchy's Theorem) *If G is a finite group, p is a prime, and $p \mid o(G)$, then G contains at least one element of order p.*

In fact, G contains at least $p - 1$ elements of order p. In the case $p = 2$, there may be a unique element of order 2, for instance, in the quaternion (dicyclic) group Q_2 (page 119) or the special linear group $SL_2(3)$ (page 172). Note that Cauchy's Theorem applies to all finite groups.

Proof We begin by introducing a new action. Let

$$X = \{(g_1, \ldots, g_p) : g_i \in G, \text{ for } i = 1, \ldots, p, \text{ and } g_1 \cdots g_p = e\},$$

that is, X is the set of all *ordered* p-tuples of elements of G whose product is the neutral element. We prove the theorem by applying an action of the group $\mathbb{Z}/p\mathbb{Z}$ to X, and counting orbits. Note first

$$o(X) = o(G)^{p-1}. \tag{6.1}$$

For arbitrary $g_1, \ldots, g_{p-1}$ in G, if we take $g_p = g_{p-1}^{-1} \cdots g_1^{-1}$, then $g_p \in G$, the p-tuple $(g_1, \ldots, g_p)$ belongs to X, and g_p is unique by Theorem 2.5.

We define an action of $\mathbb{Z}/p\mathbb{Z}$ (the integers with addition modulo p) on X as follows. If $(g_1, \ldots, g_p) \in X$ and $a \in \mathbb{Z}/p\mathbb{Z}$, then

$$(g_1, \ldots, g_p)\backslash a = (g_{a+1}, \ldots, g_p, g_1, \ldots, g_a).$$

The action of a on an element of X cyclically permutes its entries by a places modulo p. The set X is closed under this action because, if $(g_1, \ldots, g_p) \in X$ (and so $g_1 \cdots g_p = e$), then by associativity

$$g_{a+1} \cdots g_p g_1 \cdots g_a = (g_1 \cdots g_a)^{-1}(g_1 \cdots g_a)(g_{a+1} \cdots g_p)(g_1 \cdots g_a) = e.$$

Also it is easily seen that the action axioms (5.1) are satisfied.

By the Orbit–Stabiliser Theorem (Theorem 5.7), an orbit of this action has order 1 or p (the divisors of $o(\mathbb{Z}/p\mathbb{Z})$). Suppose there are r orbits of order 1, and s orbits of order p. An element of an orbit of order 1 has the form $(g, \ldots, g)$ for some $g \in G$ with $g^p = e$ (as all cyclic permutations give the same p-tuple). Now $r > 0$ because there is at least one orbit of this type, namely the orbit of the p-tuple $(e, \ldots, e)$. If this was the only orbit of order 1, then $r = 1$, and by (6.1),

$$1 + sp = o(X) = o(G)^{p-1},$$

as X is the disjoint union of its orbits. But this is impossible because $p \mid o(G)$ by hypothesis, and so p divides the right-hand side of the equation above whilst it clearly does not divide the left-hand side. Therefore, $r > 1$ (in fact, r is a positive multiple of p), and so there must exist at least one $g \in G$ satisfying $g \neq e$ and $o(g) = p$. □

We give a second proof of this theorem using the Class Equation. As noted above the Abelian case must be established first, see Problem 4.15.

Second proof—Non-Abelian case We use induction on $o(G)$. Choose $a \in G$ such that $a \notin Z(G)$, it exists because G is not Abelian. This implies that $o(\mathcal{C}\ell_G\{a\}) > 1$, and so by Theorem 5.19 $[G : C_G(a)] > 1$, that is, $C_G(a)$ is a proper subgroup of G. If $p \mid o(C_G(a))$, then the theorem follows by induction because $o(C_G(a)) < o(G)$ and so by the inductive hypothesis $C_G(a)$ has an element of order p. Hence we may suppose $p \nmid o(C_G(a))$ for all $a \in G\backslash Z(G)$. But as $p \mid o(G)$ by hypothesis, this implies

$$p \mid [G : C_G(a)],$$

for all $a \in G\backslash Z(G)$ by Lagrange's Theorem (Theorem 2.27). Applying this to the second Class Equation (Theorem 5.20) gives $p \mid o(Z(G))$. But $Z(G)$ is Abelian, and so by Problem 4.15, $Z(G)$ contains an element of order p, and hence so does G. □

We can now show that our two definitions of a finite p-group are equivalent.

Theorem 6.3 *If G is finite and p is prime, then G is a p-group if and only if $o(G) = p^r$ for some non-negative integer r.*

Proof There is nothing to prove if $r = 0$. Suppose G is a p-group (Definition 6.1) of order $q^t n$ where q is a prime which does not divide n, $t > 0$, and $q \neq p$. By Theorem 6.2, G contains an element of order q, which contradicts the hypothesis. Therefore, $q = p$ and $n = 1$. The converse follows from Lagrange's Theorem and Definition 2.19. $\square$

Next we show that the property of being a p-group is preserved by taking both subgroups and factor groups, *and* vice versa. Very few properties are preserved in this way. Abelianness is not one, but it is true for finiteness and, as we shall show later, it is also true for solubility (Chapter 11).

Theorem 6.4 (i) *Subgroups and factor groups of p-groups are p-groups.*
(ii) *If $K \lhd G$, and K and G/K are both p-groups, then G is also a p-group.*

Proof (i) The first part follows from the definition. For the second part, suppose $g \in G$ and $o(g) = p^r$, then using coset multiplication we have $K = eK = g^{p^r} K = (gK)^{p^r}$. Hence the order of gK is a divisor of p^r, and so the order of every element of G/K is a power of p.

(ii) If $g \in G$, then $(gK)^{p^r} = K$ for some integer r by the second hypothesis, hence $g^{p^r} \in K$. But K is a p-group (the first hypothesis), and so $o(g^{p^r}) = p^s$ for some non-negative integer s, so the order of g is a divisor of p^{r+s}. The result follows as this holds for all $g \in G$. $\square$

The p-groups have many important properties, for example, they are reverse Lagrange. In fact, the following stronger result follows from Lemma 5.21.

Theorem 6.5 *If G is a group with order p^r, then G is a finite p-group (by Theorem 6.3) and it has subgroups $G_0, \ldots, G_r$ satisfying*

(a) $G_0 = \langle e \rangle$, $G_r = G$;
(b) *for $i = 1, \ldots, r$, $G_i \lhd G$ and $G_{i-1} \lhd G_i$;*
(c) *for $i = 0, \ldots, r$, $o(G_i) = p^i$.*

Proof The proof is by induction on r. If $r = 1$ there is nothing to prove, and if $r = 2$ the result follows from Lemma 5.21 and Theorem 5.22, for then G is Abelian. Hence we may assume that the result holds for all p-groups of order less than p^r with $r > 1$. By Lemma 5.21 again, $Z(G) \neq \langle e \rangle$, and so $Z(G)$ contains an element y which satisfies $o(y) = p^t$ for some positive t, and if we put $z = y^{p^{t-1}}$, then $z \in Z(G)$ and $o(z) = p$. Set

$$G_1 = \langle z \rangle,$$

we have $G_1 \lhd G$ because $G_1 \leq Z(G) \lhd G$, see Problem 2.14(ii).

Let $H = G/G_1$, by Lagrange's Theorem (Theorem 2.27), $o(H) = p^{r-1}$. The inductive hypothesis provides a sequence of subgroups $H_0, \ldots, H_{r-1}$ which satisfies (a), (b) and (c) with $r - 1$ for r, H for G, and H_{i-1} for G_i.

Let θ be the natural homomorphism $G \to H \simeq G/G_1$, then the Correspondence Theorem (Theorem 4.16(iii) and (iv)) gives $G_i = H_{i-1}\theta^{-1}$ which are subgroups of G containing G_1, for $i = 1, \ldots, r-1$. Applying this theorem again for $i = 0, \ldots, r-1$, we have

$$G_{i+1} \lhd G \quad (\text{as } H_i \lhd H) \quad \text{and} \quad G_i \lhd G_{i+1} \quad (\text{as } H_{i-1} \lhd H_i).$$

Parts (a) and (b) follow, and (c) follows by Lagrange's Theorem. $\square$

The groups G_i in Theorem 6.5 are in many cases not unique. If r_i denotes the number of subgroups of G with order p^i, both normal and non-normal, then $r_i \equiv 1 \pmod{p}$. We shall prove this fact for 'Sylow' subgroups in the next section, and the general result is given in Rotman (1994).

There is an extension to Theorem 6.5 which applies when K is a maximal subgroup of G. In this case, K has prime index and is normal in G (*cf.* Problem 2.19(i)). We derive these properties now.

Theorem 6.6 *Suppose G is a finite p-group, and K is a maximal subgroup.*

(i) $K \lhd G$.
(ii) $[G : K] = p$.

Proof (i) By induction on r where $o(G) = p^r$, the result clearly holds when $r = 1$. By Lemma 4.14(ii), $K \leq KZ(G) \leq G$ (the second inequality holds as $Z(G) \lhd G$), and so by the maximality of K we have

$$KZ(G) = G \quad \text{or} \quad KZ(G) = K,$$

both are possible. If $KZ(G) = G$, there exists $g \in Z(G) \backslash K$, but $g \in N_G(K)$ (because $g \in Z(G)$); therefore, K is a proper subgroup of $N_G(K)$ which implies, by the maximality of K, that $N_G(K) = G$. In turn, this implies that $K \lhd G$ by Lemma 5.25(ii).

The second possibility is $KZ(G) = K$, then $Z(G) \lhd K$ by Problem 2.14 and Lemma 2.31. As K is a maximal subgroup of G, we also have $K/Z(G)$ is a maximal subgroup of $G/Z(G)$ by Problem 4.13(ii). By Lemma 5.21 and as G is a *p*-group, we have $o(Z(G)) > 1$, so

$$o\big(G/Z(G)\big) < o(G).$$

Using the inductive hypothesis this gives $K/Z(G) \lhd G/Z(G)$, and the Correspondence Theorem shows that $K \lhd G$. Hence in both cases $K \lhd G$.

(ii) This follows easily from (i) and Theorem 6.5, see Problem 6.1. $\square$

Theorems 10.6 and 10.7 on pages 213 and 214 give a second proof of (i).

A large number of p-groups exist with order p^r if r is large, many of which differ only slightly from one another. If $v(n)$ denotes the number of (isomorphism classes of) groups of order n, then the following data has been established for powers of 2 (see Besche *et al.* 2001):

$$v(4) = 2, \quad v(8) = 5, \quad v(16) = 14, \quad v(32) = 51,$$
$$v(64) = 267, \quad v(128) = 2328, \quad v(256) = 56092,$$
$$v(512) = 10494213 \quad \text{and} \quad v(1024) = 49487365422;$$

the second of these equations is established below and in Chapter 7. No exact formula is known for $v(n)$ in general, but the following estimate has been given by Higman and Sims:

$$v(p^n) = p^{2n^3/27} + O\left(n^{8/3}\right).$$

On the other hand, if n has a large number of prime factors and is square-free, then $v(n)$ can be quite small. For all integers m, there are infinitely many integers n, with m prime factors, that satisfy $v(n) = 1$; this is a corollary of Dirichlet's Theorem on Primes in Arithmetic Progressions; see, for instance, Rose (1999). As examples we have $v(15) = 1$, $v(105) = v(3 \cdot 5 \cdot 7) = 1$, and $v(5865) = v(3 \cdot 5 \cdot 17 \cdot 23) = 1$; see the table in Appendix D.

Non-Abelian Groups of Order 8

The smallest non-Abelian p-groups have order 8 (Problem 2.20), we consider these now. Groups of order p^3, $p > 2$, can be treated similarly (Problem 6.5), and Abelian p-groups will be discussed in Chapter 7. Let G be a non-Abelian group of order 8. If G contains an element of order 8, then G is cyclic, also if every non-neutral element has order 2, then G is Abelian (Corollary 2.20), hence we may assume that G contains an element a, say, of order 4 and, by Problem 2.19(i), $\langle a \rangle \lhd G$ (this also follows from Theorem 6.6). Choose $b \in G \backslash \langle a \rangle$, then the elements of G are

$$e, a, a^2, a^3, b, ab, a^2b, \text{ and } a^3b.$$

The reader should check that no two of these elements are equal—for example, if $a = ab$ then by cancellation $b = e$. Now as $b^2 \in G$, it follows that b^2 equals one of the eight elements of G listed above. If $b^2 = a^rb$, then $b = a^r$ but by definition $b \notin \langle a \rangle$, and if $b^2 = a$ or a^3, then $o(b) = 8$, and G is cyclic. Hence as we are considering the non-Abelian case, $b^2 = e$ or a^2; both are possible.

Also, as $\langle a \rangle \lhd G$ we have

$$b^{-1}ab = a^s \in \langle a \rangle,$$

for $s = 0, 1, 2$, or 3. If $s = 0$ then $a = e$, if $s = 1$ then G is Abelian, and if $s = 2$ then $e = a^4 = (b^{-1}ab)^2 = b^{-1}a^2b$, which implies that $a^2 = e$. Hence there is only one possibility: $s = 3$, that is, $b^{-1}ab = a^3$. This gives $ba^3 = ab$, and so $ba^t = a^{4-t}b$ for

$t = 1, 2, 3$. These calculations show that there are at most two (isomorphism classes of) non-Abelian groups of order 8:

$$D_4 = \langle a, b \mid a^4 = b^2 = e, b^{-1}ab = a^3 \rangle,$$

$$Q_2 = \langle a, b \mid a^4 = e, b^2 = a^2, b^{-1}ab = a^3 \rangle,$$

see pages 58 and 59. The first group, which is isomorphic to the dihedral group of the square, has five elements of order 2. The second is the *quaternion group* (sometimes called the first dicyclic group), it contains one element of order 1, one element of order 2, and six elements of order 4. As these groups have different numbers of elements of order 2 they are not isomorphic, and so there are exactly two isomorphism classes of non-Abelian groups of order 8. The group Q_2 also has one property that it shares with very few others: it is not Abelian *and* all of its subgroups are normal (groups with this property are called *Hamiltonian*), see Problem 7.13.

Historical note In the middle of the nineteenth century, the Irish mathematician W.R. Hamilton (1805–1865) was looking for fields which extend the complex numbers, he discovered the quaternions which have some remarkable properties but do not (quite) form a field. They can be defined in a similar manner to the complex numbers (that is, as a vector space, now of dimension 4, over $\mathbb{R}$ with a vector multiplication) except that the number i is replaced by three entities i, j and k which satisfy

$$i^2 = j^2 = k^2 = -1 \quad \text{and} \quad ij = k = -ji,$$

and a general quaternion is an entity of the form $x + iy + zj + tk$ where $x, y, z, t \in \mathbb{R}$. All field axioms are satisfied except one, commutativity of multiplication fails as can be seen above (a system of this type is known as a *division algebra*). The *octonions* which are sometimes called *Cayley numbers* extend this construction to dimension 8, but in this case both commutativity and associativity fail; some brief details are given in Rose (2002), page 176. The ring of integral quaternions has a similar definition to that for the quaternions given above except now $x, y, z, t \in \mathbb{Z}$. The quaternion group Q_2 forms the group of units (divisors of 1) of this ring when we map $a \to i$ and $b \to j$, for then $ab \to ij$ where $ij = k$.

6.2 Sylow Theory

We have seen earlier that the converse of Lagrange's Theorem is false—if $m \mid o(G)$ it does not follow that G has a subgroup of order m. For instance, A_5 has no subgroup of order 15, see page 101. But there is a partial converse if we restrict m to be a prime power—if $p^r \mid o(G)$, then G does have a subgroup of order p^r, and this is the starting point for the Sylow theory. As for Cauchy's Theorem, we shall give a number of proofs of this important (existence) result. The first uses a new action and is due to Wielandt, and the second uses the Class Equations but as with the second proof of Cauchy's result, it has the disadvantage that the Abelian case must be dealt with separately. The third proof uses some matrix theory.

We begin with the main existence theorem first proved by the Norwegian mathematician Ludwig Sylow (1832–1918) in 1872.

Theorem 6.7 (Sylow Theorem, Part 1) *If G is finite group with order $p^r m$, where p is prime and $p \nmid m$, then G contains a subgroup of order p^r.*

Proof There is nothing to prove if $r = 0$, so we may assume that $r > 0$. The following fact concerning binomial coefficients will be used:

$$\text{The binomial coefficient } \binom{p^r m}{p^r} \text{ is not divisible by } p. \qquad (6.2)$$

(The numerator of $\binom{p^r m}{p^r}$ has the following factors (p^r in total):

$$(p^r m)(p^r m - 1) \cdots (p(p^{r-1}m - 1))(p^r m - (p+1)) \cdots (p^2(p^{r-2}m - 1))$$
$$\cdots (p(p^{r-1}m - (p^{r-1} - 1))) \cdots (p^r m - (p^r - 1))$$

and when this expression is divided by the denominator of the binomial coefficient, that is $(p^r)!$, all factors of the form p^i are removed by cancellation.)

We define a new action. Let $\mathcal{X}$ denote the set of all *unordered subsets with no repetitions* of the underlying set of G which contain exactly p^r elements. As there are $\binom{p^r m}{p^r}$ ways of choosing these subsets, we have by (6.2)

$$p \nmid o(\mathcal{X}). \qquad (6.3)$$

The action of G on $\mathcal{X}$ is defined by right multiplication: If $X_1 \in \mathcal{X}$ then

$$X_1 \backslash g = \{xg : x \in X_1\} \quad \text{for } g \in G.$$

Using cancellation we have $o(X_1 \backslash g) = o(X_1) = p^r$, and so $X_1 \backslash g \in \mathcal{X}$, for all $g \in G$. The action axioms (5.1) follow. Let $\mathcal{X}_1, \mathcal{X}_2, \ldots$ denote the orbits of this action, as $\mathcal{X}$ is a disjoint union of its orbits, we see by (6.3) that there is at least one orbit $\mathcal{X}_j$, say, with the property

$$p \nmid o(\mathcal{X}_j). \qquad (6.4)$$

Let $Y \in \mathcal{X}_j$ where $e \in Y$ (if $Y = \{y_1, \ldots\} \in \mathcal{X}_j$, then $Y \backslash y_1^{-1} = \{e, \ldots\}$ is in the same orbit as Y by definition), note that $o(Y) = p^r$. Now J is defined by $J = \text{stab}_G(Y)$. By Lemma 5.6, $J \leq G$, hence if we can show that $o(J) = p^r$, the result will follow. We do this by proving both $p^r \leq o(J)$, and $p^r \geq o(J)$. First, by the Orbit–Stabiliser Theorem (Theorem 5.7), we have

$$o(\mathcal{X}_j) = [G : \text{stab}_G(Y)] = [G : J] = p^r m / o(J),$$

using Lagrange's Theorem for the final equation. Now the last entity in this sequence of equations is an integer not divisible by p (by (6.4)), and so

$$p^r \mid o(J), \quad \text{which implies } p^r \leq o(J). \tag{6.5}$$

For the reverse, note that $J \subseteq YJ = Y$ (as $e \in Y$ and J is the stabiliser of Y). Hence $o(J) \leq o(Y) = p^r$, which with (6.5) proves the theorem. $\qquad \square$

Example We illustrate the above construction with the group $D_3 \simeq \langle a, b \mid a^3 = b^2 = (ab)^2 = e \rangle$ which has order 6, and we look for a subgroup of order 3. Here $p^r = 3$, $m = 2$ and so the binomial coefficient $\binom{6}{3} = 20$, and $3 \nmid 20$. Hence in this example $\mathcal{X}$ has 20 elements—the underlying set of D_3 has 20 unordered 3-element subsets, and using the action defined above these split into the four orbits given by:

$$\{e, a, a^2\}, \{b, ab, a^2 b\};$$

$$\{e, a, b\}, \{e, a^2, ab\}, \{e, b, ab\}, \{a, a^2, a^2 b\}, \{a, b, a^2 b\}, \{a^2, ab, a^2 b\};$$

$$\{e, a, ab\}, \{e, a^2, a^2 b\}, \{e, ab, a^2 b\}, \{a, a^2, b\}, \{a, b, ab\}, \{a^2, b, a^2 b\};$$

$$\{e, a, a^2 b\}, \{e, a^2, b\}, \{e, b, a^2 b\}, \{a, a^2, ab\}, \{a, ab, a^2 b\}, \{a^2, b, ab\}.$$

The first orbit has an order not divisible by 3, and we can take its triple which contains e, that is, $\{e, a, a^2\}$ for Y and J. Now this stabiliser provides the subgroup of order 3 that we were looking for. Reader, find a subgroup of order 2.

Second proof of Theorem 6.7 We use induction on $o(G) = p^r m$, there is nothing to prove if $o(G) = 1$. We treat the cases of G non-Abelian, and G Abelian, separately. The first of these has two subcases.

Subcase 1.1 $Z(G) < G$ and $p \mid o(Z(G))$.

By Cauchy's Theorem (Theorem 6.2) and as p divides $o(Z(G))$, $Z(G)$ contains an element of order p, and so it contains a subgroup H of order p. By Problem 2.14(ii), $H \lhd G$, hence we can form the factor group G/H which has order $p^{r-1}m$. By the inductive hypothesis, this factor group has a subgroup J/H of order p^{r-1}, and the Correspondence Theorem now shows that $J < G$ and $o(J) = p^r$ as required.

Subcase 1.2 $Z(G) < G$ and $p \nmid o(Z(G))$.

We use the second Class Equation (Theorem 5.20). As G is not Abelian, there is at least one $y_i \in G$ whose conjugacy class has order larger than 1, that is at least one term of the form $[G : C_G(y_i)]$ occurs in the Class Equation. Now p divides $o(G)$ but not $o(Z(G))$ in this subcase. Hence there must exist at least one y_i with the properties (for the inequality use Theorem 5.19)

$$p \nmid [G : C_G(y_i)] = o(G)/o\big(C_G(y_i)\big) \quad \text{and} \quad o\big(C_G(y_i)\big) < o(G).$$

From this it follows that $o(C_G(y_i)) = p^r n$ for some n satisfying $1 \le n < m$. The inductive hypothesis implies that $C_G(y_i)$ has a subgroup L of order p^r, and $L < G$ by Corollary 2.14 as required.

Case 2 G is Abelian.

We can argue as in the first subcase. By Cauchy's Theorem and as G is Abelian, G contains a normal subgroup K of order p and $o(G/K) = p^{r-1}m$. By the inductive hypothesis, G/K has a subgroup L/K of order p^{r-1}, and the Correspondence Theorem shows that $L < G$ and $o(L) = p^r$. Note that this case also follows directly from Lemma 7.13. These three cases complete the proof.   □

The *third proof* is as follows. By Problem 4.17, G is isomorphic to a subgroup of $GL_n(p)$ for some suitably chosen n. Also, by Problem 3.15(iii), we have $IT_n(p)$ is a Sylow p-subgroup of $GL_n(p)$. Reader, check that $o(IT_n(p)) = p^{n(n-1)/2}$ (there are p choices for each entry above the main diagonal) and use Theorem 3.15. The result follows by Problem 6.11(iv) if in this problem we put $G = GL_n(p)$, $H = IT_n(p)$ and $J = G$.

For our first application of this result, we reprove Cauchy's Theorem (Theorem 6.2), note that we did not use Cauchy's result in the first (or third) proof of Theorem 6.7. We have if $o(G) = p^r m$ where $p \nmid m$ and $r > 0$, then by the First Sylow Theorem, G has a subgroup P of order p^r, and by Lagrange's Theorem (Theorem 2.27), all elements of P have order a power of p. If $h \in P$ and $o(h) = p^s$, then $h^{p^{s-1}}$ has order p, as required by Cauchy's result.

We shall discuss some more applications and examples below, but first we make

Definition 6.8 Let $o(G) = p^r m$ where p is prime and $p \nmid m$. A subgroup of G with order p^r is called a *Sylow p-subgroup* of G. A *Sylow subgroup* is a Sylow p-subgroup for some unspecified prime p.

Note that a Sylow subgroup is only specified by its order. By Theorem 6.7, for every prime p, G has a Sylow p-subgroup P; if $p \nmid o(G)$ then $P = \langle e \rangle$. Also if $m = 1$, that is, G is a p-group, then G has only one Sylow p-subgroup—itself. We shall use the symbols P or Q for Sylow subgroups. If $o(G) = p^r m$ with $p \nmid m$ and $r > 0$ as above, then by Theorem 6.7, G always has at least one Sylow p-subgroup, and so by Theorem 6.5, G also has subgroup(s) of orders p^s, for all s satisfying $1 \le s \le r$. In the next theorem, we prove the converse, that is, every p-subgroup of G is contained in a Sylow p-subgroup of G. This theorem and its successor also provide more information and a number of further properties.

Examples (a) If $G = C_n$ where $n = p_1^{r_1} \cdots p_k^{r_k}$, then G has a unique normal (see Theorem 6.10(iii) below) Sylow p_i-subgroup of order $p_i^{r_i}$, for each $i = 1, \ldots, k$, as all subgroups are normal in a cyclic group. This follows from Theorem 4.20.

(b) Consider the group A_5 which has order $60 = 2^2 \cdot 3 \cdot 5$. Theorem 6.7 implies that A_5 has Sylow subgroups of orders 3, 4 and 5. In fact, A_5 has ten subgroups of order 3, five of order 4, and six of order 5 (Problems 3.3 and 6.13). But note that A_5 has no subgroups of order 15, 20 or 30, see the example on page 101.

This second example shows that in the general finite case there is no extension of Sylow's First Theorem to non-prime power divisors of the group order. There is an extension if the group is 'soluble', see Section 11.2.

We come now to the remaining Sylow results. Let n_p denote the number of Sylow p-subgroups of G; by Theorem 6.7, $n_p \geq 1$. The first of these theorems below gives the main properties including: For a fixed prime p, all Sylow p-subgroups of a group G are conjugate in G, and so in particular they are isomorphic.

Theorem 6.9 (Sylow Theorem, Part 2) *Let G be a finite group.*

(i)   *Each p-subgroup of G is a subgroup of some Sylow p-subgroup of G.*
(ii) *All Sylow p-subgroups of G are conjugate in G.*
(iii) $n_p \equiv 1 \pmod{p}$.

Proof We use a similar method for all three parts of this theorem. Let P be a Sylow p-subgroup of G and let $\mathcal{X}$ denote the set of subgroups conjugate to P in G, that is, $\mathcal{X} = \{g^{-1}Pg : g \in G\}$. Further, let R be some p-subgroup of G and suppose R acts on $\mathcal{X}$ by conjugation:

$$g^{-1}Pg \backslash x = x^{-1}(g^{-1}Pg)x = (gx)^{-1}P(gx) \quad \text{for } g \in G \text{ and } x \in R. \quad (6.6)$$

As R is a p-group, the Orbit–Stabiliser Theorem (Theorem 5.7) shows that the order of an orbit of this action is a power of p, possibly $p^0 = 1$. Now

$$o(\mathcal{X}) = [G : N_G(P)] \quad \text{and} \quad p \nmid [G : N_G(P)].$$

The equation here is given by Lemma 5.25(iii), and the non-divisibility property follows because P is a Sylow p-subgroup of both G and $N_G(P)$ (they have the same (maximal) p-power order). Hence $p \nmid o(\mathcal{X})$ and so, as $\mathcal{X}$ is a disjoint union of its orbits (all of order a power of p), there must be at least one orbit of order $1 (=p^0)$.

Suppose $\{P_1\}$ is an orbit of the action (6.6) with order 1. Then $x^{-1}P_1x = P_1$ for all $x \in R$, so $P_1R = RP_1$, and $RP_1 \leq G$ (Problem 2.19(v)). But $P_1 \leq RP_1$, hence $o(P_1) \leq o(RP_1)$. Applying Theorem 5.8 we have

$$o(RP_1)o(R \cap P_1) = o(R)o(P_1) \quad \text{or} \quad o(RP_1) = o(P_1)[R : R \cap P_1],$$

by Lagrange's Theorem. As both P_1 and R are p-groups (by definition and Problem 2.23), it follows that $o(RP_1)$ is a power of p, and so RP_1 is a p-subgroup of G. As P_1 is a Sylow p-subgroup of G, and so its order is the maximum possible power of p. Hence

$$o(RP_1) = o(P_1) \quad \text{so} \quad RP_1 = P_1, \quad \text{which gives} \quad R \leq P_1.$$

This proves (i) because R is an arbitrary p-subgroup of G.

(ii) If R in the argument above is a Sylow p-subgroup of G, then $o(R) = o(P_1)$, so $R = P_1$ where P_1 is conjugate to P in G by the action defined above. This proves (ii), and it also shows that $o(\mathcal{X}) = n_p$.

(iii) Suppose there is a second single-element orbit of the action (6.6), say $\{P_2\}$. Then, taking $R = P_2$, the first argument gives $P_1 = P_2$, that is there can only be one single-element orbit. As all other orbits have orders which are positive powers of p, (iii) follows because (by (ii))

$$o(\mathcal{X}) = n_p \equiv 1 \pmod{p}. \qquad \square$$

Notes Proposition (ii) in Theorem 6.9 is important for it implies that once we have found one Sylow p-subgroup P of a group, then all remaining Sylow p-subgroups will be isomorphic to P. This is not true for p-subgroups of smaller orders. For example, the group S_4 (note $o(S_4) = 2^3 \cdot 3$) has three Sylow 2-subgroups of order 8 each isomorphic to D_4, but it has subgroups of order 4 isomorphic to both C_4 and T_2; see Chapter 8. Proposition (iii) implies that $n_p \neq 0$, and so Sylow p-subgroups exist, but we cannot use this to reprove the main theorem (Theorem 6.7) because we began the above proof by assuming the existence of a Sylow p-subgroup P. Proposition (i) holds for infinite groups, but note that there exist infinite groups (a) with non-isomorphic Sylow p-subgroups, and (b) with isomorphic Sylow p-subgroups which are not conjugate; see, for example, Suzuki (1982), page 191.

We collect together the remaining Sylow properties in the next theorem, they all follow easily from the results above.

Theorem 6.10 (Sylow Theorem, Part 3) *Suppose $o(G) = p^r m$ where $p \nmid m$, and P is a Sylow p-subgroup of G.*

(i)   *P is a normal Sylow p-subgroup of $N_G(P)$.*
(ii)   *$n_p = [G : N_G(P)]$.*
(iii) *$P \lhd G$ if and only if P is the unique Sylow p-subgroup of G, and $P \lhd G$ implies P char G, see Problem 4.22.*
(iv)   *$n_p \mid m$.*

Proof (i) By Lemma 5.25(i), $P \lhd N_G(P)$. But P is also a Sylow p-subgroup of $N_G(P)$ as the orders of both groups have the same exponent of p in their prime factorisations, this gives (i).

(ii) This follows directly from Theorem 6.9(ii) and Lemma 5.25(iii), or from the Orbit–Stabiliser Theorem (Theorem 5.7).

(iii) If P is the unique Sylow p-subgroup of G, then by Theorem 6.9, all conjugates of P equal P, that is $P \lhd G$. Conversely, if $P \lhd G$ then $N_G(P) = G$, and the result follows from (ii) and Theorem 6.9(ii). The second part is left as an exercise.

(iv) We have

$$m = [G : P] = [G : N_G(P)][N_G(P) : P],$$

by Lagrange's Theorem and Problem 2.15, now use (ii). $\qquad \square$

We now restate the five main Sylow results which we shall often refer to as "Sylow 1", ..., and "Sylow 5" (the numbering is not standard).

Five Main Sylow Theorems *Suppose $o(G) = p^r m$ where p is prime, and $p \nmid m$.*

1. *G has at least one subgroup of order p^r called a Sylow p-subgroup.*
2. *All Sylow p-subgroups of G are conjugate in G.*
3. *A Sylow p-subgroup P is unique if and only if $P \lhd G$.*
4. *If n_p denotes the number of Sylow p-subgroups of G, then*

$$n_p \geq 1, \quad n_p \mid m \quad \text{and} \quad n_p \equiv 1 \ (\text{mod } p).$$

5. *Each p-subgroup of G is contained in a Sylow p-subgroup of G.*

Example We calculate the Sylow subgroups of S_5; some more examples are given in Problem 6.12 and Chapter 8.

We note first that $o(S_5) = 5! = 2^3 \cdot 3 \cdot 5$, and so S_5 has Sylow subgroups of order 3, 5 and 8. Further, using Sylow 4 we have

$$n_2 \equiv 1 \ (\text{mod } 2) \quad \text{and} \quad n_2 \mid 15,$$

$$n_3 \equiv 1 \ (\text{mod } 3) \quad \text{and} \quad n_3 \mid 40,$$

$$n_5 \equiv 1 \ (\text{mod } 5) \quad \text{and} \quad n_5 \mid 24.$$

By Problem 3.3 and Theorem 2.29(ii), a non-neutral normal subgroup must have order at least 12, for by Problem 3.3 the conjugacy classes not containing the neutral element have orders between 10 and 24; and so n_2, n_3 and n_5 are all larger than one by Sylow 3. Hence the third congruence above shows that $n_5 = 6$, that is S_5 has six (cyclic) subgroups of order 5. Note that S_5 possesses 24 elements of order 5 and each subgroup isomorphic to C_5 has four elements of order 5 and e; see Problem 2.5(iii). Secondly, the middle congruence above implies that $n_3 = 4, 10$ or 40. By counting the number of elements of order 3 in S_5 (there are 20), we see that $n_3 = 10$ and the group has 10 (cyclic) subgroups of order 3.

For the prime 2 we argue as follows. If we consider the subgroup generated by the permutations $(1, 2, 3, 4)$ and $(1, 3)$, we obtain an isomorphic copy of D_4 in S_5 (for a more detailed discussion of this point, see page 166). We have $o(D_4) = 8$, hence this is a Sylow 2-subgroup and all Sylow 2-subgroups of S_5 are isomorphic to D_4. The first congruence above gives $n_2 = 3, 5$ or 15. The correct value is 15 as we show now. The argument above can also be applied if we replace the two generators by either $(1, 2, 4, 3)$ and $(1, 4)$, or $(1, 3, 2, 4)$ and $(1, 2)$, this gives three copies of D_4 in S_5. Further, there are five ways of choosing four elements from five, and so the above can be repeated another four times giving a total of 15 Sylow 2-subgroups. The reader should list all these Sylow subgroups and also convince him(her)self that every subgroup of order 2 or 4 is contained in a Sylow 2-subgroup (Sylow 5). Note that S_5 has 156 subgroups in 19 conjugacy classes, how many can you find?

6.3 Applications

Here we give a number of applications of the Sylow theory, see the comment at the end of this section and Web Section 6.5.

Groups whose Orders have at most Three Prime Factors

As a first application of the Sylow theorems, we consider groups whose orders have a small number of (not necessarily distinct) prime factors. If $o(G)$ is prime, then G is cyclic and has no non-neutral proper subgroups (Theorems 2.34 and 4.20). We show that if $o(G)$ has two or three prime factors, then G has at least one proper non-neutral normal subgroup, and so is not simple. It can be shown that if $o(G)$ has four prime factors, then G is simple only if $o(G) = 60$ and $G \simeq A_5$ (Section 3.2), and if it has five prime factors then G is simple in a few well-known cases. (They are $L_2(7)$, $L_2(11)$ and $L_2(13)$ with orders 168, 660 and 1092, respectively, see Chapter 12 and the ATLAS 1985.) We begin by considering the two-prime case.

Theorem 6.11 *If $o(G) = pq$, p and q are prime, and $p \le q$, then G has a normal (cyclic) subgroup of order q. If $p \ne q$ it is unique.*

It is also unique if $p = q$ and G is cyclic, but it is not unique if G is elementary Abelian.

Proof If $p = q$ this follows by Theorem 6.5, hence we may suppose $p < q$. By Sylow 1 and 4, G has a subgroup P of order q which is cyclic by Theorem 2.34, and the number of such subgroups, n_q, satisfies

$$n_q \equiv 1 \pmod{q} \quad \text{and} \quad n_q \mid p.$$

These conditions imply $n_q = 1$, so by Sylow 3, $P \lhd G$ and P is unique. $\square$

Corollary 6.12 *If $o(G) = 2p$ and p is an odd prime, then G is isomorphic to either C_{2p} or D_p.*

Proof By Theorem 6.11, G has a normal cyclic subgroup P of order p. Suppose $P = \langle a \rangle$, and so $a^p = e$. By Cauchy's Theorem (Theorem 6.2), G also has an element b of order 2. As $P \lhd G$, we have

$$b^{-1}ab = bab = a^s,$$

for some positive integer s less than p. If r is also a positive integer, then $ba^r b = (bab)(bab) \cdots (bab) = a^{rs}$ (as $b^2 = e$) where there are r copies of bab in the second term. This proposition gives

$$a = b^2 a b^2 = b(bab)b = ba^s b = a^{s^2},$$

and so $s^2 \equiv 1 \pmod{p}$ because a is an element of order p. This congruence has exactly two solutions: 1 and $p - 1$ (Appendix B). If $s = 1$, then G is Abelian, and so cyclic, see Chapter 7. If $s = p - 1$, then G is isomorphic to the pth dihedral group:

$$G \simeq D_p = \langle a, b \mid a^p = b^2 = e, b^{-1}ab = a^{p-1} \rangle. \qquad \square$$

The pq case, where p and q are both odd, is discussed in Problems 3.6, 6.14, 7.21, and 9.14, and in Web Section 14.3. For the three prime case we have

Theorem 6.13 *If $o(G)$ has exactly three prime factors, then G is not simple.*

Proof There are three cases to consider.

Case 1. $o(G) = p^3$, p prime.

In this case, the result follows by Theorem 6.5.

Case 2. $o(G) = p^2q$, p and q prime, and $p \neq q$.

We give two proofs. In the first, we show that either n_p or n_q equals 1 using Sylow 3 and 4. Hence suppose both $n_p > 1$ and $n_q > 1$. As $n_p \mid q$ and q is prime, we have $n_p = q$. But $n_p \equiv 1 \pmod{p}$, and so

$$q > p. \tag{6.7}$$

We also have $n_q \equiv 1 \pmod{q}$ and $n_q \mid p^2$, so (6.7) and our supposition show that $n_q = p^2$. This implies that G has p^2 cyclic subgroups each of order q (as q is prime), and so G has $p^2(q - 1)$ elements of order q, see Problem 2.5(iii). This further implies that G has exactly

$$p^2q - p^2(q - 1) = p^2$$

elements whose orders are not equal to q. But this is impossible because we are assuming that G also has q subgroups each having order p^2, and $q \geq 3$ by (6.7). Hence our supposition is false, either $n_p = 1$ or $n_q = 1$ (or both), and so G contains at least one normal subgroup by Sylow 3.

Second proof—this method has applications elsewhere. If $p^2 < q$ use the same proof as that for Theorem 6.11. So assume that $p^2 > q$ and $n_p > 1$. This implies that G has distinct Sylow p-subgroups P and Q. Let $R = P \cap Q$. By Theorem 5.8, we have $o(R) = p$ (as $o(PQ) \leq p^2q$ and $P \neq Q$). Now R is a p-group and a maximal subgroup of both P and Q. Using Problem 5.13(vi), this implies that

$$P, Q \leq N_G(R), \quad \text{which in turn gives} \quad \langle P, Q \rangle \leq N_G(R).$$

But P is a maximal subgroup of G because it has prime index in G, this further implies that $\langle P, Q \rangle = N_G(R) = G$, and hence $R \triangleleft G$. This is impossible if G is simple, and so this case is established.

All three possibilities occur, for example, with groups of order 12, see Section 7.3. The cyclic group C_{12} has normal subgroups of order 3 and 4 ($n_2 = n_3 = 1$), D_6 has a normal subgroup of order 3 ($n_2 = 3$ and $n_3 = 1$), and A_4 has a normal subgroup of order 4 ($n_2 = 1$ and $n_3 = 4$); see Problem 6.13(i).

Case 3. $o(G) = pqr$, p, q and r are prime, and $p < q < r$.

As above suppose n_p, n_q and n_r are all larger than one. By Sylow 4, $n_r \equiv 1 \pmod{r}$ and $n_r \mid pq$, hence $n_r = pq$ as $r > q$. Secondly, $n_q \equiv 1 \pmod{q}$ and $n_q \mid pr$, so $n_q \geq r$ as $p < q$, and, arguing similarly we have $n_p \geq q$. Now as in Case 2, the group G has $n_p(p - 1)$ elements of order p, $n_q(q - 1)$ elements of order q, and $n_r(r - 1)$ elements of order r. Hence counting the neutral element and these only we have

$$o(G) = pqr \geq 1 + n_p(p - 1) + n_q(q - 1) + n_r(r - 1)$$
$$\geq 1 + q(p - 1) + r(q - 1) + pq(r - 1)$$
$$= pqr + (q - 1)(r - 1),$$

which is impossible. Therefore, at least one of n_p, n_q or n_r is one, showing that G has at least one normal (Sylow) subgroup by Sylow 3. $\qquad\qquad\square$

Using this theorem the structure of all groups whose orders have at most three prime factors follows: For groups of order $p^2 q$, see Burnside's classic 1897 (2nd edition, 1911 reprinted in 2004) textbook—the first on group theory to be published in English. For groups of order pqr, see Theorem 6.18 (their Sylow subgroups are necessarily cyclic) and Problem 6.20. Groups of order 12 will be discussed at the end of the next chapter.

Frattini Argument and Nilpotent Groups

To prove our next result on so-called nilpotent groups, we need the *Frattini Argument*, a simple procedure which has important applications in the theory; see Section 10.2. It also has one of the most elegant proofs in all mathematics!

Lemma 6.14 (Frattini Argument) *If G is a finite group, $K \lhd G$, and P is a Sylow subgroup of K, then*

$$G = N_G(P)K.$$

Proof For $g \in G$ we have

$$g^{-1}Pg \subseteq g^{-1}Kg = K,$$

as $P \leq K \lhd G$. Hence both P and $g^{-1}Pg$ are Sylow subgroups of K (they have the same order, Problem 2.23), and so by Sylow 2 they are conjugate in K. Therefore, we can find $k \in K$ to satisfy

$$k^{-1}(g^{-1}Pg)k = (gk)^{-1}P(gk) = P.$$

By Definition 5.24, this shows that $gk \in N_G(P)$, and so

$$g \in N_G(P)k^{-1} \subseteq N_G(P)K \subseteq G.$$

The result follows because this argument applies to all $g \in G$. $\square$

Example Suppose $G = S_5$, $K = A_5$ and $P = \langle(1, 2, 3)\rangle$; see the example on page 125. Now P is a Sylow 3-subgroup of K and $K \triangleleft G$, and so Frattini's Argument gives $S_5 = N_G(P)A_5$ which shows that $N_G(P)$ must contain an odd permutation. In fact, $N_G(P) = \langle(1, 2, 3), (2, 3), (4, 5)\rangle \simeq D_6$, a group of order 12 (look for the permutations in S_5 which commute with $(1, 2, 3)$); see Section 7.3.

As an illustration of this result we prove the following results, two of many dealing with properties of the normaliser; see Problem 6.10.

Corollary 6.15 *If P is a Sylow p-subgroup of a finite group G, and $N_G(P) \leq H \leq G$, then H is self-normalising, that is, $H = N_G(H)$.*

Note this shows that $N_G(P)$ is itself self-normalising.

Proof We have by hypothesis and Theorem 6.10(i)

$$P \triangleleft N_G(P) \leq H \triangleleft N_G(H) \leq G,$$

and P is a Sylow p-subgroup of each group in this sequence. Applying the Frattini Argument with $N_G(H)$ for G, and H for K, we have

$$N_G(H) = N_{N_G(H)}(P)H.$$

But $N_{N_G(H)}(P) \leq N_G(P) \leq H$, as $N_G(H) \subseteq G$, and so $N_G(H) = H$. $\square$

A finite group is called *nilpotent* if all of its Sylow subgroups are normal. A second definition is: Every maximal subgroup of G is normal in G (*cf.* Theorem 6.6). Here we use the Frattini Argument to show that this second property implies the first. In the next chapter (page 144), we prove the reverse implication and introduce a third equivalent definition involving direct products; a full discussion of nilpotency is given in Chapter 10.

Theorem 6.16 *If every maximal subgroup of a finite group G is normal in G, then every Sylow subgroup of G is normal in G; that is, G is nilpotent.*

Proof Suppose P is a non-normal Sylow subgroup of G. By Lemma 5.25,

$$P \leq N_G(P) < G.$$

Hence there is a maximal subgroup H of G (possibly $N_G(P)$ itself) which satisfies

$$P \leq N_G(P) \leq H < G.$$

Now P is a Sylow subgroup of H, and $H \lhd G$ by hypothesis. By the Frattini Argument (Lemma 6.14), this gives

$$G = N_G(P)H \leq H,$$

which is impossible because maximal subgroups are always proper, and H is maximal by definition. Therefore, $P \lhd G$. This applies to all Sylow subgroups of G and so the result follows.   □

Frattini introduced his argument to show that the subgroup of a group formed by the intersection of all of its maximal subgroups, the so-called Frattini subgroup, is nilpotent; we prove this result in Chapter 10 where further applications are given.

Web Section 6.5 gives some more applications of Sylow's theorems. Two are quite important and are used in a few of the problems, hence we shall state them here; their proofs (the second is basically a consequence of the first) are given in the web section. The first—one of Burnside's most important—concerns normal complements and is as follows.

Theorem 6.17 (Burnside's Normal Complement Theorem) *Suppose G is a finite group with Sylow subgroup P. If $C_G(P) = N_G(P)$, then P has a normal complement K in G, that is, G has a normal subgroup K with the property $G = PK$.*

In Problem 5.17(ii), we showed that the centraliser/normaliser condition in this theorem is equivalent to the condition $P \leq Z(N_G(P))$.

The second application, which is due to Hölder, Burnside and Zassenhaus, characterises groups with cyclic Sylow subgroups.

Theorem 6.18 *Suppose G is a finite group all of whose Sylow subgroups are cyclic, then G has a presentation in the form*

$$\langle a, b \mid a^m = b^n = e, b^{-1}ab = a^r \rangle$$

where m is odd, $(m, n(r - 1)) = 1$, $0 \leq r < m$ and $r^n \equiv 1 \pmod{m}$.

Groups satisfying these conditions are called *Frobenius* or *metacyclic*, they are denoted by $F_{m,n}$, or $F_{m,n,r}$ if for fixed m and n there are several values of r (and so several non-isomorphic groups). For example, there are six non-isomorphic groups of order 42, all Frobenius; see Web Section 6.5 and Problem 8.3.

6.4 Problems

Problem ♦ 6.1 (i) Complete the proof of Theorem 6.6 using Cauchy's (Theorem 6.2) and the Correspondence (Theorem 4.16) Theorems.

(ii) Using Theorem 4.20, show that every non-generator in a cyclic p-group is a p-th power. (Hint. See Problem 2.13(ii).)

(iii) Use the Class Equations (Theorem 5.20) to show that if G is a finite p-group, $K \triangleleft G$ and $o(K) = p$, then $K \leq Z(G)$. This result can also be derived using Problem 5.3.

Problem 6.2 Given a group G, let $G_p = \{g \in G : g^p = e\}$.

(i) Prove that G_p char G if G is an Abelian p-group; see Problem 4.22.

(ii)* Let G be a p-group and, for this problem, let $\mathrm{Exp}(G)$ denote the subgroup generated by the elements of the set G_p. Note that the inverse of an element of order p is itself an element of order p. Use Problem 4.6 and one of the commutator properties given in Problem 2.17 to show that (a) $\mathrm{Exp}(G) = G_p$ if $p > 2$, and (b) $G/\mathrm{Exp}(G)$ is Abelian.

Problem 6.3 For each group G of order 8, see Section 6.1, list their subgroups, find $Z(G)$, and determine the isomorphism type of $G/Z(G)$ using direct calculation and some results from this chapter and its predecessor. Also list their normal subgroups and draw subgroup lattice diagrams.

Problem 6.4 (An Example on 2-Groups) Let $G_1 = \langle a, b \mid b^2 = e, bab = a^{-3} \rangle$ and $G_2 = \langle a, b \mid b^2 = e, bab = a^3 \rangle$.

(i)   Show that $a^8 = e$ in each case, and so prove that both G_1 and G_2 have order 16.
(ii) Find the subgroups of G_1 and G_2, and determine which are normal.
(iii) Verify all parts of Theorems 6.5 and 6.6 for both G_1 and G_2.
(iv) Find the centres and the derived subgroups.
(v)   Show that the subgroup lattice diagrams for G_1 and $C_8 \times C_2$, see Chapter 7, are identical except for the 'top' group. Hence prove that non-isomorphic groups of the same order can have identical proper subgroup lattices.

The group G_2 is called *semi-dihedral*, another representation is given in Problem 6.23. The reader should note the similarities and differences between this group and the dihedral group with the same order, that is D_8, and see Problem 8.12. Also note that the group G_1 can be represented by the semi-direct product $C_2 \rtimes C_8$ which is discussed in Chapter 7.

Problem 6.5 (Extra-Special Groups) For more details on these groups, see Aschbacher (1994), Chapter 2.

(i)   Let G be a non-Abelian group of order p^3 where p is an odd prime, see Problem 5.22(iii). By considering $G/Z(G)$, show that there are at most two isomorphism classes for G, and so deduce that G has one of the following presentations

$$ES_1(p) = \langle a, b \mid a^{p^2} = b^p = e, ab = ba^{1+p}\rangle,$$
$$ES_2(p) = \langle a, b, c \mid a^p = b^p = c^p = e, c = [a,b], ca = ac, cb = bc\rangle.$$

Each of these groups is called *extra-special*, that is, they satisfy the conditions: $Z(G)$ is cyclic and $Z(G) = G' = \Phi(G)$; see Section 10.2. Note that for the second group every non-neutral element has order p, but it is not elementary Abelian if $p > 2$. (If $p = 2$ the group is elementary Abelian, see Corollary 2.20.)

(ii) Show that $ES_2(p)$ is represented by the matrix group $IT_3(p)$; see Problem 3.15. We shall give a representation of $ES_1(p)$ in Problem 7.22.

(iii) Find the subgroups of these groups when $p = 3$, and indicate which are normal and which are maximal.

Problem 6.6* Suppose G is a finite p-group with a single subgroup of order p, and if $p = 2$ suppose also G is Abelian. Show that G is cyclic using the following method. Use induction on n where $o(G) = p^n$. For $n = 2$ use Theorem 5.22, and for $n = 3$ use the previous problem and Section 7.2. In the general case, let $K \leq G$ with $o(K) = p$, and suppose G/K has two subgroups H/K and J/K both with order p. For more details, see Doerk and Hawkes (1992), page 204.

Show finally that the extra condition in the case $p = 2$ is necessary.

Problem 6.7 (The group C_{p^∞}—An Infinite Group all of whose Proper Subgroups are Finite) Fix a prime p. The group C_{p^∞} has the following presentation. The generators are $a_0, a_1, \ldots$, and the relations are given by

$$a_i a_j = a_j a_i, \quad \text{for all } i \text{ and } j,$$
$$a_0^p = e, \quad a_1^p = a_0, \quad \ldots, \quad a_{n+1}^p = a_n, \quad \ldots$$

(i)   Prove that C_{p^∞} is an infinite Abelian p-group.
(ii) Show that if $H \leq C_{p^\infty}$ and $a_m \in H$, then $a_n \in H$ for all $n < m$.
(iii) Using (ii), prove that all proper subgroups of C_{p^∞} are finite and cyclic.

Ol'shanskii (1983) has given an example, usually called the 'Tarski monster' (it applies some model theory based on the work of Alfred Tarski and involves the use of large primes—hence the name) of a countably infinite non-Abelian (simple) group all of whose subgroups are finite.

Problem 6.8 (i) For the given groups and primes determine the isomorphism classes of the Sylow p-subgroups of (a) S_6 with $p = 3$, (b) A_6 with $p = 2$, and (c) $SL_2(5)$ again with $p = 2$. (Hint. Use Problem 3.19(i) for (c).)

(ii) Prove that if n is odd, then all Sylow subgroups of D_n are cyclic. Is this true if n is even?

(iii) Show that Sylow subgroups can intersect both neutrally and non-neutrally. To be more precise, there are groups G with distinct Sylow p-subgroups P_1, P_2 and P_3 (with p fixed) that satisfy:

$$P_1 \cap P_2 = \langle e \rangle \quad \text{and} \quad P_1 \cap P_3 > \langle e \rangle.$$

An example can be found in a subgroup H of S_6 of permutations in which the elements of the subsets $\{1, 2, 3\}$ and $\{4, 5, 6\}$ are permuted amongst themselves but there is no mixing; that is, $H \simeq S_3 \times S_3$, see Problem 3.9 and Section 7.1.

Problem ♦ 6.9 (Properties of Sylow Subgroups 1) (i) Suppose $H \leq G$, P is a Sylow p-subgroup of H, and Q is a Sylow p-subgroup of G with the property $P \leq Q$, prove that $P = Q \cap H$.

This result can be extended to: If $H \leq G$ and G has a Sylow p-subgroup P, then for some $g \in G$ we have $H \cap g^{-1}Pg$ is a Sylow p-subgroup of H, and this extension can be used as the starting point for another proof of the First Sylow Theorem (Theorem 6.7). See also Problem 6.10(iv).

(ii) Show that if K is a normal p-subgroup of a finite group G, then $K \leq P$ for all Sylow p-subgroups P of G.

(iii) Let H be a non-Sylow p-subgroup of G, so $o(H)$ is a power of p and $p \mid [G : H]$. Using a coset action and Problem 5.3 show that $H < N_G(H)$.

(iv) If G is finite, $K \triangleleft G$, $(o(K), p) = 1$, and P is a Sylow p-subgroup of G with the property $P = C_G(P)$, prove that $o(K) \equiv 1 \pmod{p}$. (Hint. Use Problem 5.3 and see Problem 5.11(ii), or use Theorem 6.4 and Problem 5.13(iii).)

(v) Using Problem 4.22, show that if P is a unique Sylow p-subgroup of G, then P char G.

(vi) The subgroup $O_p(G) = \bigcap_r P_r$, where the intersection is taken over all Sylow p-subgroups of G, is called the p-*radical* of G; see Section 10.2. Using Problem 4.22 again, show that (a) $O_p(G) \triangleleft G$, and (b) using (ii), it is the largest subgroup of G with this property.

Problem ♦ 6.10 (Properties of Sylow Subgroups 2) Suppose G is finite and P is a Sylow p-subgroup.

(i)   Show that if $N_G(P) \leq H \leq G$, then $[G : H] \equiv 1 \pmod{p}$.
(ii) Show that if g and h belong to $C_G(P)$ and are conjugate in G, then they are also conjugate in $N_G(P)$—a useful fact.
(iii) Using Theorem 5.8, prove that if $K \triangleleft G$, then $K \cap P$ is a Sylow subgroup of K, and PK/K is a Sylow subgroup of G/K.
(iv) Give an example to show that if J is a non-normal subgroup G, then $J \cap P$ need not be a Sylow subgroup of J; see (iii).
(v)   Use (iii) to show that if $K \triangleleft G$, then $n_p(G/K) \leq n_p(G)$.

Problem 6.11 (Properties of Sylow Subgroups 3) Let $H \leq G$.

(i)   Let $\{p_1, \ldots, p_k\}$ be a list of the prime divisors of $o(G)$ and let P_i be a Sylow
      p_i-subgroup of G for $i = 1, \ldots, k$. Show that $G = \langle P_1, \ldots, P_k \rangle$ and $G = P_1 P_2$,
      if $k = 2$. See also Condition 6 on page 242.
(ii) If $K_1, K_2 \lhd G$ and P is a Sylow subgroup of G, show that

$$K_1 K_2 \cap P = (K_1 \cap P)(K_2 \cap P) \quad \text{and} \quad K_1 P \cap K_2 P = (K_1 \cap K_2)P.$$

(iii) Let P and Q be distinct Sylow p-subgroups of H, and let $P_1 \geq P$ and $Q_1 \geq Q$
      where P_1 and Q_1 are Sylow p-subgroups of G. Prove that P_1 and Q_1 are also
      distinct. Use this to show that $n_p(H) \leq n_p(G)$.
(iv) Let $J \leq G$ and let H be a Sylow p-subgroup of G. Show how to find an
      element $a \in G$ with the property: $J \cap a^{-1} H a$ is a Sylow p-subgroup of J.
      (Hint. Use Problem 2.29.)

Problem 6.12 (i) If P is a Sylow p-subgroup of the symmetric group S_p, show that
$o(N_{S_p}(P)) = p(p-1)$; see Problems 3.11 and 6.19.
  (ii)* Let G be a transitive subgroup of S_p (that is for each pair of integers
r and s in $\{1, \ldots, p\}$ there exists a permutation in G which maps r to s; see
page 94), let n_p equal the number of Sylow p-subgroups in G as usual, and let P,
a Sylow p-subgroup of G, be one of these subgroups. Show that if $n_p > 1$, then
P is not self-normalising (that is, $P < N_G(P)$). One method is as follows. Let
$P = \langle (1, 2, \ldots, p) \rangle$ and $t = [N_G(P) : P]$. Show that $o(G) = p n_p t$, and deduce using (i) the properties: $t \mid p - 1$, $0 < t < p$ and $t \equiv o(G)/p \pmod{p}$. Now assume
that $t = 1$, count the number of elements of G with no fixed points to give an estimate of $o(\text{stab}_G(j))$ where $j \in \{1, \ldots, p\}$, and so show that $n_p = 1$ which gives the
required contradiction.

Problem 6.13 (i) Show that if $o(G) = 12$, then G has a normal Sylow subgroup of
order 3 or 4.
  (ii) Suppose m is a proper divisor of 60 which is itself divisible by 5. Prove that
if $o(G) = m$, then G has a unique Sylow 5-subgroup.
  (iii) List the Sylow subgroups of A_5.

Problem 6.14 Let G be a non-Abelian group of order pq where p and q are primes,
and $p \mid q - 1$. Using Theorem 6.11, (a) find a presentation for G with two generators, (b) give a formula in terms of the generators of the product of one element
by another, and (c) show that up to isomorphism there is only one group satisfying
these conditions. See also Problems 3.6, 7.21 and 9.14, and Web Section 15.3.

Problem ♦ 6.15 By applying the results proved in this chapter and its predecessor
including Theorem 5.15, show that the smallest non-Abelian simple group has order
60. Note that the next non-Abelian simple group has order 168, try to prove this;
some parts are covered in Problem 6.17.

Problem 6.16 (i) Using the method indicated below, or otherwise, show that if G has order 60 and possesses more than one Sylow 5-subgroup, then it is simple. Hence show that A_5 is simple. Method: suppose $K \lhd G$ and $5 \mid o(K)$. Use Problems 6.13(ii) and 4.22 to show this is impossible. Deduce $5 \mid o(G/K)$, and so prove that $o(K) = 12$. Obtain the final contradiction using Problem 6.13(i). (Hint. You need to show that if H is a Sylow 5-subgroup of G, then $HK = G$.)

(ii)* Prove that a simple group G of order 60 is isomorphic to A_5. One method is as follows. (a) Show that G has a subgroup of index 5 by considering Sylow 2-subgroups and their intersections, supposing the contrary, and counting elements of order a power of 2. (b) Apply Theorem 5.15 to show that G is isomorphic to a subgroup of S_5, and then use Theorem 5.8. Deduce $A_5 \simeq L_2(4) \simeq L_2(5)$ using Theorems 3.11 and 12.7. Note that $L_2(q) = SL_2(q)/Z(SL_2(q))$, see Section 12.2.

Problem 6.17 Using one of the methods suggested, or otherwise, show that there are no simple groups with the following orders.

(i) 90. Method—by the Sylow theory if the group G is simple, then $n_3 = 10$ and $n_5 = 6$. Firstly, use a counting argument to show that each pair of the ten Sylow 3-subgroups cannot have neutral intersection. Secondly, if P and Q are distinct Sylow 3-subgroups of G with non-neutral intersection, $R = P \cap Q$ and $S = N_G(R)$, show that $o(S) = 18, 45$ or 90 applying the method used in the second proof of Theorem 6.13, Part 2. Finally, show that each of these cases is impossible if G is simple. Using Problem 3.7(ii), this fact can also be established by defining an injection from the supposed simple group into A_6, and arguing that A_6 has no subgroups of index 4.

(ii) 108. Method—use Theorem 5.15. Similar methods apply for groups of order $p^r(p+1)$ if p is prime and $r > 1$.

(iii) 112*. The method suggested works for groups G of order $p^n q$. Assuming that G simple, show first that $n_p = q$. Prove that if all pairs of distinct Sylow p-subgroups P and Q have neutral intersection, then using a counting argument, deduce $n_q = 1$ and so obtain a contradiction. Now suppose $R = P \cap Q$ where $o(R)$ is of maximal size which we assume to be larger than 1, and let $S = N_G(R)$. By Problem 5.13(vi), note that R is a proper subgroup of both $N_P(R)$ and $N_Q(S)$. Secondly, suppose S has a single Sylow p-subgroup T, say, and let U be the Sylow p-subgroup of G containing T. Show that $P = U = Q$, and so deduce T is not unique. Use this to show that S has q Sylow p-subgroups, and by applying Problem 6.10(iii) deduce that all of these subgroups contain R. Now use Problem 6.10(iii) again, and the supposed simplicity of G, to show that $R = \langle e \rangle$. As in the case of groups of order 90, this non-simplicity result (for a group of order 112) can also be established by considering injections into the alternating group A_7.

(iv) 132. Method—use the Sylow theory and a counting argument.

(v) 144. Method—assume that $n_3 > 1$, so $n_3 = 4$ or 16. Apply Theorem 5.15 in the first case, and in the second case, use a counting argument to show that if each pair of Sylow 3-subgroups has neutral intersection, then $n_2 = 1$. If not, apply the first method suggested in (i) treating the normalisers of orders 18, 36, 72, and 144 separately.

Problem 6.18 Prove that $\operatorname{Aut} Q_2 \simeq S_4$; one method is as follows. Noting that automorphisms map maximal subgroups to maximal subgroups (this is part of the Correspondence Theorem (Theorem 4.16)), use Theorem 5.15 to construct a homomorphism $\phi : \operatorname{Aut} Q_2 \to S_3$, and show that ϕ is surjective (these automorphisms interchange elements of order 4). Secondly, use Corollary 5.27 to show that there are four inner automorphisms (each of these permute the elements of the maximal subgroups but leave the subgroups themselves unaltered); thirdly, deduce $o(\operatorname{Aut} Q_2) = 24$, and so prove the result using Section 8.1.

Problem 6.19 Working in S_5, let $\sigma = (1,2,3,4,5)$, $P = \langle \sigma \rangle \leq S_5$ and $K = N_{S_5}(P)$. Show that $o(K) = 20$ and $K \simeq \langle \sigma, \tau \rangle$ where $\tau = (2,3,5,4)$; see Problems 3.11 and 6.12.

Problem ◆ 6.20 (i) Using the Frattini Argument (Lemma 6.14), show that if $H \triangleleft G$, P is a Sylow subgroup of G and $P \triangleleft H$, then $P \triangleleft G$; note that normality is not transitive in general.

(ii)* Suppose G has square-free order. You are given: G is not simple, this can be proved using Burnside's Normal Complement Theorem (Theorem 6.17). Show that if p^* is the largest prime factor of $o(G)$, then G has a normal p^*-subgroup by applying the following method: use induction on the number of factors of $o(G)$, Theorem 6.11, and Problem 6.10(iii). If H is the given normal subgroup of G, treat the cases (a) $p^* \mid o(H)$, and (b) $p^* \nmid o(H)$ separately, then use (i). See also Theorem 6.18.

Problem 6.21 (i) Let $K_1, K_2 \triangleleft G$ have the property: $K_1 K_2 = G$, and let $H \leq G$. Show that $H = (K_1 \cap H)(K_2 \cap H)$. (Hint. Use Problem 2.18.)

(ii) Secondly, show by an example that the result in (i) is false if either K_1 or K_2 is not normal; but see (iv) below. In one example, $G = S_4$.

(iii)* Suppose $K \triangleleft G$, P is a p-subgroup of G, and $(o(K), p) = 1$. Show that

$$N_{G/K}(P/K) = N_G(P)/K.$$

(Hint. Use the Frattini Argument (Lemma 6.14).)

(iv)* Now assume that H is a complement of K in G, so $G = HK$ where $H \leq G$. Using (i) and the Second Isomorphism Theorem (Theorem 4.15) show that

$$N_G(P) = \big(K \cap N_G(P)\big)\big(H \cap N_G(P)\big).$$

Problem 6.22 Suppose $P_1, \ldots, P_m$ is a list of the Sylow p-subgroups of a finite group G. Note that $m \equiv 1 \pmod p$. Further, let $S_{(P)}$ denote the group of permutations of the set $\{P_1, \ldots, P_m\}$, and define a map $\theta : G \to S_{(P)}$ so that $g\theta$ is the permutation that maps P_i to $g^{-1}P_i g$, for $i = 1, \ldots, m$.

(i) Show that θ is a homomorphism and determine its kernel.

(ii) Find the kernel of θ when $G = D_n$, n is odd, and $p = 2$. Hence show that D_n is isomorphic to a subgroup of S_n when n is odd.

Problem 6.23 (Project—Subgroups of $GL_2(3)$) In this project, you are asked to find all 55 subgroups of the matrix group $GL_2(3)$ some of which are *semi-dihedral*, see Problem 6.4. This project is a continuation of the one given in Problem 3.24 where you were asked to show that the group $GL_2(3)$ has the following presentation

$$\langle a, b, c \mid a^8 = b^2 = c^3 = e, bab = a^3, bcb = c^2, c^2 a^2 c = ab, c^2 abc = aba^2 \rangle.$$

You were further asked to find matrix and permutation representations. As noted there the work can be done by hand, but a computer package which includes matrix calculations modulo a prime (3 in this case) would be an asset.

Method: To find the subgroups, use some results from Section 3.3 to write out all elements of the group with their orders. Secondly, using Problem 5.19 show that the elements of the centre together with those of order 4 form a normal subgroup K isomorphic to the quaternion group Q_2. Thirdly, show that the subset of elements of the group whose orders are a power of 2 define the Sylow 2-subgroup(s) and, using Problem 6.9(ii), show that K is a subgroup of each of them. Fourthly, determine these Sylow 2-subgroups. Now use Problem 5.19 again to find the subgroups of order 12, note there are four of them, and using suitably chosen elements of the Sylow 2-subgroup(s) find the remaining subgroups (seven in all) with order 8. Finally, using the methods applied in the last part of Section 6.1, find the cyclic and remaining dihedral subgroups of the group, indicate which subgroups are normal, and find the centre and the derived subgroup.

Chapter 7
Products and Abelian Groups

This chapter can be read before Chapters 5 and 6, see the note on page 91.

Suppose $(G_i, \odot_i)$, $i = 1, 2, \ldots$, are groups, and H is the (set-theoretic) Cartesian product of their underlying sets (page 278)

$$H = G_1 \times G_2 \times \cdots = \prod_i G_i.$$

A number of new groups can be formed using H as the underlying set. The operations of these new groups are constructed using the operations $\odot_i$ of G_i, $i = 1, 2, \ldots$. For example, if $G_1 = C_2$ and $G_2 = C_3$, then H is the set of pairs (a, b) where $a \in C_2$ and $b \in C_3$, and we define a new operation $\odot$ by

$$(a_1, b_1) \odot (a_2, b_2) = (a_1 \odot_1 a_2, b_1 \odot_2 b_2).$$

It is easily seen that this construction defines a group which we denote by $C_2 \times C_3$. It is called the *direct product* of C_2 and C_3; direct because there is no mixing between the first and second arguments. In the first argument, we have elements $a_i \in G_1$ and we use the operation $\odot_1$, and in the second we have $b_i \in G_2$ and use the operation $\odot_2$. In other products, some mixing is allowed, the terms in one or both arguments of the product are constructed using a_i, b_i, $\odot_1$ and $\odot_2$, see Section 7.3.

In this chapter, we discuss direct products, and two types of product where some mixing is allowed, they are called *semi-direct* and *wreath*; the latter, a special case of the former, is only considered briefly. We have seen some examples in the previous chapters; there is no circularity here because the prerequisites for most of the work in this chapter were given in Chapter 2, and Sections 4.1 and 4.2. We shall also show that all finite Abelian groups can be represented as direct products of cyclic groups—an important result with applications to other areas of mathematics. Some examples of groups built up using the semi-direct construction will be given in the final section. A brief account of the basic facts about infinite Abelian groups can be found in Web Section 7.5.

H.E. Rose, *A Course on Finite Groups*,
Universitext,
DOI 10.1007/978-1-84882-889-6_7, © Springer-Verlag London Limited 2009

7.1 Direct Products

Given sets G_i, $i = 1, \ldots, n$, the underlying sets of the groups G_i, we can form the (set-theoretic) Cartesian product of ordered n-tuples of elements of G_1 to G_n by

$$G_1 \times \cdots \times G_n = \{(a_1, \ldots, a_n) : a_i \in G_i, i = 1, \ldots, n\},$$

see Appendix A. This will be used as the starting point for the construction of new groups. The (finite) direct product is given by

Definition 7.1 For each $i = 1, \ldots, n$, let $(G_i, \odot_i)$ be a group with neutral element e_i. On the set $G = G_1 \times \cdots \times G_n$ as given above, define the operation $\odot$ by

$$(a_1, \ldots, a_n) \odot (b_1, \ldots, b_n) = (a_1 \odot_1 b_1, \ldots, a_n \odot_n b_n),$$

where $a_i, b_i \in G_i$, for $i = 1, \ldots, n$. We denote the set G with the operation $\odot$ by $G_1 \times \cdots \times G_n$, it is a group called the *direct product* of $G_1, \ldots, G_n$; see Theorem 7.2. Each G_i is called a *factor* of the product group G.

Note that we use the same symbol $\times$ for the set-theoretic Cartesian product and the group-theoretic direct product.

Theorem 7.2 *If G_i are groups for $i = 1, \ldots, n$, then the set $G_1 \times \cdots \times G_n$, with the operation $\odot$ given in Definition 7.1, also forms a group.*

Proof The operation $\odot$ is clearly closed and associative because the constituent operations $\odot_i$ have these properties. The neutral element is

$$(e_1, \ldots, e_n),$$

and the inverse operation is given by

$$(a_1, \ldots, a_n)^{-1} = (a_1^{-1}, \ldots, a_n^{-1}),$$

using the inverse operations in the constituent groups. □

The group constructed in Theorem 7.2 is sometimes called the *external direct product* of the groups G_i. We are not restricted to finite products, products with an infinite number of factors can be defined similarly; see Web Section 7.5. We shall give some examples later, but first we derive the basic properties. To ease the notation, we give the results for the product of two groups first, they can easily be extended to products with a finite number of factors; see Lemma 7.5 and Theorem 7.6.

Lemma 7.3 *Suppose H and J are groups.*

(i) *If $o(H), o(J) < \infty$ then, $o(H \times J) = o(H) \cdot o(J)$.*

(ii)   *H and J are Abelian if and only if $H \times J$ is Abelian.*
(iii) $H \times \langle e \rangle \lhd H \times J, \langle e \rangle \times J \lhd H \times J, H \times \langle e \rangle \simeq H$, *and* $\langle e \rangle \times J \simeq J$.
(iv) $(H \times J)/(H \times \langle e \rangle) \simeq J$ *and* $(H \times J)/(\langle e \rangle \times J) \simeq H$.

Proof (i) and (ii) These are immediate consequences of the definitions.
   (iii) and (iv) Define a map $\phi : H \times J \to J$ by

$$(h, j)\phi = j \quad \text{for } h \in H, \ j \in J.$$

It is easily checked that this defines a surjective homomorphism with kernel $H \times \langle e \rangle$. The first parts of both (iii) and (iv) follow by the First Isomorphism Theorem (Theorem 4.11), and the second parts follow in the same way. Similar arguments can be applied to establish the third and fourth parts of (iii), the reader should write them out.   □

Secondly, we consider these products from the opposite point of view: Under what conditions is a group isomorphic to a direct product of two or more of its subgroups? For example, we constructed the group $C_2 \times C_3$ on page 139, but in this case we obtained nothing new, for this group is isomorphic to C_6; that is, C_6 can be treated as a direct product of its two subgroups isomorphic to C_2 and C_3, see Theorem 4.20. The next result gives necessary and sufficient conditions in the two-subgroup case.

Theorem 7.4 *Suppose H and J are subgroups of the group G, $H \lhd G, J \lhd G$, $H \cap J = \langle e \rangle$, and $HJ = G$.*

(i)   *If $g = hj$ where $g \in G, h \in H$ and $j \in J$, then h and j are uniquely determined by g.*
(ii) *$hj = jh$, for all $h \in H$ and $j \in J$.*
(iii) *$G \simeq H \times J \simeq J \times H$.*

In this case, we say that G is the *internal direct product* of its subgroups H and J.

Proof Throughout we assume that $h, h', h_1, h_2 \in H$ and $j, j', j_1, j_2 \in J$.
   (i) As $G = HJ$, for each $g \in G$ we can find $h \in H$ and $j \in J$ to satisfy

$$g = hj.$$

Proposition (i) follows for if $g = hj = h'j'$, then $h'^{-1}h = j'j^{-1}$ where $h'^{-1}h \in H$ and $j'j^{-1} \in J$. But $H \cap J = \langle e \rangle$, and so $h'^{-1}h = j'j^{-1} = e$ which gives $h = h'$ and $j = j'$.
   (ii) As $H \lhd G$, we have $j^{-1}h^{-1}j \in H$ which shows that $j^{-1}h^{-1}jh \in H$; also as $J \lhd G$, we have $j^{-1}h^{-1}jh \in J$. But as above $H \cap J = \langle e \rangle$, hence $j^{-1}h^{-1}jh = e$, or $hj = jh$, which gives (ii).

(iii) We define a map $\psi : G \to H \times J$ by

$$\text{if } g = hj \quad \text{then } g\psi = (h, j).$$

The uniqueness condition proved in (i) shows that ψ is well-defined, and it is clearly surjective (as $(hj)\psi = (h, j)$). To prove injectivity we argue as follows. Suppose $g_1\psi = g_2\psi$, where $g_i = (h_i, j_i)$, for $i = 1, 2$. This gives $(h_1, j_1) = (h_2, j_2)$, and so

$$(e, e) = (h_1, j_1)(h_2, j_2)^{-1} = \left(h_1 h_2^{-1}, j_1 j_2^{-1}\right),$$

which is only possible if $h_1 = h_2$ and $j_1 = j_2$. But then $g_1 = g_2$ proving injectivity. For the homomorphism property, we have

$$g_1\psi g_2\psi = (h_1 j_1\psi)(h_2 j_2\psi) = (h_1, j_1)(h_2, j_2)$$

$$= (h_1 h_2, j_1 j_2) = h_1 h_2 j_1 j_2\psi$$

$$= h_1 j_1 h_2 j_2\psi = g_1 g_2\psi,$$

using the definition of ψ and (ii) for the last but one equation (that is, $h_2 j_1 = j_1 h_2$). Putting these properties together, we see that ψ is an isomorphism. The second isomorphism in (iii) is an direct consequence of (ii). $\qquad\square$

The converse of Theorem 7.4 is given by Lemma 7.3.

The n-subgroup versions of Lemma 7.3 and Theorem 7.4 follow, their proofs will be left as exercises for the reader, see Problem 7.1. Note that (iii) and (iv) show that the direct product is both associative and commutative.

Lemma 7.5 *Suppose $H_1, \ldots, H_n$ are groups.*

(i) *If $o(H_i) < \infty$ for $i = 1, \ldots, n$, then $o(H_1 \times \cdots \times H_n) = o(H_1) \cdots o(H_n)$.*
(ii) *$H_1, \ldots, H_n$ are all Abelian if and only if $H_1 \times \cdots \times H_n$ is Abelian.*
(iii) *If J, K and L are direct products of groups (with one or more factors), then $J \times (K \times L) \simeq (J \times K) \times L$.*
(iv) *If $\{i_1, \ldots, i_n\}$ is a permutation of $\{1, \ldots, n\}$, $H_{i_1} \times \cdots \times H_{i_n} \simeq H_1 \times \cdots \times H_n$.*
(v) *For each i, let H_i^* denote the group $\langle e \rangle \times \cdots \times \langle e \rangle \times H_i \times \langle e \rangle \times \cdots \times \langle e \rangle$ with n factors and where H_i is in the ith place, then*

$$H_i^* \lhd H_1 \times \cdots \times H_n, \quad \text{and}$$

$$(H_1 \times \cdots \times H_n)/H_i^* \simeq H_1 \times \cdots \times H_{i-1} \times H_{i+1} \times \cdots \times H_n.$$

Theorem 7.6 *If $H_1, \ldots, H_n$ are subgroups of a group G, then $G \simeq H_1 \times \cdots \times H_n$ if and only if the following three conditions hold:*

(i) *$H_i \lhd G$, for $i = 1, \ldots, n$,*
(ii) *$G = H_1 \cdots H_n$,*
(iii) *$H_i \cap H_1 \cdots H_{i-1} H_{i+1} \cdots H_n = \langle e \rangle$, for $i = 1, \ldots, n$.*

Condition (iii) in this theorem cannot be replaced by

$$H_i \cap H_j = \langle e \rangle, \quad \text{for } 1 \le i < j \le n,$$

as the following example shows.

Example Let $G = \langle a^t b^u : t = 0, 1, u = 0, 1 \rangle$, the 4-*group* T_2. It has three subgroups of order 2: $H_1 = \langle a \rangle$, $H_2 = \langle b \rangle$ and $H_3 = \langle ab \rangle$; and $H_i \lhd G$ for $i = 1, 2, 3$, $G = H_1 H_2 H_3$, and $H_1 \cap H_2 = H_1 \cap H_3 = H_2 \cap H_3 = \langle e \rangle$. But $G \not\simeq H_1 \times H_2 \times H_3$ because G has order 4 whilst $H_1 \times H_2 \times H_3$ has order 8.

Uniqueness of Representation

Suppose G is represented as a direct product of (some of) its subgroups $H_1, \ldots, H_m$. We can ask under what conditions is this representation unique? Clearly, it is not unique if some of the subgroups H_i are themselves direct products. So we ask: If

$$G \simeq H_1 \times \cdots \times H_m,$$

and each H_i is *indecomposable*, that is, it cannot be represented as a direct product with non-neutral factors then, apart from the order of the terms, is this representation unique? The answer is yes in the finite and some infinite cases. This result was first conjectured by J.H.M. Wedderburn (1882–1948), and it was proved by him, Remak, Krull and Schmidt. For the proof, and details of the infinite case, the reader should consult Suzuki (1982), page 127*ff*. We shall consider the Abelian case in the next section.

For our first application of the results above, we return to the topic of cyclic groups; see Section 4.3. We have

Theorem 7.7 (i) *If* $n = p_1^{s_1} \ldots p_k^{s_k}$ *is the prime factorisation of* n, *then*

$$C_n \simeq C_{p_1^{s_1}} \times \cdots \times C_{p_k^{s_k}}.$$

(ii) *If* $(m, n) = 1$ *then* $C_{mn} \simeq C_m \times C_n$.

Proof (i) Let $n_i = n / p_i^{s_i}$, for $i = 1, \ldots, k$, and let g be a generator of the cyclic group C_n. Then g^{n_i} generates a subgroup $P_i = \langle g^{n_i} \rangle$ of C_n of order $p_i^{s_i}$. (Incidentally, P_i is the unique Sylow p_i-subgroup of C_n; one proof of Sylow's First Theorem reduces the general case to this example; see Alperin and Bell 1995, page 64.) By unique factorisation of n, we have $C_n = P_1 \cdots P_k$. Also $P_i \cap P_1 \cdots P_{i-1} P_{i+1} \cdots P_k = \langle e \rangle$, for $i = 1, \ldots, k$ as P_i only contains elements of order a power of p_i. Now apply Theorem 7.6 as all subgroups are normal.

(ii) This is an immediate consequence of (i) and Lemma 7.5. $\square$

For our second application, we consider subgroups of direct product groups. We have the following useful

Lemma 7.8 *If* $G \simeq H_1 \times \cdots \times H_k$, $(o(H_i), o(H_j)) = 1$ *when* $i \neq j$, *and* $J \leq G$, *then*

$$J \simeq (H_1 \cap J) \times \cdots \times (H_k \cap J).$$

Proof We treat the case $k = 2$, the general case follows by induction. The second hypothesis gives $(H_1 \cap J) \cap (H_2 \cap J) = \langle e \rangle$. Also, using direct product properties, and Problems 2.5 and 2.14, we have $H_i \cap J \triangleleft J$ for $i = 1, 2$. Hence we can construct the direct product $(H_1 \cap J) \times (H_2 \cap J)$ inside J. By the first hypothesis, if $j \in J$, then $j = h_1 h_2$ where $h_i \in H_i$ for $i = 1, 2$. The result will follow if we show that $h_1, h_2 \in J$ as j is an arbitrary element of J.

As $(o(h_1), o(h_2)) = 1$, and h_1 and h_2 commute (Theorem 7.4(i)), we have $o(h_1 h_2) = o(h_1) o(h_2)$, and by Theorem 7.7(ii), $\langle h_1 h_2 \rangle \simeq \langle h_1 \rangle \times \langle h_2 \rangle$. Hence $\langle j \rangle = \langle h_1 h_2 \rangle \simeq \langle h_1 \rangle \times \langle h_2 \rangle$, and so $h_1, h_2 \in \langle j \rangle \leq J$. The lemma follows. $\square$

Examples (a) Let $G = C_3 \times D_4$, see Appendix C. The group G has a subgroup J isomorphic to C_{12}, and so the lemma gives $J \simeq J \cap C_3 \times J \cap D_4$, that is, $J \cap C_3 \simeq C_3$, $J \cap D_4 \simeq C_4$, and $J \simeq C_{12} \simeq C_3 \times C_4$.

(b) Referring back to the example given below Theorem 7.6, suppose G, H_1 and H_2 are as defined there, and we set $J = H_3$. Now $J \leq G$ and $H_1 \cap J = H_2 \cap J = \langle e \rangle$, and so the conclusion of Lemma 7.8 is false in this case. But $o(H_1) = o(H_2) = 2$, and so this shows that second condition in the lemma above is essential.

We introduced finite nilpotent groups in Section 6.3 and we shall discuss general groups of this type in Chapter 10. In the finite case, there are a number of equivalent definitions one of which involves direct products (Theorem 10.9), and so we prove the following result now.

Theorem 7.9 *Suppose G is a finite group. The statements below are equivalent*:

(i)   *G is the direct product of its non-neutral Sylow subgroups.*
(ii) *All maximal subgroups of G are normal in G.*
(iii) *All Sylow subgroups of G are normal in G.*

Note that if G is a p-group, the result follows from Theorem 6.6. Also, using the Frattini Argument (Lemma 6.14), we have shown previously that (ii) implies (iii), see Theorem 6.16. It is also of interest to note the range and number of subsidiary results that are used in the proof given below.

Proof Let $o(G) = p_1^{s_1} \cdots p_k^{s_k}$ and, for $i = 1, \ldots, k$, let P_i be a Sylow p_i-subgroup of G. We need to show that (i) implies (ii), and (iii) implies (i).

(i) implies (ii) By (i) we have

$$G \simeq P_1 \times \cdots \times P_k.$$

By Theorem 7.6, this shows that $P_i \lhd G$, for $i = 1, \ldots, k$, and so P_i is the unique Sylow p_i-subgroup of G by Theorem 6.10. Let H be a maximal subgroup of G. As $(o(P_i), o(P_j)) = 1$ if $i \neq j$, we can apply Lemma 7.8 to obtain, for some j satisfying $1 \leq j \leq k$,

$$H \simeq (P_1 \cap H) \times \cdots \times (P_j \cap H) \times \cdots \times (P_k \cap H)$$
$$\simeq P_1 \times \cdots \times (P_j \cap H) \times \cdots \times P_k,$$

where $P_j \cap H < P_j$. To see why this second isomorphism follows, we argue as follows. Clearly, $P_i \cap H \leq P_i$ for all i. Suppose, for instance, both $P_1 \cap H < P_1$ and $P_2 \cap H < P_2$ (that is, the first inequality above holds in at least two cases). Then we would have

$$H < (P_1 \cap H) \times P_2 \times \cdots \times P_k < G,$$

but this is impossible as it contradicts the maximality of H in G. For the same reason, $P_j \cap H$ is a maximal subgroup of P_j; and so by Theorem 6.6 and as P_j is a p_j-group

$$P_j \cap H \lhd P_j.$$

We can now apply Lemma 4.14(iv). Above we noted that $P_1 \lhd G$, and so this lemma gives

$$P_1(P_j \cap H) \lhd P_1 P_j;$$

if $j = 1$ replace P_1 by P_2. We also have $P_2 \lhd G$, and so we can repeat this argument to obtain $P_1 P_2(P_j \cap H) \lhd P_1 P_2 P_j$. Continuing we have finally

$$H \simeq P_1 \cdots (P_j \cap H) \cdots P_k \lhd P_1 \cdots P_j \cdots P_k = G.$$

This gives (ii).

(iii) implies (i) First we prove

$$P_1 \cdots P_j \lhd G \quad \text{and} \quad o(P_1 \cdots P_j) = o(P_1) \cdots o(P_j), \tag{7.1}$$

for $j = 1, \ldots, k$, by induction on j. For $j = 1$ this follows by hypothesis, so suppose it is true for j. As $P_1 \cdots P_j \lhd G$ (by the inductive hypothesis) and $P_{j+1} \lhd G$ by (iii), Lemma 4.14(iii) gives $P_1 \cdots P_{j+1} \lhd G$. Secondly

$$o(P_1 \cdots P_{j+1}) = o(P_1 \cdots P_j)o(P_{j+1})$$

by Theorem 5.8 as $P_{j+1} \cap P_1 \cdots P_j = \langle e \rangle$ and P_{j+1} is a Sylow p_{j+1}-subgroup. This shows that (7.1) holds for all $j \leq k$, and Condition (ii) of Theorem 7.6 follows if we put $j = k$. This last conclusion also justifies Condition (iii) in Theorem 7.6, and so the result follows. $\qquad \square$

Example Referring to Appendix C, we note that the group

$$Q_2 \times C_3$$

is an example of a nilpotent group (of order 24) satisfying the conditions of the theorem. This group has unique Sylow subgroups isomorphic to Q_2 and C_3 (of orders 8 and 3, respectively) which are therefore normal, it is a direct product, and all of its maximal subgroups are normal. These subgroups are three copies of C_{12} (of index 2, and so normal; see Problem 2.19), and the unique Sylow 2-subgroup is Q_2 (normal by Theorem 6.10). The group $D_4 \times C_3$ is another non-Abelian example listed in Appendix C, as an exercise the reader should describe its maximal subgroups.

7.2 Finite Abelian Groups

We now consider finite Abelian groups in more detail. When studying Abelian groups, some authors use an additive notation. We use the multiplicative notation throughout to 'keep the work in context'.

With cyclic groups as factors, many Abelian groups can be constructed using the direct product construction described in the previous section, see Lemma 7.5(ii). Remarkably, the opposite is also true in the finite case, that is, *all* finite Abelian groups are isomorphic to direct products of cyclic groups. This is not true for infinite groups, for example, the rational numbers $\mathbb{Q}$ with addition cannot be expressed as a direct product of cyclic groups; see Web Section 7.5.

Several direct products of cyclic groups can be constructed having order n when n is composite, but they may not all be distinct (non-isomorphic). Previously we showed that C_6 and $C_2 \times C_3$ are isomorphic; note that 2 and 3 are coprime. On the other hand, consider the case $n = 8$. We have three possibilities

$$C_8, \quad C_4 \times C_2 \quad \text{and} \quad C_2 \times C_2 \times C_2,$$

and no two are isomorphic—the first contains an element of order 8 whilst the others do not, and the first two contain elements of order 4 whilst the last is an elementary Abelian 2-group. These two examples are typical. If G is a group of order n, and n is a product of distinct primes then the group is completely determined provided it is Abelian, but if high powers occur in the factorisation of n then many Abelian groups are possible, see the note at the end of this section. The *Fundamental Theorem of Finite Abelian Groups* characterises these groups completely; it is in two parts. The first part, called the *Basis Theorem*, describes the essential structure. We give two proofs, the first is relatively short and describes the basic facts, but it does not establish the full picture as the prime factorisation of the group order is involved only indirectly, the second gives more information. The remaining part of the Fundamental Theorem determines the isomorphism classes, that is, it provides conditions under which two finite Abelian groups are isomorphic, its proof will be given in the problem section (Problem 7.15).

A number of proofs of the Basis Theorem use the Euclidean Algorithm (Theorem B.2). Our first proof relies on the following matrix result which is a consequence of this algorithm.

An $n \times n$ matrix A is called *unimodular* if every entry is an integer, and $\det A = 1$ (and so A is non-singular); in this case, A^{-1} is also unimodular. We have

Lemma 7.10 *If $r_1, \ldots, r_n$ are integers having no positive common factor except 1, then there exists an $n \times n$ unimodular matrix with first column $(r_1, \ldots, r_n)^\top$, where $\top$ denotes the transpose.*

For a proof of this result, see Rose (1999), page 165; the 2×2 case is given by the Euclidean Algorithm, and the general case follows by induction.

We give now the first proof of the Basis Theorem, it relies on the following

Lemma 7.11 *If $\{g_1, \ldots, g_n\}$ is a generating set for an Abelian group G, and $r_1, \ldots, r_n$ are coprime integers (that is, they have no common factor larger than 1), then G has a second generating set one of whose elements is*

$$h_1 = g_1^{r_1} \cdots g_n^{r_n}.$$

Proof Let $A = (s_{ij})$ be one of the unimodular matrices with first column $(r_1, \ldots, r_n)^\top$ given by Lemma 7.10 where

$$r_i = s_{i1} \quad \text{for } i = 1, \ldots, n,$$

and let

$$h_i = g_1^{s_{1i}} \cdots g_n^{s_{ni}} \quad \text{for } i = 1, \ldots, n.$$

Note that $h_i \in G$ for all i, as each $s_{ij} \in \mathbb{Z}$. Now let $A^{-1} = (t_{ij})$, then $t_{ij} \in \mathbb{Z}$ for all i and j, and we have $s_{i1}t_{1j} + \cdots + s_{in}t_{nj}$ equals 1 if $i = j$, and it equals 0 if $i \neq j$ (as $AA^{-1} = I_n$). Hence

$$g_j = g_1^{s_{11}t_{1j} + \cdots + s_{1n}t_{nj}} \cdots g_j^{s_{j1}t_{1j} + \cdots + s_{jn}t_{nj}} \cdots g_n^{s_{n1}t_{1j} + \cdots + s_{nn}t_{nj}}$$

$$= \left(g_1^{s_{11}} \cdots g_j^{s_{j1}} \cdots g_n^{s_{n1}}\right)^{t_{1j}} \cdots \left(g_1^{s_{1n}} \cdots g_j^{s_{jn}} \cdots g_n^{s_{nn}}\right)^{t_{nj}}$$

$$= h_1^{t_{1j}} \cdots h_n^{t_{nj}},$$

as G is Abelian. This shows that the set $\{h_1, \ldots, h_n\}$ can be used as a generating set for G, and h_1 has the required property. $\square$

Using this lemma, we give the first proof of

Theorem 7.12 (Basis Theorem for Finite Abelian Groups) *Every finite Abelian group G is isomorphic to a direct product of cyclic groups.*

In some cases, the product has only one term, for instance, when $G = C_p$, see also Theorem 7.7(ii).

Proof By induction on $o(G)$. The result is clearly true for the neutral group; and so, using the inductive hypothesis, we may assume that it holds for all Abelian groups of order less than $o(G)$.

As G is finite, it has a finite generating set. Let n be the *least* positive integer such that G has a generating set with n elements; as noted above if $n = 1$ there is nothing to prove. So we have

every subset of G with at most $n - 1$ elements fails to generate G.   (∗)

Secondly, amongst elements of all n-element generating sets for G choose g_1 of *least* order t, say (note $t > 1$), and let the remaining elements of the generating set containing g_1 be $g_2, \ldots, g_n$. Let H be the subgroup of G generated by $g_2, \ldots, g_n$. By (∗) and the inductive hypothesis, H is a proper subgroup of G, and it is isomorphic to a direct product of cyclic groups using Lemma 7.5(iv). We prove the theorem by showing

$$G \simeq \langle g_1 \rangle \times H.$$

By Theorem 7.4, this will follow if we can prove that

$$\langle g_1 \rangle \cap H = \langle e \rangle,   \tag{7.2}$$

because $G = \langle g_1 \rangle H$ by definition, and all subgroups are normal.

Suppose (7.2) is false, that is, a non-neutral element $x \in G$ exists satisfying $x \in \langle g_1 \rangle$ and $x \in H$. It follows that integers $r_1, \ldots, r_n$ exist with

$$x = g_1^{r_1} = g_2^{r_2} \cdots g_n^{r_n},   \tag{7.3}$$

and

$$0 < r_1 < t,   \tag{7.4}$$

where t was defined above, $r_1 \neq 0$ because $x \neq e$, and $r_1 < t$ because $\langle g_1 \rangle$ is a cyclic group of order t. Also, at least one $r_{i+1} \neq 0$ (again as $x \neq e$). Suppose the GCD of $r_1, \ldots, r_n$ is d. As $d \leq r_1$, we have $d < t$ by (7.4). Let

$$y = g_1^{-r_1/d} g_2^{r_2/d} \cdots g_n^{r_n/d}.   \tag{7.5}$$

We know that $\{g_1, \ldots, g_n\}$ is a generating set for G and, by Lemma 7.10 and as the GCD of the set of integers $\{-r_1/d, r_2/d, \ldots, r_n/d\}$ equals 1, the element of G on the right-hand side of (7.5) is a member of a generating set for G with n elements. But by (7.3)

$$y^d = e,$$

that is, $o(y) \leq d < t$. This is impossible because, by construction, the minimum order of an element of a generating set for G with n members is t. Hence our assumption (that $x \neq e$) is false, and the result follows.   □

We come to the second proof of Theorem 7.12, as noted above it is longer but it provides more information. In the first part (Lemma 7.13), we show that a finite Abelian group G is isomorphic to a direct product of p_i-subgroups where $p_i \mid o(G)$.

Lemma 7.13 *Suppose G is a finite Abelian group and $o(G) = p_1^{s_1} \cdots p_k^{s_k}$, where $p_1, \ldots, p_k$ are distinct primes, and $s_i \geq 1$, $i = 1, \ldots, k$. For $i = 1, \ldots, k$, let H_{p_i} denote the subset of all those elements of G whose orders are a power of p_i.*

(i) $H_{p_i} \leq G$.
(ii) $G \simeq H_{p_1} \times \cdots \times H_{p_k}$.

Proof (i) If $a, b \in H_{p_i}$, $o(a) = p_i^j$, $o(b) = p_i^k$, and if l equals the maximum of j and k, then, as G is Abelian, $o(a^{-1}b) \mid p_i^l$. Hence (i) follows.

(ii) The first proof was given in Theorem 7.9—the subgroups H_{p_i} given in (i) are the Sylow p_i-subgroups of G, and they are normal because the group is Abelian. For a second more direct proof, we argue as follows using Theorem 7.6. Let $m_i = o(G)/p_i^{s_i}$. The integers $m_1, \ldots, m_k$ have no common factor larger than one by definition, so using the Euclidean Algorithm (Theorem B.2), integers $r_1, \ldots, r_k$ exist satisfying

$$r_1 m_1 + \cdots + r_k m_k = 1.$$

Now if $x \in G$, then for each $i = 1, \ldots, k$, the integer $o(x^{m_i})$ is a power of p_i (as $o(x) \mid o(G)$), and so $x^{m_i} \in H_{p_i}$. Hence

$$x = x^{m_1 r_1 + \cdots + m_k r_k} = (x^{m_1})^{r_1} \cdots (x^{m_k})^{r_k} \in H_{p_1} \cdots H_{p_k}.$$

As this holds for all $x \in G$, Condition (ii) in Theorem 7.6 is satisfied.

For the third condition, suppose $y \in H_{p_i} \cap H_{p_1} \cdots H_{p_{i-1}} H_{p_{i+1}} \cdots H_{p_k}$. So there exist integers s_i satisfying

$$o(y) = p_i^{s_i} = p_1^{s_1} \cdots p_{i-1}^{s_{i-1}} p_{i+1}^{s_{i+1}} \cdots p_k^{s_k}.$$

But this is only possible if $s_1 = \cdots = s_k = 0$ (the primes p_i are distinct), in which case $y = e$ and so Condition (iii) of Theorem 7.6 is also satisfied. Condition (i) is automatically satisfied, so the proof is complete. $\square$

Continuing the second proof of Theorem 7.12, by Lemma 7.13 we now need to prove the result in the case when G is a p-group, and so we can make use of the work on these groups given in Section 6.1.

Proof of Theorem 7.12 in the case when G is a finite Abelian p-group, (and so by Theorem 6.3, when $o(G)$ is a power of p) We use induction on $o(G)$, there is nothing to prove if $o(G) \leq p$.

Let H be a maximal subgroup of G. By Theorem 6.6, $o(G/H) = p$ and, by the inductive hypothesis, we can express H as a direct product of j, say, cyclic p-groups H_i where $o(H_i) = p^{r_i}$, $i = 1, \ldots, j$; that is, by the inductive hypothesis we have

$$H \simeq H_1 \times \cdots \times H_j \quad \text{and} \quad r_1 \geq \cdots \geq r_j \geq 1. \tag{7.6}$$

This holds by rearranging the H_i; see Lemma 7.5(iv).

Choose $a \in G \backslash H$ and then $a^p \in H$, there are p cosets $a^t H$, one for each t in the range $0, \ldots, p - 1$. Hence we can choose $h_i \in H_i$ to satisfy

$$a^p = h_1 \cdots h_j. \tag{7.7}$$

We may assume that each h_i is either a generator of H_i, or equals e. For by Problem 6.1(ii), a non-generator of H_i is a pth power, and if $h_i = b_i^p$ for some $b_i \in H_i$, then using (7.6) and as G is Abelian we have

$$\left(ab_i^{-1}\right)^p = a^p b_i^{-p} = a^p h_i^{-1} = h_1 \cdots h_{i-1}h_{i+1} \cdots h_j.$$

But $ab_i^{-1} \notin H$ by definition of a and b_i; and so this contradiction establishes our assumption. Now if $a^p = e$, then G is isomorphic to a direct product of $\langle a \rangle$, $H_1, \ldots, H_{j-1}$ and H_j, and the theorem follows in this case. Hence we may assume that $h_i \neq e$ for some i satisfying $1 \leq i \leq j$. Let i' equal the smallest such i, so $h_{i'} \neq e$ and $a^p = h_{i'} \cdots h_j$ by (7.7).

Now by (7.7) again, $o(a^p) = p^{r_{i'}}$ (order is LCM of $o(h_{i'}), \ldots, o(h_j)$), hence $o(\langle a \rangle) = p^{r_{i'}+1}$. Let $J = H_1 \times \cdots \times H_{i'-1} \times \langle e \rangle \times H_{i'+1} \times \cdots \times H_j$. Clearly, $o(J) = o(H)/p^{r_{i'}}$, and the theorem will follow if we can show that

$$\langle a \rangle \cap J = \langle e \rangle, \tag{7.8}$$

for then $\langle a \rangle J$ is a direct product with order equal to $o(G)$.

Suppose (7.8) is false, and so $a^t \in J$ for some t. Now $p \mid t$ by definition. So if $t = sp$ for some s satisfying $0 \leq s < p^{r_{i'}}$, then by (7.7)

$$a^t = \left(a^p\right)^s = h_{i'}^s \cdots h_j^s \quad \text{and} \quad h_{i'}^s \neq e \tag{7.9}$$

by definition of s. This shows that the unique (we have a direct product) representation (7.9) of a^t given in $H = H_1 \times \cdots \times H_j$ has a non-neutral element in its i'th place. But this cannot belong to J because J has e here. This establishes (7.8), and the main result follows.   □

The second proof of Theorem 7.12 is now complete. For, by Lemma 7.13, a finite Abelian group G is isomorphic to a direct product of p_i-subgroups where $p_i \mid o(G)$, and each of these is itself a direct product of cyclic p_i-subgroups.

Both of these proofs rely on an argument involving induction. We apply the inductive hypothesis to G with one of its smallest terms H, say, removed, to obtain a direct product K, and then show that we can form the direct product of H and K. Other proofs work by removing the largest factor.

Example Consider the case $n = 72$. We have $72 = 2^3 \cdot 3^2$, and each Abelian group of order 72 is a direct product of a group H of order 8 and a group J of order 9 (Lemma 7.13). There are three possibilities for H: C_8, $C_4 \times C_2$ and $C_2 \times C_2 \times C_2$, and two for J: C_9 and $C_3 \times C_3$. Hence there are six (isomorphism classes of) Abelian groups of order 72:

$$C_8 \times C_9, \quad C_8 \times C_3 \times C_3, \quad C_4 \times C_2 \times C_9, \quad C_4 \times C_2 \times C_3 \times C_3,$$

$$C_2 \times C_2 \times C_2 \times C_9, \quad \text{and} \quad C_2 \times C_2 \times C_2 \times C_3 \times C_3.$$

This list is complete; for example, $C_{18} \times C_4$ is included because $C_{18} \simeq C_2 \times C_9$, and so $C_{18} \times C_4 \simeq C_4 \times C_2 \times C_9$. Note also that the groups T_n, for $n = 1, 2, \ldots$, introduced in Section 2.2 can now be defined by $T_n = C_2 \times \cdots \times C_2$ with n copies of C_2.

In general, if $n > 1$ and part(n) denotes the number of *partitions* of n (that is, the number of ways of writing n as a sum of equal or smaller positive integers; for example, $4 = 3 + 1 = 2 + 2 = 2 + 1 + 1 = 1 + 1 + 1 + 1$, and so part$(4) = 5$), then there are part$(n)$ distinct (non-isomorphic) Abelian groups of order p^n. The integer part(n) increases exponentially with n; for instance, there are 42 (isomorphism classes of) Abelian groups of order 1024, and 627 of order 2^{20}. For further details on partitions, see Andrews (1976).

7.3 Semi-direct Products

As noted earlier, given two groups H and J, a number of new groups can be constructed on the underlying set $H \times J$. One is the direct product, but in many cases others are possible. The new group is, in a sense to be made precise, either an extension of H by J, or of J by H; see Chapter 9. Contrary to our discussion of direct products given in Section 7.1, we begin our work with the *internal* semi-direct product, this will lead to the 'right' definition for the *external* product. We take the same underlying set as in the direct case and define a different operation, and we begin with (*cf.* Theorem 7.4)

Definition 7.14 Let G be a group with a subgroup A and a normal subgroup K which satisfy

$$G = AK \quad \text{and} \quad A \cap K = \langle e \rangle. \tag{7.10}$$

In this case, G is called an (*internal*) *semi-direct product* of K by A, and we write

$$G \simeq A \rtimes K.$$

For example, the dihedral group $D_n = \langle a, b \mid a^n = b^2 = e, bab = a^{n-1} \rangle$ can be represented as a semi-direct product if we take $\langle b \rangle$ for A and $\langle a \rangle$ for K, and then $D_n \simeq \langle b \rangle \rtimes \langle a \rangle$.

Notes (a) In symbols we always place the normal subgroup K second next to the triangular part of the $\rtimes$ sign. The subgroup A is called a *complement* of K in G, and G is sometimes called a *split extension* of K by A. Some authors put the normal subgroup K first and write $K \ltimes A$. The ATLAS (1985) uses $K : A$ for our $A \rtimes K$ (and $K.A$ for a general extension of K by A).

(b) If G is the direct product of A and K, then clearly it is also a semi-direct product of K by A, and of A by K.

(c) G can only be simple if either A or K is the neutral subgroup, also we shall see below (Lemma 7.15) that G is not Abelian when the product is not direct even if both A and K are Abelian.

(d) It is important to note that given A and K, the group G is *not uniquely determined*; for example, both D_3 and C_6 are semi-direct products of C_3 by C_2. Note also that some groups can be expressed as a semi-direct product in several different ways; see page 183, for example.

(e) There are non-simple groups that cannot be expressed as semi-direct products, for example, the quaternion group Q_2. This group has a number of normal subgroups K_i with corresponding complements (subgroups) A_i, but (7.10) fails in each case; see Problem 7.17. Direct and semi-direct products do provide a number of interesting groups, but many more constructions need to be considered if all groups are to be described.

(f) A number of authors have considered the question: when can a group G be expressed as a product of two of its proper subgroups (with no restriction on these subgroups except that they are both proper), that is, when is G *factorisable*; see, for example, Scott (1964), Chapter 13. Clearly C_p is not factorisable, but Scott gives other instances—one is the simple group $L_2(13)$, see Section 12.2. He also gives some positive results which we shall discuss in Section 11.2.

Note (d) above suggests that something more is needed. The following lemma provides this missing link.

Lemma 7.15 *Suppose $G = A \rtimes K$.*

(i)   *If G is finite, $o(G) = o(A)o(K)$.*
(ii) *If $g \in G$, then g can be uniquely represented by $g = ak$ where $a \in A$ and $k \in K$.*
(iii) *If $g_i = a_i k_i$ where $a_i \in A$ and $k_i \in K$, for $i = 1, 2$, then $g_1 g_2$ has the unique representation as $g_1 g_2 = ak$, where $a = a_1 a_2 \in A$ and $k = a_2^{-1} k_1 a_2 k_2 \in K$.*
(iv) *There is a homomorphism $\gamma : A \to \operatorname{Aut} K$ with the property that the product $g_1 g_2$ given in (iii) can be defined by*

$$g_1 g_2 = (a_1 k_1)(a_2 k_2) = (a_1 a_2)(k_1 (a_2 \gamma))k_2.$$

(v)   *The homomorphism γ given in (iv) is the trivial map (see page 69) if and only if the product is direct.*
(vi) *If the homomorphism γ given in (iv) is not the trivial map, then G is not Abelian.*

Proof For $g_i \in G$ we write $g_i = a_i k_i$ throughout where $a_i \in A$ and $k_i \in K$; see Definition 7.14.

(i) By the Second Isomorphism Theorem (Theorem 4.15) and (7.10) we have $A \simeq A/A \cap K \simeq AK/K = G/K$, and so the result follows by Lagrange's Theorem (Theorem 2.27).

(ii) Suppose $g = a_1 k_1 = a_2 k_2$, $a_i \in A$ and $k_i \in K$, for $i = 1, 2$. Then $a_2^{-1} a_1 = k_2 k_1^{-1} = e$ by (7.10), that is $a_1 = a_2$ and $k_1 = k_2$.

(iii) We have $g_1 g_2 = a_1 k_1 a_2 k_2 = a_1 a_2 a_2^{-1} k_1 a_2 k_2$, where $a_1 a_2 \in A$ and, as $K \lhd G$, $a_2^{-1} k_1 a_2 \in K$, and so $a_2^{-1} k_1 a_2 k_2 \in K$. Uniqueness follows from (ii).

(iv) Define the map $\gamma : A \to \operatorname{Aut} K$ using conjugation as follows. If $a \in A$ and $k \in K$, let

$$k(a\gamma) = a^{-1} k a.$$

By definition, $a\gamma$ is an automorphism of K for each $a \in A$, and so $k(a\gamma) \in K$ also for each $a \in A$. Now γ is a homomorphism because $a_1 \gamma \circ a_2 \gamma = (a_1 a_2)\gamma$, this follows because $(a_1 a_2)^{-1} k (a_1 a_2) = a_2^{-1}(a_1^{-1} k a_1) a_2$.

(v) First, γ is the trivial map on A if, and only if, $a\gamma$ equals the identity map (automorphism) on K for all $a \in A$, that is $k(a\gamma) = k$ for all $a \in A$ and $k \in K$. If the product is direct, then $ak = ka$ for all $a \in A$ and $k \in K$, and so γ is the trivial map. Conversely, if γ is the trivial map, then for all k we have $k(a\gamma) = k$, but $k(a\gamma) = a^{-1} k a$, and so $ak = ka$ for all $a \in A$ and $k \in K$. This shows that $A \lhd G$, hence the product is direct by Theorem 7.4.

(vi) By (iv), if $\gamma : A \to \operatorname{Aut} K$ is not trivial, there is an $a \in A$ such that $a\gamma$ is not the identity map, hence there exist $a \in A$ and $k \in K$ such that $k(a\gamma) \neq k$. But by (iv) $k(a\gamma) = a^{-1} k a$, that is, a and k do not commute. $\square$

These results suggest that to define an *external* semi-direct product of K by A we need to consider the homomorphism γ; in many cases different γ will give rise to different groups even if both A and K are fixed.

Definition 7.16 Given groups A and K, and a homomorphism $\gamma : A \to \operatorname{Aut} K$, the *(external) semi-direct product* $A \rtimes_\gamma K$ of K by A relative to γ is the group with underlying set $A \times K$, and operation

$$(a_1, k_1)(a_2, k_2) = \big(a_1 a_2, (k_1 (a_2 \gamma)) k_2\big), \tag{7.11}$$

where $a_i \in A$ and $k_i \in K$, for $i = 1, 2$.

Note that $a_2 \gamma$ is an automorphism of K for each $a_2 \in A$, and so $k_1(a_2 \gamma) \in K$, hence we also have $(k_1(a_2 \gamma))k_2 \in K$ when $k_1, k_2 \in K$.

Before stating the next result (Theorem 7.17) we consider the following

Example Let $A = \langle a \rangle \simeq C_2$ and $K = \langle k \rangle \simeq C_3$, see (d) opposite. By Theorem 4.23, $\operatorname{Aut} C_3 \simeq C_2$, that is C_3 has two automorphisms η_1 and η_2 as follows:

(i) η_1 is the identity map which satisfies $a^t \eta_1 : k^u \mapsto k^u$ for all $k \in K$ and $t, u \in \mathbb{Z}$.

(ii) η_2 is the map which satisfies, for all integers u, $e\eta_2 : k^u \mapsto k^u$ and $a\eta_2 : k^u \mapsto k^{2u}$, and so η_2^2 is the identity map η_1 on K.

Combining the two statements in (ii) we obtain: for $t = 0$ or 1, and for all u

$$a^t \eta_2 : k^u \mapsto k^{(t+1)u}. \tag{7.12}$$

We can define two groups using (i) or (ii), respectively. In the first case, $k^s(a^t \eta_1) = k^s$ for all s and t, which gives for $0 \le r, t \le 1, 0 \le s, u \le 2$,

$$\left(a^r, k^s\right)\left(a^t, k^u\right) = \left(a^{r+t}, \left(k^s\left(a^t \eta_1\right)\right)k^u\right) = \left(a^{r+t}, k^{s+u}\right),$$

the direct product of A and K isomorphic to C_6.

For the second group, we use (7.12). We have, for r, s, t and u in the same ranges as those listed above,

$$\left(a^r, k^s\right)\left(a^t, k^u\right) = \left(a^{r+t}, \left(k^s\left(a^t \eta_2\right)k^u\right)\right) = \left(a^{r+t}, k^{(t+1)s+u}\right).$$

The reader should check that this gives a group isomorphic to the dihedral group D_3 where the usual generators a and b are replaced by k and a, respectively.

The next result shows that the semi-direct product defines a group.

Theorem 7.17 (i) *The set with operation in Definition 7.16 forms a group.*
(ii) *The map γ given in Definition 7.16 is defined by conjugation.*

To make this proof easier to follow we have reintroduced the $\odot$ symbol to denote the group operation in K, that is, (7.11) now reads

$$(a_1, k_1)(a_2, k_2) = \left(a_1 a_2, \left(k_1(a_2\gamma)\right) \odot k_2\right).$$

Proof (i) The operation is well-defined by definition. The neutral element is (e, e) for

$$(a, k)(e, e) = \left(ae, \left(k(e\gamma)\right) \odot e\right) = (a, k)$$

as $e\gamma$ is the identity map. The inverse of (a, k) is $(a^{-1}, k^{-1}(a^{-1}\gamma))$ for

$$(a, k)\left(a^{-1}, k^{-1}(a^{-1}\gamma)\right) = \left(aa^{-1}, \left(k(a^{-1}\gamma)\right) \odot \left(k^{-1}(a^{-1}\gamma)\right)\right)$$
$$= \left(e, \left(k \odot k^{-1}\right)(a^{-1}\gamma)\right) = (e, e)$$

because $a^{-1}\gamma$ is a homomorphism (automorphism) which maps e to e. For associativity, we proceed as follows: The square brackets in the expressions below are not strictly necessary (they enclose products in K) but are inserted to aid clarity. We have, where $a_i \in A$ and $k_i \in K$ for $i = 1, 2, 3$,

$$((a_1, k_1)(a_2, k_2))(a_3, k_3) = (a_1 a_2, [(k_1(a_2\gamma)) \odot k_2])(a_3, k_3)$$
$$= (a_1 a_2 a_3, ([(k_1(a_2\gamma)) \odot k_2](a_3\gamma)) \odot k_3),$$

and

$$(a_1, k_1)((a_2, k_2)(a_3, k_3)) = (a_1, k_1)(a_2 a_3, [(k_2(a_3\gamma)) \odot k_3])$$
$$= (a_1 a_2 a_3, (k_1((a_2 a_3)\gamma)) \odot [(k_2(a_3\gamma)) \odot k_3])$$
$$= (a_1 a_2 a_3, [(k_1((a_2 a_3)\gamma)) \odot (k_2(a_3\gamma))] \odot k_3),$$

by associativity in K. The expressions on the right-hand sides of these equations are equal because

$$(k_1((a_2 a_3)\gamma)) \odot (k_2(a_3\gamma))$$
$$= (k_1(a_2\gamma \circ a_3\gamma)) \odot (k_2(a_3\gamma)) \qquad \text{as } \gamma \text{ is a homomorphism}$$
$$= [(k_1(a_2\gamma))(a_3\gamma)] \odot (k_2(a_3\gamma)) \qquad \text{by definition of composition}$$
$$= [(k_1(a_2\gamma)) \odot k_2](a_3\gamma) \qquad \text{as } a_3\gamma \text{ is an automorphism.}$$

This proves (i).

(ii) The set $\{(a, e) : a \in A\}$ forms an isomorphic copy of A in G, similarly the set $\{(e, k) : k \in K\}$ forms an isomorphic copy of K in G. Now using the equations above we have

$$(a, e)^{-1} = (a^{-1}, e(a^{-1}\gamma)) = (a^{-1}, e),$$

because $a^{-1}\gamma$ is a homomorphism, and

$$(e, k)(a, e) = (ea, (k(a\gamma)) \odot e) = (a, k(a\gamma)).$$

Hence

$$(a, e)^{-1}(e, k)(a, e) = (a^{-1}, e)(a, (k(a\gamma)))$$
$$= (a^{-1}a, (e(a\gamma)) \odot (k(a\gamma))) = (e, k(a\gamma))$$

for all $a \in A$ and $k \in K$, and as $e(a\gamma) = e$. Therefore, the homomorphism γ is given by conjugation of an element in K by a element in A. $\qquad\square$

As an exercise the reader should show directly (using the same methods as above) that $(e, e)(a, k) = (a, k)$ and $(a^{-1}, k^{-1}(a^{-1}\gamma))(a, k) = (e, e)$; note these equations also follow from Theorems 7.17 and 2.5.

Wreath Product

Before providing some examples we give a brief introduction to a particular type of semi-direct product called a *wreath product*, a number of groups with special properties can be defined using it. Suppose G and H are groups, $X = \{1, 2, \ldots, n\}$, H acts on X, and G^* is the direct product of n copies of G. We define an action of H on G^* by

$$(g_1, \ldots, g_n)\backslash h = \left(g_{1\backslash h^{-1}}, \ldots, g_{n\backslash h^{-1}}\right) \quad \text{for } h \in H.$$

It is easy to see that the action axioms (5.1) hold, and

$$(g_1, \ldots, g_n)\backslash h\left(g_1', \ldots, g_n'\right)\backslash h = \left(g_1 g_1', \ldots, g_n g_n'\right)\backslash h,$$

gives a homomorphism. The *wreath product* of G by H is defined as the semi-direct product $H \times_\phi G^*$ where ϕ is the homomorphism given above. The product is denoted by either G wr H or $G \wr H$. We have (a) the product is associative, and (b) $o(G \text{ wr } H) = o(H) \cdot o(G)^n$. An example is: C_2 wr $C_2 \simeq D_4$. We mention two applications.

(a) If G is soluble with derived length n (page 234), then G wr C_2 has derived length $n + 1$, hence there exist soluble groups with arbitrarily high derived length; for a proof, see the first reference given below.
(b) The Sylow subgroups of the symmetric groups S_{p^m} are isomorphic to iterated ($m - 1$ times) wreath products of the cyclic group C_p, and the Sylow subgroups of general symmetric groups can be constructed using direct products of these iterated wreath products; this is proved in the last reference below.

For more details and proofs, the reader should see Rose (1978), page 219 *ff*. Rotman (1994), page 173 *ff*, Suzuki (1982), page 268 *ff*, and Cameron (1999) should also be consulted for (b).

Groups of Order 12

As an application of the theorems above, we consider the five (isomorphism classes of) groups of order 12; in fact, all of these groups can be treated as semi-direct products. By Theorem 7.12, there are two (isomorphism classes of) Abelian groups of order 12:

$$C_{12} \simeq C_4 \times C_3 \quad \text{and} \quad C_6 \times C_2 \simeq C_3 \times C_2 \times C_2;$$

these are direct products (and so are special cases of semi-direct products). For non-Abelian groups, we have the following preliminary

Lemma 7.18 *If G is a non-Abelian group of order* 12, *then G is either* (i) *isomorphic to A_4, or* (ii) *it contains a normal cyclic subgroup of order* 6.

Proof Let P be a Sylow 3-subgroup of G. As $o(P) = 3$, P is cyclic, and so we can assume that it has the form $\langle a \rangle$ where $a^3 = e$. We also have $[G : P] = 4$, and so by Theorem 5.15, there is an injective homomorphism

$$\theta : G \Big/ \bigcap_{h \in G} h^{-1} P h \to S_4.$$

Now $\bigcap_{h \in G} h^{-1} P h$ equals $\langle e \rangle$ or P (as $o(P)$ is prime). If the first case applies then θ is an injective homomorphism of G into S_4, and so G is isomorphic to a normal (as the index is 2) subgroup of S_4. Problem 3.3 now shows that $G \simeq A_4$. Note that A_4 has no elements of order 6.

For the second case, we have $\bigcap_{h \in G} h^{-1} P h = P$, and so $P \triangleleft G$. This implies that G has only one (Sylow) subgroup of order 3 (Sylow 3), and so G has only two elements of order 3: a and a^2. This further implies that a can only have one or two conjugates in G as the order of a conjugate of a equals the order of a. Theorem 5.19 now gives $[G : C_G(a)] = 1$ or 2, where $C_G(a)$ denotes the centraliser of a in G. Hence $o(C_G(a)) = 12$ or 6. By Cauchy's Theorem (Theorem 6.2), in either case $C_G(a)$ contains an element b, say, of order 2, and this element commutes with a. Therefore, $ab \in G$, $o(ab) = 6$, and so $\langle ab \rangle$ is a normal (as its index in G is 2) subgroup of G of order 6. $\qquad\square$

Continuing the argument given in the second part of the above proof, let $ab = c$, then $o(\langle c \rangle) = 6$ and $\langle c \rangle \triangleleft G$. Suppose $d \in G \backslash \langle c \rangle$, then

$$d^{-1} c d = c^t \quad \text{for some } t = 0, \dots, 5.$$

If $t = 0$ then $c = e$, if $t = 1$ then G is Abelian, if $t = 2$ then $e = (d^{-1}cd)^3 = d^{-1}c^3d$ which implies $c^3 = e$, and if $t = 3$ or 4 a similar argument shows that $c^2 = e$ or $c^3 = e$. All of these contradict the properties of c given above, and so $t = 5$ is the only possibility. Secondly, as the elements c^r and $c^r d$, for $r = 0, \dots, 5$, are all distinct in G (the reader should check this), and $o(G) = 12$, we have $d^2 \in \langle c \rangle$. If $d^2 = c$ or c^5 then $o(d) = 12$ and G is cyclic, hence we may exclude these possibilities. If $d^2 = e$ then

$$G \simeq D_6 = \langle a, b \mid a^6 = b^2 = e, bab = a^5 \rangle.$$

Secondly, if $d^2 = c^2$ then, if we replace d by $d_1 = c^2 d$, we have $d_1^2 = (c^2 d)^2 = e$ and $d_1^{-1} c d_1 = d^{-1} c^7 d = c^5$, and so again $G \simeq D_6$ now with generators c and d_1. Thirdly, if $d^2 = c^4$, arguing similarly we again have $G \simeq D_6$. The final possibility, that is, $d^2 = c^3$, gives the dicyclic group Q_3 of order 12 (Section 3.4) where

$$Q_3 = \langle c, d \mid c^6 = e, d^2 = c^3 \text{ and } d^{-1} c d = c^5 \rangle.$$

To recap, we have shown that there are at most three (isomorphism classes of) non-Abelian subgroups of order 12: A_4, D_6 and Q_3. As an exercise the reader should write down the Sylow subgroups of these groups (Problem 7.23). We show now that each of these groups can be defined using the semi-direct product construction, that is,

$$A_4 \simeq C_3 \rtimes T_2, \quad D_6 \simeq C_2 \rtimes C_6, \quad \text{and} \quad Q_3 \simeq C_4 \rtimes C_3.$$

Case 1: A_4.

The group A_4 (treated as a permutation group on $\{1, 2, 3, 4\}$) has a normal subgroup K_1 of order 4 containing the three products of two 2-cycles and e, and so it is isomorphic to the 4-group T_2. Taking $B_1 = \langle (1, 2, 3) \rangle$, we have

$$B_1 \leq A_4, \quad K_1 \lhd A_4, \quad B_1 \cap K_1 = \langle e \rangle, \quad \text{and} \quad A_4 = B_1 K_1.$$

Reader, why does $B_1 K_1$ have twelve elements?

Hence $A_4 \simeq B_1 \rtimes K_1$, and we need to construct the homomorphism $\gamma_1 : B_1 \to \text{Aut } K_1$. We have $\text{Aut } K_1 \simeq S_3$ (see Section 4.4, note that K_1 is an elementary Abelian 2-group), and so the only non-trivial homomorphism from B_1 (of order 3) to $\text{Aut } K_1$ (of order 6) is an injection of B_1 onto the (unique) subgroup of S_3 generated by its 3-cycles and isomorphic to C_3. For example, using (7.11) on page 153 with $a_1 = (1, 2, 3), k_1 = (1, 2)(3, 4), k_2 = (1, 3)(2, 4)$ and a_2 ranging over the elements $e, (1, 2, 3)$ and $(1, 3, 2)$ of B_1 (they are underlined below) we have

$$(1, 2, 3)(1, 2)(3, 4) \cdot \underline{e}(1, 3)(2, 4) = \underline{(1, 2, 3)} \cdot (1, 2)(3, 4)(1, 3)(2, 4),$$

$$(1, 2, 3)(1, 2)(3, 4) \cdot \underline{(1, 2, 3)}(1, 3)(2, 4) = \underline{(1, 2, 3)}^2 \cdot (1, 4)(2, 3)(1, 3)(2, 4),$$

$$(1, 2, 3)(1, 2)(3, 4) \cdot \underline{(1, 2, 3)}^2(1, 3)(2, 4) = \underline{e} \cdot (1, 3)(2, 4)(1, 3)(2, 4).$$

This illustrates the fact that we can define the image of $(1, 2)(3, 4)$ under γ_1 using (7.11) by:

$$\underline{e}\gamma_1 : (1, 2)(3, 4) \mapsto (1, 2)(3, 4),$$

$$\underline{(1, 2, 3)}\gamma_1 : (1, 2)(3, 4) \mapsto (1, 4)(2, 3), \quad \text{and}$$

$$\underline{(1, 3, 2)}\gamma_1 : (1, 2)(3, 4) \mapsto (1, 3)(2, 4),$$

with similar expressions for the images of $e, (1, 3)(2, 4)$ and $(1, 4)(2, 3)$ which the reader should write out. See Problem 3.10 for further properties on A_4.

Case 2: D_6.

By Lemma 7.18 and the discussion given below its proof, the group D_6 contains a normal cyclic subgroup (of order 6) $K_2 \simeq \langle c \rangle$, and an element d of order 2. If we let $B_2 = \langle d \rangle$, we have $D_6 = B_2 K_2$ and $B_2 \cap K_2 = \langle e \rangle$, as required for a semi-direct product. As above we need to construct the homomorphism $\gamma_2 : B_2 \to \operatorname{Aut} K_2$. By Theorem 4.23, and Problems 4.19 and 7.10, $\operatorname{Aut} K_2 \simeq C_2$, and the only non-identity automorphism of K_2 maps c to c^5; that is, $e\gamma_2$ is the identity automorphism on K_2, and $d\gamma_2$ is the automorphism that maps c^s to c^{5s}. Hence we have for $r = 0$ or 1, and $s, u \in \{0, \ldots, 5\}$,

$$d^r c^s \cdot d^0 c^u = d^{r+0} \big(c^s (e\gamma_2) \big) c^u = d^r c^{s+u},$$

$$d^r c^s \cdot d c^u = d^{r+1} \big(c^s (d\gamma_2) \big) c^u = d^{r+1} c^{5s+u},$$

which gives a representation of D_6. Note that D_6 can also be represented by semi-direct product $T_2 \rtimes C_3$, this is a consequence of the fact that D_6 is isomorphic to $C_2 \times D_3$.

Case 3: Q_3.

The group Q_3 contains a normal cyclic subgroup $K_3 = \langle c^2 \rangle$ of order 3, and a second non-normal cyclic subgroup $B_3 = \langle d \rangle$ of order 4. Clearly, we have $B_3 \cap K_3 = \langle e \rangle$ and $B_3 K_3 = Q_3$, and so Q_3 is a semi-direct product of K_3 by B_3. By Theorem 4.23, $\operatorname{Aut} K_3 \simeq C_2$, and the only non-trivial homomorphism γ_3 of $B_3 = \langle d \rangle$ to $\operatorname{Aut} K_3$ associates the even powers of d with the identity automorphism, and the odd powers of d with the automorphism that maps c^s to c^{2s}. This is mirrored in Q_3 for we have in this group

$$d^r c^s \cdot d^{2t} c^u = d^{r+2t} \big(c^s (d^{2t} \gamma_3) \big) c^u = d^{r+2t} c^{s+u},$$

$$d^r c^s \cdot d^{2t+1} c^u = d^{r+2t+1} \big(c^s (d^{2t+1} \gamma_3) \big) c^u = d^{r+2t+1} c^{2s+u}$$

for $0 \le r < 4$, $0 \le t < 2$ and $0 \le s, u < 3$. The reader should check that this construction gives a group isomorphic to the dicyclic group Q_3 defined on page 59. It can also be represented as a metacyclic group, see Theorem 6.18.

7.4 Problems

Problem 7.1 (i) Give proofs of Lemma 7.5 and Theorem 7.6.

(ii) Show that if G is a product of its subgroups $H_1, \ldots, H_n$, where each $H_i \lhd G$ and $(o(H_i), o(H_j)) = 1$ if $i \ne j$, then $G \simeq H_1 \times \cdots \times H_n$.

Problem ◆ 7.2 (i) If G, H, J and K are groups, show that $[G \times H, J \times K] \simeq [G, J] \times [H, K]$, Problem 2.17.

(ii) If $H \lhd G$ and $J \lhd K$, show that $H \times J \lhd G \times K$, and

$$(G \times K)/(H \times J) \simeq G/H \times K/J.$$

(iii) Prove that if $J, K \lhd G$ and $G = JK$, then $J/(J \cap K) \times K/(J \cap K) \simeq G/(J \cap K)$.

Problem 7.3 Suppose $G = H \times J$.

(i) Show that $H \simeq J$ if and only if a subgroup D of G can be found which satisfies $G = HD = JD$ and $H \cap D = J \cap D = \langle e \rangle$.

(ii) If $H \leq L \leq G$ show that $L \simeq H \times (J \cap L)$, see Lemma 7.8.

Problem 7.4 Suppose $G = H_1 \times \cdots \times H_n$. Prove the following results.

(i) $Z(G) \simeq Z(H_1) \times \cdots \times Z(H_n)$.

(ii) $G' \simeq H_1' \times \cdots \times H_n'$.

(iii) If K is a perfect normal subgroup of G (Problem 4.8), then $K \simeq (H_1 \cap K) \times \cdots \times (H_n \cap K)$ (cf. Lemma 7.8).

Problem 7.5 Let G be a finite Abelian group, and let $\mathrm{ex}(G)$ denote its exponent.

(i) Show that there exists $g \in G$ with the property $\mathrm{ex}(G) \mid o(g)$.

(ii) Using (i) show that if $\mathrm{ex}(G) = o(G)$, then G is cyclic.

(iii) Use (ii) and the fact that in a field a polynomial of degree r has at most r roots (Theorem B.13) to show that the multiplicative group of a finite field is cyclic.

Problem 7.6 (i) Suppose $J \leq G \times H$, show that J is Abelian, or J intersects one of the factors G or H non-neutrally.

(ii) By considering an Abelian group G with order p^5 and three of its subgroups H_1, H_2 and H_3, each with order p^2, show that the following properties can all hold for a suitably chosen group G.

(a) $G = H_1 H_2 H_3$,

(b) $H_1 H_2$, $H_2 H_3$ and $H_3 H_1$ are all proper subgroups of G,

(c) $H_1 \cap H_2 = H_2 \cap H_3 = H_3 \cap H_1 = \langle e \rangle$, and

(d) G is isomorphic to a proper subgroup of $H_1 \times H_2 \times H_3$.

Problem ♦ 7.7 If G is Abelian, $o(G) = n$, and $m \mid n$, show that G has a subgroup of order m. First, reduce to the case when n is a prime power.

Problem 7.8 (i) How many Abelian groups (up to isomorphism) are there of order 385 or 432?

(ii) You are given: there are 14 (isomorphism classes of) groups of order 81. Using the Sylow and direct product theories count the number of (isomorphism classes of) groups of order (a) 891, and (b)* 405.

Problem 7.9 Suppose G is a finite group, and all of its maximal subgroups are both simple and normal. Show that G is Abelian, and $o(G) = 1, p, p^2$ or pq where p and q are prime.

Problem 7.10 Show that if G and H are finite groups and $(o(G), o(H)) = 1$, then $\mathrm{Aut}(G \times H) \simeq \mathrm{Aut}\,G \times \mathrm{Aut}\,H$.

Problem 7.11 Suppose G is a finite Abelian group with proper subgroup H; see also Problem 7.5.

(i)   Choose $g \in G$ with the largest possible order n. Prove that $h^n = e$ for all $h \in G$.
(ii)   Show that the result given in (i) is false in general for non-Abelian groups, that is, the exponent (Definition 2.19) can be larger than n.
(iii) Suppose J is maximal subject to the conditions $J \leq G$ and $H \cap J = \langle e \rangle$. If there exists $g \in G$ such that $g^p \in J$ for some prime p, show that $G = HJ$. (Hint. Consider the equation $h = jg^r$ where $h \in H$, $j \in J$ and $r \in \mathbb{Z}$.)
(iv)   With J as in (iii), show that $G \simeq H \times J$ if and only if for all primes p, and for all $g \in G$, $h \in H$ and $j \in J$ with $g^p = hj$, there exists $h_1 \in H$ satisfying $h = h_1^p$.
(v)   Finally, with g as in (i), show that $\langle g \rangle$ is a direct factor of G, that is, a subgroup $J \leq G$ exists with the property $G \simeq \langle g \rangle \times J$. (Hint. Use (i) and (iii) and consider the cases $p \mid n$ and $p \nmid n$ separately.)

Problem 7.12 A group G is called *characteristically simple* CS if its only characteristic subgroups are $\langle e \rangle$ and G itself. For the definition of characteristic, see Problem 4.22. Let G be a finite Abelian group. Show that G is CS if and only if it is elementary. A definition of 'elementary' is given in Problem 4.18, the note at the bottom of page 235 is also relevant. One method is as follows. First, suppose G is CS and $p \mid o(G)$. Let $H = \{g \in G : g^p = e\}$. Show that H is CS, and so deduce that $H = G$. For the converse, use the definition of an automorphism, and Problem 4.18.

Problem 7.13 (Hamiltonian Groups) A finite non-Abelian group is called *Hamiltonian* if all of its proper subgroups are normal; such groups are named after Hamilton, the discoverer of the quaternion groups, see page 119. R. Dedekind (1831–1916) proved that G is Hamiltonian if and only if it can be expressed as a direct product:

$$G \simeq Q_2 \times T \times O,$$

where Q_2 is the quaternion group (of order 8), T is an elementary Abelian 2-group (so $T \simeq T_n$ for some n, see Problems 4.18), and O is an arbitrary finite Abelian group of odd order. You are asked to show that if G has this form, then it is Hamiltonian. For the converse, see Robinson (1982), page 138. (Hint. Apply Lemma 7.8, consider the cases when the subgroup does, or does not, intersect Q_2, and use Problem 2.16(iii).)

Problem ◆ 7.14 Let G be a finite Abelian group, and let $m_i \in \mathbb{Z}$ for $i = 1, \ldots, k$. Elements $a_1, \ldots, a_k \in G$ are said to be *independent* if $a_1^{m_1} \cdots a_k^{m_k} = e$ implies $a_i^{m_i} = e$, for all i satisfying $1 \leq i \leq k$.

(i)   Show that the elements $a_1, \ldots, a_k$ are independent if and only if $\langle a_1, \ldots, a_k \rangle \simeq \langle a_1 \rangle \times \cdots \times \langle a_k \rangle$.
(ii)   Deduce $G \simeq \mathbb{Z}/p\mathbb{Z} \times \cdots \times \mathbb{Z}/p\mathbb{Z}$, if G has exponent p. Groups of this type are called *elementary Abelian*, see Problem 4.18.

Problem 7.15 (Isomorphism Theorem for Finite Abelian Groups) In this problem, assume that all groups are finite Abelian p-groups unless stated otherwise. For a finite Abelian group G, let $d(G)$ denote the minimum number of generators of G.

(i)   Show that if G and H are elementary, then $d(G \times H) = d(G) + d(H)$.

(ii) Suppose G has the decomposition $G \simeq C_1 \times \cdots \times C_k$ where each C_i is cyclic. Show that the number of factors C_i with order larger than p^n is $d(G^{p^n}/G^{p^{n+1}})$ where as usual $G^m = \{g^m : g \in G\}$.

(iii) For $n \geq 0$, let

$$ u_p(n, G) = d\big(G^{p^n}/G^{p^{n+1}}\big) - d\big(G^{p^{n+1}}/G^{p^{n+2}}\big). $$

Show that $u_p(n, G)$ depends on G but is independent of the particular decomposition, this gives the number of cyclic factor groups of order p^{n+1} in all decompositions of G.

(iv) Deduce G and H are isomorphic if and only if $u_p(n, G) = u_p(n, H)$ for all $n \geq 0$.

(v)   Give a condition to determine whether two general finite Abelian groups are isomorphic.

Problem ◆ 7.16 Suppose $G = A \rtimes K$ and $K \leq J \leq G$. Show that $J = (A \cap J) \rtimes K$.

Problem 7.17 Which of the following groups can be represented as semi-direct products: (i) C_{15}, (ii) S_4, (iii) Q_2? Give reasons.

(iv) Show that $GL_2(\mathbb{Q})$ can be represented as a semi-direct product of $SL_2(\mathbb{Q})$ by $\mathbb{Q}^*$.

(v)   Give an example of a group which can be represented in two distinct ways as a semi-direct product.

Problem 7.18* Let H and J be groups where H is cyclic, and let ϕ and ψ be injective homomorphisms from H to $\operatorname{Aut} J$ subject to the condition $H\phi = H\psi$. Show that $H \rtimes_\phi J \simeq H \rtimes_\psi J$. Applications of this result are given in Problems 7.21 and 7.22.

Problem 7.19* (i) Suppose G is a group and S_G is the full permutation group defined on the elements of G. Further, for $a \in G$, let $\xi_a : G \to G$ be the map given by $g\xi_a = ga$ (Cayley's Theorem (Theorem 4.7)), and let $G^* = \{\xi_a : a \in G\}$. Note that $G^* \leq S_G$. The *holomorph* of G, denoted by $\operatorname{Hol}(G)$, is defined as the group generated by G^* and $\operatorname{Aut} G$. Show that $\operatorname{Hol}(G) = \operatorname{Aut} G \rtimes G^*$. (Hint. If $\phi \in \operatorname{Aut} G$, then $\phi^{-1}\xi_a\phi = \xi_{a\phi}$.)

   (ii) Find the holomorphs of the cyclic groups C_n when $n = 1, \dots, 6$; for C_5 use Problem 6.19.

Problem 7.20 Investigate the semi-direct products of the cyclic group C_4 with itself. You should determine how many (isomorphism classes) there are, describe them, and list their elements, subgroups and normal subgroups.

Problem 7.21 Let G be a group of order pq where p and q are primes and $p < q$. Show that there is exactly one (isomorphism class) group of this type if G is Abelian (and so cyclic) or if $p \nmid q - 1$, and two if $p \mid q - 1$ one of which is not Abelian; see Problems 3.6, 6.14 and 9.14, and Web Section 14.3. (Hint. Consider semi-direct products, and use the Sylow theory and Problem 7.18*. In the non-Abelian case, only one 'semi-direct' homomorphism is possible.)

Problem 7.22 List the subgroups of the groups of order 12 discussed at the end of this chapter.

Problem 7.23 Suppose G is a non-Abelian group of order p^3 which contains an element of order p^2; see Problem 6.5. Using the method suggested below show that $G \simeq C_p \rtimes C_{p^2}$, and that all such semi-direct products are isomorphic. The method is as follows.

(a) Choose $a, b \in G$ with $o(b) = p^2$ and $a \notin \langle b \rangle$. If $a^p = e$ we have the desired semi-direct product, why?
(b) Suppose $a^p \neq e$. Show that $o(Z(G)) = p$, and so $Z(G) < \langle b \rangle$. Further show that $G/Z(G) \simeq C_p \times C_p$ (use Problem 4.16(ii)), and so deduce that $a^p = b^{tp}$ for some t with $1 \leq t < p$.
(c) Replace b by b^{-t}, and prove $(ab)^p = e$ using Problems 6.5 and 2.17(ii).
(d) Replace a by ab to obtain the required semi-direct product.
(e) Noting that $\operatorname{Aut} C_{p^2}$ has order $p(p-1)$, use the Sylow theory and Problem 7.18* to obtain uniqueness.

Problem 7.24 (Project—A Group of Order 84) Suppose G is a group with $o(G) = 84$ having the maximum possible number (28) of Sylow 3-subgroups. You are asked to determine G; there is only one isomorphism class; see also Scott (1964), page 218.

(i) Using the Burnside's Normal Complement Theorem (Theorem 6.17) show that G contains a normal subgroup J of order 28.
(ii) Show that G has a normal Sylow 7-subgroup K, and that $C_G(K) = J$ using the N/C-theorem (Theorem 5.26) and (i).
(iii) Show that G has a unique Sylow 2-subgroup H which is isomorphic to $C_2 \times C_2$. (Hint. Use (i).)
(iv) Using (i), (ii) and (iii) deduce $J \simeq C_2 \times C_2 \times C_7$.
(v) Next show that G can be treated as a semi-direct product involving J. Ideas related to holomorphs will be useful here, see Problem 7.19*.
(vi) Finally show that G has a presentation in the form

$$G \simeq \langle a, b, c, d \mid a^2 = b^2 = c^7 = d^3 = e, ab = ba, bc = cb, ca = ac,$$
$$d^2 ad = b, d^2 bd = ab, d^2 cd = c^2 \rangle.$$

Chapter 8
Groups of Order 24
Three Examples

In Chapter 1, we commented that both theorems *and* examples are important when studying groups, and so as an interlude before we introduce our next major theoretical topics we shall consider three groups of order 24 in detail to illustrate the material from the previous six chapters and motivate the work on series, simple groups, and (on the web) representation and character theory to come. 'Getting to know' the structure of these groups 'in full' will, we are sure, help the reader in his (her) general understanding of the theory as a whole. Because our work on character theory is being given in the Web Sections, we shall give the character tables for the three groups discussed in this chapter in Web Section 14.1. Some facts about the (isomorphism classes of the) remaining twelve groups of order 24 are tabulated in the Appendix C, these include details about their subgroups (including special subgroups), factor groups, automorphisms, and some more specialised properties. The reader should use this chapter to experiment with various hypotheses. For example, what is the relation between the centre and the derived subgroup of a group? Whilst reading this book similar questions will arise, many of which can be answered by considering the properties of the groups discussed here. *In this chapter a number of facts are not fully justified, the reader is asked to fill in the details.*

8.1 Symmetric Group S_4

For our first example, we consider the symmetric group S_4, and use the results concerning these groups proved in Chapter 3. Our main approach will be via permutations of the set $X = \{1, 2, 3, 4\}$ although, as we shall show later, the group has several distinct representations; see page 170. We have $o(S_4) = 24$ because there are 4! ways of ordering the set X. Also in a symmetric group, elements are conjugate if and only if they have the same cyclic structure (Theorem 3.6). In the table overleaf which lists the elements of S_4, the conjugacy classes are given by the rows or extended rows.

H.E. Rose, *A Course on Finite Groups*,
Universitext,
DOI 10.1007/978-1-84882-889-6_8, © Springer-Verlag London Limited 2009

Number of elements	Cyclic type	List of elements
1	1-cycle	e
6	2-cycle	$(1,2), (1,3), (1,4), (2,3), (2,4), (3,4)$
8	3-cycle	$(1,2,3), (1,3,2), (1,2,4), (1,4,2), (1,3,4),$ $(1,4,3), (2,3,4), (2,4,3)$
6	4-cycle	$(1,2,3,4), (1,2,4,3), (1,3,2,4), (1,3,4,2),$ $(1,4,2,3), (1,4,3,2)$
3	2-cycle $\times$ 2-cycle	$(1,2)(3,4), (1,3)(2,4), (1,4)(2,3)$

Subgroups of S_4

The group S_4 has thirty subgroups in eleven conjugacy classes. First, we list the cyclic subgroups. Each element of order 2 generates a cyclic subgroup of order 2, hence

> S_4 has nine subgroups isomorphic to C_2 in two conjugacy classes: $\langle (1,2) \rangle, \langle (1,3) \rangle, \langle (1,4) \rangle, \langle (2,3) \rangle, \langle (2,4) \rangle,$ and $\langle (3,4) \rangle$; and $\langle (1,2)(3,4) \rangle, \langle (1,3)(2,4) \rangle$ and $\langle (1,4)(2,3) \rangle$.

There are also eight elements of order 3, and so (Problem 2.5(iii))

> S_4 has four subgroups isomorphic to C_3 in one conjugacy class: $\langle (1,2,3) \rangle, \langle (1,2,4) \rangle, \langle (1,3,4) \rangle$ and $\langle (2,3,4) \rangle$;

see the Sylow entry below. The six elements of order 4 give

> S_4 has three subgroups isomorphic to C_4 in one conjugacy class: $\langle (1,2,3,4) \rangle, \langle (1,2,4,3) \rangle$ and $\langle (1,3,2,4) \rangle$.

There are no other cyclic subgroups because S_4 has no elements of order larger than 4.

Sylow Subgroups

As with all groups of order 24, S_4 has Sylow subgroup(s) of order 3 and 8. The second item above lists the four Sylow 3-subgroups. (Note that in this case the Sylow theory states: $n_3 \equiv 1 \pmod 3$ and $n_3 \mid 8$, and so $n_3 = 1$ or 4.) As S_4 has more than one Sylow 3-subgroup, none is normal, but they form a single conjugacy class by Sylow 2.

For the second case, we have $n_2 \equiv 1 \pmod 2$ and $n_2 \mid 3$. Therefore, $n_2 = 1$ or 3, and S_4 has one or three Sylow 2-subgroup(s) of order 8. First, we need to determine the type, see Theorem 6.10 (Sylow 4) and the example on page 125. This subgroup cannot be C_8 because S_4 has no elements of order 8. Also it cannot be $C_2 \times C_2 \times C_2$

because such a subgroup would contain seven of the nine elements of order 2 in S_4, and these would generate S_4; see Problem 3.1. Further, it cannot be Q_2 because such a subgroup would contain the six elements of order 4, and a single element of order 2; but this is impossible because the conjugate of a 2-cycle by a 4-cycle is another 2-cycle. Hence the Sylow 2-subgroup(s) are isomorphic to either $C_4 \times C_2$ or D_4. Suppose the former, then this subgroup would contain four 4-cycles and three 2-cycles by 2-cycles (the square of a 4-cycle is of this second type). But this is impossible because $C_4 \times C_2$ is Abelian whilst 4-cycles and 2-cycles by 2-cycles do not commute. Therefore, the Sylow 2-subgroup(s) are isomorphic to D_4 (see also (b) on page 156), and it is easily seen that

S_4 has three Sylow 2-subgroups isomorphic to D_4 in one conjugacy class: $\langle(1,2,3,4),(1,3)\rangle$, $\langle(1,2,4,3),(1,4)\rangle$ and $\langle(1,3,2,4),(1,2)\rangle$.

These subgroups are not normal (there is more than one, see Theorem 6.10).

Remaining Subgroups

Next we look for possible subgroups of order 12. If one exists, it would be normal by Problem 2.19. By Theorem 3.11, we have

the set of even permutations in S_4 forms a (normal) subgroup of order 12 which is isomorphic to A_4.

This is the only subgroup of order 12. A normal subgroup must include the neutral element, and be a union of conjugacy classes (Theorem 2.29), and in this case there are no other possibilities. This can be checked using the table opposite.

By Lagrange's Theorem (Theorem 2.27) possible subgroup orders for S_4 are 1, 2, 3, 4, 6, 8, 12, and 24. We have treated the cases 2, 3, 8, and 12; and, as with all groups,

S_4 has (normal) subgroups $\langle e \rangle$ and S_4.

Hence we need to consider possible non-cyclic subgroups of orders 4 and 6. Clearly, the permutations in S_4 which fix the symbol 1 (or 2, or 3, or 4) form a subgroup isomorphic to $S_3(\simeq D_3)$ (Problem 2.20), hence

S_4 has four subgroups isomorphic to S_3: $\langle(1,2,3),(1,2)\rangle$, $\langle(1,2,4),(1,2)\rangle$, $\langle(1,3,4),(1,3)\rangle$, and $\langle(2,3,4),(2,3)\rangle$.

A non-cyclic group of order 6 is isomorphic to S_3 (Problem 2.20), and so this is the only possibility. Finally, we note that S_4 has nine elements of order 2, and so, potentially, it has a number of subgroups of the type $C_2 \times C_2 \simeq T_2$. First, we note that e and the three elements in the second conjugacy class form a normal subgroup which we denote by V, see page 19, so

S_4 has a normal subgroup isomorphic to:

$$C_2 \times C_2 : \langle(1,3)(2,4),(1,4)(2,3)\rangle = V. \tag{8.1}$$

No other full symmetric group (apart from S_4) has a proper non-neutral normal subgroup which is not an alternating group. After some further checking, we have, noting that $(i, j)(i, k) = (i, j, k)$ if i, j and k are distinct,

S_4 has three non-normal subgroups isomorphic to $C_2 \times C_2$ which form a single conjugacy class: $\langle (1, 2), (3, 4) \rangle$, $\langle (1, 3), (2, 4) \rangle$ and $\langle (1, 4), (2, 3) \rangle$.

As we have considered all possibilities our list of subgroups of S_4 is complete.

Centre As $Z(S_4)$ is a normal Abelian subgroup of S_4 (Lemma 2.31), it follows that the centre can only be V or $\langle e \rangle$, see (8.1) and the diagram opposite. But

$$(1, 2)(3, 4) \cdot (1, 3) = (1, 2, 3, 4) \neq (1, 4, 3, 2) = (1, 3) \cdot (1, 2)(3, 4),$$

for instance, and so at least one element of V does not belong to the centre. Therefore, $Z(S_4) = \langle e \rangle$, and because of this the group S_4 is called *centreless*.

Derived Subgroups The conjugate of a j-cycle is another j-cycle (Theorem 3.6), and so the commutator of two elements in S_4 is an even permutation and belongs to A_4. Also we have, for distinct $i, j, k \in \{1, 2, 3, 4\}$,

$$\left[(i, j), (i, k) \right] = (i, j)(i, k)(i, j)(i, k) = (i, j, k)^2 = (i, k, j);$$

that is, every 3-cycle belongs to the derived subgroup, hence $S_4' \simeq A_4$. For the second derived subgroup, we have

$$\left[(i, j, k), (i, k, l) \right] = (i, l)(j, k).$$

By Theorem 3.12, this shows that $(S_4)'' = V$, see (8.1). The third derived subgroup is $\langle e \rangle$ because V is Abelian.

$$* \quad * \quad * \quad * \quad * \quad *$$

The diagram given on the opposite page illustrates the structure of the lattice of subgroups for S_4. The left-hand column gives the subgroup type, the second column gives the order, and the right-hand column gives the number of subgroups. The symbol 'N' indicates that the corresponding subgroup is normal in S_4. The subgroups are circled and the lines between the circles correspond to the subgroup relation working upwards; so, for example, $\langle (1, 2, 3) \rangle$ is a (maximal) subgroup of both $\langle (1, 2, 3), (1, 2) \rangle$ and A_4. Also the rows in the diagram correspond to subgroup conjugacy classes. To simplify the diagram slightly, the linkages with the six subgroups generated by single 2-cycles are indicated by arrows and the letters $A, B, \ldots, F$. For instance, the far left-hand downward arrow with A, D, E below it indicates that the subgroup generated by $(1, 2, 3)$ and $(1, 2)$ has itself three subgroups of order 2 (they are $\langle (1, 2) \rangle$ labelled E, $\langle (1, 3) \rangle$ labelled A and $\langle (2, 3) \rangle$ labelled D). The far right-hand upward arrow labelled F indicates that $\langle (3, 4) \rangle$ is a subgroup of two of the subgroups isomorphic to S_3 on the left-hand page ($\langle (1, 3, 4), (1, 3) \rangle$ and $\langle (2, 3, 4), (3, 4) \rangle$ again with the label F). Note that $\langle (3, 4) \rangle$ is also a subgroup of $\langle (1, 2), (3, 4) \rangle$ with the label E, F below it.

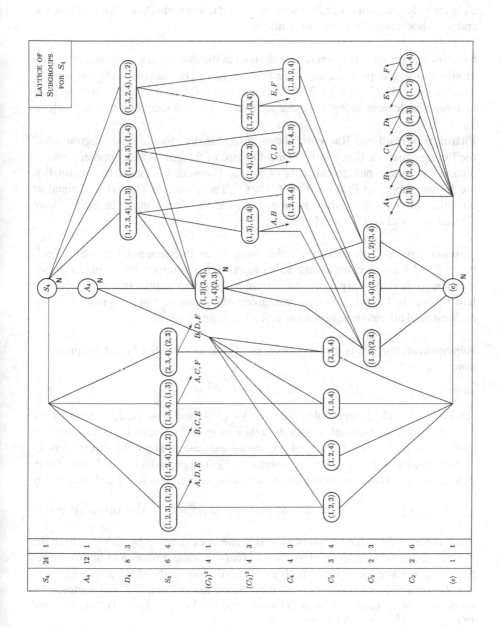

LATTICE OF SUBGROUPS FOR S_4

Central Series The lower $\mathcal{D}_i(G)$ and upper $\mathcal{Z}_i(G)$ central series of a group G are defined in Section 10.1. For S_4 the lower series has two terms: $\mathcal{D}_1 = S_4$ and $\mathcal{D}_2 \simeq S_4' = A_4$, and the upper series only has one term $\mathcal{Z}_0 = \langle e \rangle$ because S_4 is centreless. S_4 is the only group of order 24 whose lower and upper series have different lengths, and this shows that the group is not nilpotent.

Frattini Subgroup This subgroup is defined as the intersection of all maximal subgroups of the group in question, see Section 10.2. The maximal subgroups of S_4 include $\langle (2, 3, 4), (2, 3) \rangle$, $\langle (1, 2, 3, 4), (1, 3) \rangle$ and $\langle (1, 3, 2, 4), (1, 2) \rangle$; as these have no common element except e we deduce that the Frattini subgroup $\Phi(S_4) = \langle e \rangle$.

Fitting Subgroup and Radicals The group S_4 has only four normal subgroups and the Fitting subgroup (Section 10.2) must be one of these. It is also nilpotent, and so it cannot be isomorphic to either A_4 or S_4, see Theorem 7.9. Hence by maximality, the Fitting subgroup $F(S_4) = V \simeq C_2 \times C_2$. The p-radical $O_p(S_4)$ is defined as the intersection of the Sylow p-subgroups, see Problem 6.9(vi). In S_4, we have $O_2(S_4) = V$ and $O_3(S_4) = \langle e \rangle$.

Automorphisms The basic properties were given in Section 4.4 and by Corollary 5.27. As S_4 is centreless the subgroup of inner automorphisms $\text{Inn } S_4$ is isomorphic to S_4 (Corollary 5.27). The group has no outer automorphisms, see Problem 8.5, and so the full automorphism group of S_4, that is, $\text{Aut } S_4$, is isomorphic to S_4 itself, and all automorphisms are given by conjugation.

Representations It was first shown by von Dyck in 1882 that S_4 has the presentation

$$G_1 = \langle a, b \mid a^2 = b^3 = (ab)^4 = e \rangle \simeq S_4, \qquad (8.2)$$

see Problem 3.18. To prove this we set $a \mapsto (3, 4)$ and $b \mapsto (1, 2, 3)$, then $ab \mapsto (1, 2, 3, 4)$, and a, b and ab satisfy the relations in (8.2). Hence G_1 is a homomorphic image of S_4, and it is a straightforward exercise to show that it contains 24 distinct elements, therefore it is isomorphic to S_4. The group has several other easily defined presentations, one using involutions was given in Problem 3.21*. Another is

$$S_4 \simeq \langle a_1, b_1, c_1 \mid a_1^3 = b_1^3 = c_1^4 = (a_1 b_1)^2 = (b_1 c_1)^2 = (c_1 a_1)^2 = (a_1 b_1 c_1)^2 = e \rangle$$

see Problem 8.5(ii). This presentation is sometimes denoted by $G^{3,3,4}$ where the general group $G^{r,s,t}$ is defined as above except that the powers 3, 3 and 4 are replaced by r, s and t, respectively. It arose during the study of the *projective general linear group* $PGL_n(q)$, that is, the general linear group $GL_n(q)$ factored by its centre; see Coxeter and Moser (1984), page 96. We also have $A_5 \simeq G^{3,5,5}$ and $PGL_2(7) \simeq G^{3,7,8}$ which has order 336.

   The group S_4 can also be treated as the rotational symmetry group of a (regular) octahedron as follows. An octahedron has eight identical equilateral triangular faces, and so it has three types of symmetry: (i) rotation by π about a line through the centre points of opposite edges, (ii) rotation by $2\pi/3$ about a line through the centres of opposite faces and (iii) rotation by $\pi/2$ about a line through opposite vertices. To show that this group is isomorphic to S_4, use the presentation given in (8.2), and associate a rotation of type (i) with a, a rotation of type (ii) with b, and a rotation of type (iii) with ab. The reader should check that a rotation of type (i), followed by a rotation of type (ii), has the same effect on an octahedron as a rotation of type (iii). For this reason some authors call S_4 the *octahedral group*. Note also that a cube can be inscribed in an octahedron (with its vertices corresponding to the centres of the faces), and vice versa; and so S_4 can equally be treated as the *rotational* symmetry group of a cube. But the full symmetry group (that is, if reflections are also counted) of an octahedron or a cube is $S_4 \times C_2$. This can also be represented by a wreath product of the form C_2 wr S_3 (page 156).

There are a number of representations of S_4 as a matrix group. In Problem 3.12, we showed how a permutation group can be represented as a group of 'permutation' matrices, and in Problem 12.6 we prove that $S_4 < L_2(7) \simeq GL_3(2)$.

We give two more here. First, working over $\mathbb{F}_2$, the two element field, let

$$A = \begin{pmatrix} 0 & 0 & 1 \\ 1 & 1 & 1 \\ 1 & 0 & 0 \end{pmatrix}, \qquad B = \begin{pmatrix} 0 & 0 & 1 \\ 0 & 1 & 0 \\ 1 & 0 & 1 \end{pmatrix}, \qquad C = \begin{pmatrix} 1 & 0 & 1 \\ 1 & 1 & 0 \\ 0 & 0 & 1 \end{pmatrix}.$$

Remember that $1 + 1 = 0$ in $\mathbb{F}_2$. It is easily checked that $C = AB$, $A^2 = B^3 = C^4 = I_3$, the 3×3 identity matrix; and so A and B generate an isomorphic copy of S_4 in $GL_3(2)$. Second, if we take the group $GL_2(3)$ and factor out its centre (which has order 2 and is generated by the matrix $\begin{pmatrix} 2 & 0 \\ 0 & 2 \end{pmatrix}$), we obtain another group of order 24 usually denoted by $PGL_2(3)$ which is isomorphic to S_4; see page 170.

S_4 is not decomposable (it cannot be written as a direct product with non-neutral factors), but it has two representations as semi-direct products as follows:

$S_4 \simeq C_2 \rtimes A_4$ if we take $C_2 = \langle(1, 2)\rangle$ and A_4 to be the set of even permutations in S_4; we could replace $\langle(1, 2)\rangle$ by a subgroup generated by another single 2-cycle in S_4; reader, why?

$S_4 \simeq S_3 \rtimes T_2$ if, for example, we take S_3 to be the subgroup of permutations on 1, 2 and 3 only, and $T_2 = V$, see (8.1).

Finally, note that, for groups of order 24, each of the following statements characterise S_4 (these are not the only characterisations, for instance, see the paragraph on central series above): The group S_4 is the only group of order 24 (Appendix C) such that

(a) it has no normal subgroups of orders 2 or 3,
(b) it is the only group with no normal Sylow subgroups, that is, with both $n_2 > 1$ and $n_3 > 1$,
(c) it is centreless—its centre is the neutral subgroup,
(d) it has the largest derived subgroup,
(e) it occurs as a subgroup of A_6, see Problem 3.7 and Cayley's Theorem (Theorem 4.7).

The group S_4 is also the largest soluble symmetric group, and it occurs as a maximal subgroup of several simple groups; see the ATLAS (1985).

8.2 Special Linear Group $SL_2(3)$

For our second example, we consider $SL_2(3)$, the group whose initial definition is as the set of all 2×2 matrices with determinant 1 defined over $\mathbb{F}_3$, the 3-element field. By Theorem 3.15, it has 24 elements. We shall see below that this group can be defined in several ways (that is, like most groups it has several distinct representations), but we introduce the group using direct matrix calculations via a suitable computer algebra package such as GAP. The elements with their orders are tabulated below. A further series of straight-forward calculations shows that the conjugacy classes of $SL_2(3)$ are given by the rows in this table. An important point to note is that this group has only one involution. Cauchy's Theorem (Theorem 6.2), or Problem 2.8, implies that at least one involution must occur as the group has even order, in this case there are no more.

Number of elements	Element order	List of elements
1	1	$\begin{pmatrix} 1 & 0 \\ 0 & 1 \end{pmatrix}$
1	2	$\begin{pmatrix} 2 & 0 \\ 0 & 2 \end{pmatrix}$
4	3	$\begin{pmatrix} 1 & 1 \\ 0 & 1 \end{pmatrix}, \begin{pmatrix} 1 & 0 \\ 2 & 1 \end{pmatrix}, \begin{pmatrix} 2 & 1 \\ 2 & 0 \end{pmatrix}, \begin{pmatrix} 0 & 1 \\ 2 & 2 \end{pmatrix}$
4	3	$\begin{pmatrix} 1 & 0 \\ 1 & 1 \end{pmatrix}, \begin{pmatrix} 1 & 2 \\ 0 & 1 \end{pmatrix}, \begin{pmatrix} 2 & 2 \\ 1 & 0 \end{pmatrix}, \begin{pmatrix} 0 & 2 \\ 1 & 2 \end{pmatrix}$
6	4	$\begin{pmatrix} 2 & 2 \\ 2 & 1 \end{pmatrix}, \begin{pmatrix} 1 & 2 \\ 2 & 2 \end{pmatrix}, \begin{pmatrix} 2 & 1 \\ 1 & 1 \end{pmatrix}, \begin{pmatrix} 1 & 1 \\ 1 & 2 \end{pmatrix}, \begin{pmatrix} 0 & 2 \\ 1 & 0 \end{pmatrix}, \begin{pmatrix} 0 & 1 \\ 2 & 0 \end{pmatrix}$
4	6	$\begin{pmatrix} 2 & 1 \\ 0 & 2 \end{pmatrix}, \begin{pmatrix} 2 & 0 \\ 2 & 2 \end{pmatrix}, \begin{pmatrix} 0 & 1 \\ 2 & 1 \end{pmatrix}, \begin{pmatrix} 1 & 1 \\ 2 & 0 \end{pmatrix}$
4	6	$\begin{pmatrix} 2 & 0 \\ 1 & 2 \end{pmatrix}, \begin{pmatrix} 2 & 2 \\ 0 & 2 \end{pmatrix}, \begin{pmatrix} 0 & 2 \\ 1 & 1 \end{pmatrix}, \begin{pmatrix} 1 & 2 \\ 1 & 0 \end{pmatrix}$

Next we look for a presentation of the group, one is as follows. If we let

$$a \mapsto \begin{pmatrix} 2 & 1 \\ 0 & 2 \end{pmatrix} \quad \text{and} \quad b \mapsto \begin{pmatrix} 2 & 0 \\ 1 & 2 \end{pmatrix}, \tag{8.3}$$

then $ab \mapsto \begin{pmatrix} 2 & 2 \\ 2 & 1 \end{pmatrix}$, $ba \mapsto \begin{pmatrix} 1 & 2 \\ 2 & 2 \end{pmatrix}$ and

$$a^3 = b^3 = (ab)^2 \mapsto \begin{pmatrix} 2 & 0 \\ 0 & 2 \end{pmatrix},$$

the unique element of order 2.

Lemma 8.1 *If*

$$G_2 = \langle a, b \mid a^3 = b^3 = (ab)^2 \rangle,$$

then $G_2 \simeq SL_2(3)$.

Proof We begin by showing that

$$a^6 = b^6 = e. \tag{8.4}$$

By cancellation, the relations in G_2 give

$$a^2 = bab \quad \text{and} \quad b^2 = aba, \tag{8.5}$$

and using (8.5) several times we obtain

$$a^6 = \left(a^2\right)^3 = bab^2ab^2ab = ba^2ba^3ba^2b = ba^2b^5a^2b = ba^2b^6ab^2$$

$$= ba^9b^2 = b^{12} = a^{12},$$

and so (8.4) follows by cancellation. The maps in (8.3) and the table above show that G_2 is isomorphic to a factor group of $SL_2(3)$, hence to prove the lemma we need to show that $o(G_2) = 24$. This can be done directly using (8.4) and (8.5), we leave this as an exercise, see below. For instance, $(ab)^3 = ababab = a^4b = b^3ab = b^2a^2$. $\square$

This result shows that the elements of G_2 are

$e; \quad a^3; \quad a^2, b^4, b^2a, ab^2; \quad b^2, a^4, ba^2, a^2b; \quad ab, ba, a^2b^2, b^2a^2, ab^2a, ba^2b;$
$a, b^5, a^5b, ba^5; \quad b, a^5, ab^5, b^5a;$

where the orders of the elements in this list correspond to those given in the table opposite starting with e in row 1.

Subgroups of $SL_2(3)$

The group $SL_2(3)$ has fifteen subgroups in seven conjugacy classes. As in the previous example, we begin by listing the cyclic subgroups. The group has a single element of order 2, so

$SL_2(3)$ has a single cyclic subgroup of order 2: $\left\langle \left(\begin{smallmatrix} 2 & 0 \\ 0 & 2 \end{smallmatrix} \right) \right\rangle \simeq \langle a^3 \rangle$.

The group has eight elements of order 3 (use table opposite), and so

$SL_2(3)$ has four cyclic subgroups of order 3: $\left\langle \left(\begin{smallmatrix} 1 & 1 \\ 0 & 1 \end{smallmatrix} \right) \right\rangle \simeq \langle a^2 \rangle$, $\left\langle \left(\begin{smallmatrix} 1 & 0 \\ 1 & 1 \end{smallmatrix} \right) \right\rangle \simeq \langle b^2 \rangle$,
$\left\langle \left(\begin{smallmatrix} 2 & 1 \\ 2 & 0 \end{smallmatrix} \right) \right\rangle \simeq \langle b^2a \rangle$, and $\left\langle \left(\begin{smallmatrix} 2 & 2 \\ 1 & 0 \end{smallmatrix} \right) \right\rangle \simeq \langle ba^2 \rangle$,

see the Sylow entry below. Similarly, as there are six elements of order 4, the group contains three cyclic subgroups of order 4; the intersection of any two of which equals the unique 2-element cyclic group:

$SL_2(3)$ has three subgroups isomorphic to C_4: $\langle\left(\begin{smallmatrix}2&2\\2&1\end{smallmatrix}\right)\rangle \simeq \langle ab\rangle$, $\langle\left(\begin{smallmatrix}2&1\\1&1\end{smallmatrix}\right)\rangle \simeq \langle a^2b^2\rangle$, and $\langle\left(\begin{smallmatrix}0&2\\1&0\end{smallmatrix}\right)\rangle \simeq \langle ab^2a\rangle$.

Finally, as there are eight elements of order 6, we have

$SL_2(3)$ has four cyclic subgroups of order 6: $\langle\left(\begin{smallmatrix}2&1\\0&2\end{smallmatrix}\right)\rangle \simeq \langle a\rangle$, $\langle\left(\begin{smallmatrix}2&0\\1&2\end{smallmatrix}\right)\rangle \simeq \langle b\rangle$, $\langle\left(\begin{smallmatrix}0&1\\2&1\end{smallmatrix}\right)\rangle \simeq \langle a^5b\rangle$, and $\langle\left(\begin{smallmatrix}0&2\\1&1\end{smallmatrix}\right)\rangle \simeq \langle ab^5\rangle$,

and again each contain the matrix $\left(\begin{smallmatrix}2&0\\0&2\end{smallmatrix}\right)$.

Sylow and Remaining Subgroups

The four Sylow 3-subgroups are listed above. As in the S_4 case, the group has four non-normal cyclic Sylow 3-subgroups. Now $SL_2(3)$ has a single element of order 2. The group C_8 cannot be a Sylow 2-subgroup (because $SL_2(3)$ contains no elements of order 8), and of the remaining groups of order 8, Q_2 is the only one with a single involution, hence

$SL_2(3)$ has a single subgroup isomorphic to Q_2 which is generated by $\left(\begin{smallmatrix}2&2\\2&1\end{smallmatrix}\right)$ and $\left(\begin{smallmatrix}1&2\\2&2\end{smallmatrix}\right)$, that is, by ab and ba; it is normal (by Sylow 4).

No other subgroups exist apart from those listed above, $\langle e\rangle$ and $SL_3(2)$ itself. By Lagrange's Theorem (Theorem 2.27) and the Sylow theory we only need to consider non-cyclic subgroups of orders 4, 6 and 12. But each of these groups, that is, $C_2 \times C_2$, D_3, and the five groups of order 12, have more than one element of order 2, hence they cannot be isomorphic to a subgroup of $SL_2(3)$.

Centre As $a^3(= b^3)$ commutes with both a and b, we have $\langle a^3\rangle \leq Z(SL_2(3))$. In fact, we have equality because the centre is both normal and Abelian, and $SL_2(3)$ has no other normal Abelian subgroup.

Derived Subgroups Using the presentation given in Lemma 8.1, and (8.4) and (8.5), we have

$$[a, b] = a^5b^5ab = a^2b^2ab = a^2ba^2 = ab^2a,$$

and similarly $[b, a] = ba^2b$. Hence $(SL_2(3))'$ equals either $\langle ab, ba\rangle \simeq Q_2$ or the group itself. But as $SL_2(3)/Q_2$ is Abelian (it has order 3), we have $(SL_2(3))' \leq Q_2$ (Problem 4.6(ii)) and so these groups are equal. In Problem 6.3, we showed that $(SL_2(3))'' = \langle a^3\rangle$, and so the third derived subgroup is $\langle e\rangle$. Also note that $Z(SL_2(3)) = (SL_2(3))''$.

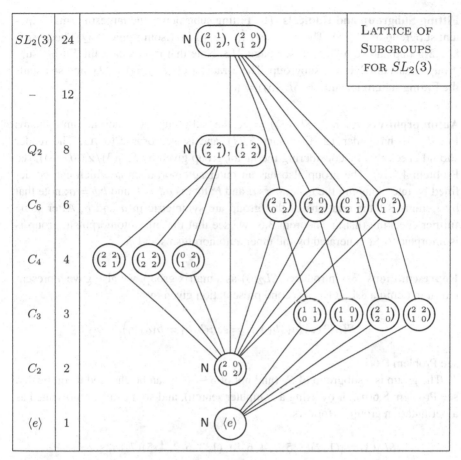

$SL_2(3)$	24	N $\left(\left(\begin{smallmatrix} 2 & 1 \\ 0 & 2 \end{smallmatrix}\right), \left(\begin{smallmatrix} 2 & 0 \\ 1 & 2 \end{smallmatrix}\right) \right)$ — LATTICE OF SUBGROUPS FOR $SL_2(3)$
–	12	
Q_2	8	N $\left(\left(\begin{smallmatrix} 2 & 2 \\ 2 & 1 \end{smallmatrix}\right), \left(\begin{smallmatrix} 1 & 2 \\ 2 & 2 \end{smallmatrix}\right) \right)$
C_6	6	$\left(\begin{smallmatrix} 2 & 1 \\ 0 & 2 \end{smallmatrix}\right)$ $\left(\begin{smallmatrix} 2 & 0 \\ 1 & 2 \end{smallmatrix}\right)$ $\left(\begin{smallmatrix} 0 & 1 \\ 2 & 1 \end{smallmatrix}\right)$ $\left(\begin{smallmatrix} 0 & 2 \\ 1 & 1 \end{smallmatrix}\right)$
C_4	4	$\left(\begin{smallmatrix} 2 & 2 \\ 2 & 1 \end{smallmatrix}\right)$ $\left(\begin{smallmatrix} 1 & 2 \\ 2 & 2 \end{smallmatrix}\right)$ $\left(\begin{smallmatrix} 0 & 2 \\ 1 & 0 \end{smallmatrix}\right)$
C_3	3	$\left(\begin{smallmatrix} 1 & 1 \\ 0 & 1 \end{smallmatrix}\right)$ $\left(\begin{smallmatrix} 1 & 0 \\ 1 & 1 \end{smallmatrix}\right)$ $\left(\begin{smallmatrix} 2 & 1 \\ 2 & 0 \end{smallmatrix}\right)$ $\left(\begin{smallmatrix} 2 & 2 \\ 1 & 0 \end{smallmatrix}\right)$
C_2	2	N $\left(\left(\begin{smallmatrix} 2 & 0 \\ 0 & 2 \end{smallmatrix}\right) \right)$
$\langle e \rangle$	1	N $\left(\langle e \rangle \right)$

The diagram above illustrates the subgroup structure of $SL_2(3)$—note there is no subgroup of order 12. The first left-hand column gives the subgroup type, and the second column gives the corresponding order. As before the symbol N indicates that the subgroup in question is normal in the group, and the rows list the members of the subgroup conjugacy classes.

* * * * * *

Central Series Both the lower and upper central series of $SL_2(3)$ have two terms and the group is not nilpotent, see Section 10.1:

Lower series: $\mathcal{D}_1\big(SL_2(3)\big) = SL_2(3)$ and $\mathcal{D}_2\big(SL_2(3)\big) = SL_2(3)' \simeq Q_2$,

Upper series: $\mathcal{Z}_0\big(SL_2(3)\big) = \langle e \rangle$ and $\mathcal{Z}_1\big(SL_2(3)\big) = Z(SL_2(3)) \simeq C_2$.

Frattini Subgroup This normal subgroup is the intersection of the maximal subgroups of $SL_2(3)$. From the diagram above we can see immediately that the Frattini subgroup $\Phi(SL_2(3))$ equals $\left\langle \left(\begin{smallmatrix} 2 & 0 \\ 0 & 2 \end{smallmatrix}\right) \right\rangle$ $(=Z(SL_2(3)))$.

Fitting Subgroup and Radicals The Fitting subgroup is the largest normal nilpotent subgroup of $SL_2(3)$. This is clearly $\left\langle \left(\begin{smallmatrix} 2 & 1 \\ 1 & 1 \end{smallmatrix} \right), \left(\begin{smallmatrix} 1 & 1 \\ 1 & 2 \end{smallmatrix} \right) \right\rangle$ (isomorphic to Q_2) because it is a 2-group and so nilpotent, see page 213. Note that in this case the Fitting subgroup equals the Sylow 2-subgroup. The 2-radical $O_2(SL_2(3)) \simeq Q_2$, and so equals the Fitting subgroup, and $O_3(SL_2(3)) = \langle e \rangle$.

Automorphisms As $o(Z(SL_2(3)) = 2$, the subgroup of inner automorphisms $\mathrm{Inn}\, SL_2(3)$, has order 12 (Corollary 5.27), and is isomorphic to A_4. The reader should check this by considering the factor group given by $SL_2(3)/Z(SL_2(3))$, see Problem 4.4(iv). The group also has an outer automorphism ϕ which can be defined by interchanging the generators a and b; that is $a\phi = b$ and $b\phi = a$; note that by Lemma 8.1 the relations of this group are symmetric in a and b. After some further checking (using Theorem 3.6) we see that the full automorphism group is isomorphic to S_4 generated by the inner automorphisms and ϕ.

Representations We introduced $SL_2(3)$ as a matrix group, we also gave a presentation in Lemma 8.1. It has a second presentation given by

$$SL_2(3) \simeq \langle a_1, b_1 \mid a_1^3 = e, a_1 b_1 a_1 = b_1 a_1 b_1 \rangle,$$

see Problem 8.6(i).

The group is a subgroup of S_8 (and not of S_7—this can be checked theoretically, see Problem 8.6(ii), or by using a computer search), and so it can be represented as a permutation group as follows:

$$SL_2(3) \simeq \langle (1,2)(3,5,7,4,6,8), (1,3,6,2,4,5)(7,8) \rangle \leq S_8.$$

Note that if the elements in the permutation representation above are denoted by a and b we have $a^3 = b^3 = (1,2)(3,4)(5,6)(7,8) = c$, an element of order 2, and $(ab)^2 = ((1,4,2,3)(5,8,6,7))^2 = c$, see Lemma 8.1. Further, $SL_2(3)$ is sometimes called the *binary tetrahedral group* because if we factor it by its centre (of order 2) we obtain a group isomorphic to A_4 which as we have seen can be represented as the symmetry group of the tetrahedron, see Problem 3.10. The group also has a representation as a semi-direct product:

$SL_2(3) \simeq C_3 \rtimes Q_2$ if we take C_3 to be one of the Sylow 3-subgroups, and Q_2 to be the unique Sylow 2-subgroup.

Finally, note $SL_2(3)$ can be described as the only group of order 24 which

(a) has no subgroup of order 12;
(b) has the property that its derived and Sylow 2-subgroups are equal;
(c) has the smallest *sockel* (product of minimal normal subgroups), see (17) on page 293.

8.3 Exceptional Group E

For our final example, we consider what we call the *exceptional group* E of order 24. It is not a member of some easily recognisable class, but it can be defined as a semi-direct product in several different ways and it has an important connection with the alternating group A_7; see the representation subsection below. It does also appear once a list of all groups of order 24 is sought (Problems 8.10* and 8.11*), and it has a kind of 'symmetry representation'. We define the group E by the presentation

$$E = \langle a, b, c \mid a^4 = b^2 = c^3 = e, bab = a^3, bcb = c, aca^3 = c^2 \rangle. \qquad (8.6)$$

As a and b generate a copy of D_4 and c generates a copy of C_3, the group E can be treated as a semi-direct product C_3 by D_4 as $\langle c \rangle \simeq C_3$ is normal in E. We shall see later that it has at least four distinct and easily defined representations as semi-direct products.

We show first that $o(E) = 24$ using the following

Lemma 8.2 *The element $a^2 \in E$ commutes with both b and c, and so belongs to the centre of E.*

Proof The last relation for E in (8.6) can be rewritten as $ac = c^2 a$, and so

$$a^2 c = aac = acca = c^2 aca = c^4 a^2 = ca^2.$$

Two applications of the first relation in (8.6) give $ba^2 = a^2 b$, and the result follows from these two equations. $\qquad\qquad\square$

Using this lemma, we can easily derive the following equations which we use when working with the group E.

$$cb = bc,$$
$$ba = a^3 b, \quad ba^2 = a^2 b, \quad ba^3 = ab,$$
$$ca = ac^2, \quad ca^2 = a^2 c, \quad ca^3 = a^3 c^2, \qquad (8.7)$$
$$c^2 a = ac, \quad c^2 a^2 = a^2 c^2, \quad c^2 a^3 = a^3 c.$$

Theorem 8.3 $o(E) = 24$.

Proof Using (8.7), we note that every element of E can be put in the form $a^r b^s c^t$. This shows that $o(E) \le 24$, and to prove equality use the fact that E is defined as a semi-direct product and in this product we have $a^r b^s c^t = e$ implies that $r = s = t = 0$. $\qquad\qquad\square$

A list of elements and conjugacy classes of E is tabulated overleaf where, as in the previous examples, the rows give the conjugacy classes, nine in all.

Number of elements	Element order	List of elements
1	1	e
1	2	a^2
2	2	$b, a^2 b$
6	2	$ab, a^3 b, abc, a^3 bc, abc^2, a^3 bc^2$
2	3	c, c^2
6	4	$a, a^3, ac, a^3 c, ac^2, a^3 c^2$
2	6	$a^2 c, a^2 c^2$
2	6	$bc, a^2 bc^2$
2	6	$bc^2, a^2 bc$

Subgroups of E

The group E has thirty subgroups in sixteen conjugacy classes. As in the previous examples (S_4 and $SL_2(3)$), we begin by listing the cyclic subgroups using the table above. We have

> E has nine cyclic subgroups of order 2 in three conjugacy classes (of sizes 1, 2 and 6): $\langle a^2 \rangle$; $\langle b \rangle$, $\langle a^2 b \rangle$; $\langle ab \rangle$, $\langle a^3 b \rangle$, $\langle abc \rangle$, $\langle a^3 bc \rangle$, $\langle abc^2 \rangle$, and $\langle a^3 bc^2 \rangle$.

As only two elements of order three occur in E, that is, c and c^2, we have

> E has a unique cyclic subgroup of order 3: $\langle c \rangle$.

This subgroup is the unique Sylow 3-subgroup, and so by Theorem 6.10, it is normal. There are six elements of order 4, and so

> E has three cyclic subgroups of order 4: $\langle a \rangle$, $\langle ac \rangle$, $\langle ac^2 \rangle$ which form a single conjugacy class.

Note that the intersection of any two of these subgroups is $\langle a^2 \rangle$.
Lastly, there are six elements of order 6, hence

> E has three cyclic subgroups of order 6: $\langle a^2 c \rangle$, $\langle bc \rangle$, $\langle a^2 bc \rangle$; the first is normal and the remaining two form a conjugacy class, use $a^3 \langle bc \rangle a = \langle a^2 bc \rangle$, *et cetera*.

Sylow and Non-cyclic Subgroups

The unique Sylow 3-subgroup was given above. For Sylow 2-subgroup(s), we can see directly from the definition of E that the subgroup generated by a and b is isomorphic to D_4, and so by the Sylow theory all Sylow 2-subgroups (and so all subgroups of order 8) are isomorphic to D_4. There are three in all, and they form a single conjugacy class by Sylow 2; hence

E has three Sylow 2-subgroups isomorphic to D_4: $\langle a, b \rangle$, $\langle ac, b \rangle$, $\langle ac^2, b \rangle$.

By Lagrange's Theorem and the Sylow theory, E could also have non-cyclic subgroups of orders 12, 6 or 4; in fact, this is true in all three cases as we show now.

Order 12 There are five (isomorphism types of) groups of order 12 (Section 7.3). Perhaps surprisingly three occur as subgroups of E; cf. the Sylow theory. The group C_{12} cannot be a subgroup because E contains no elements of order 12. Also A_4 cannot be a subgroup because it contains eight elements of order 3, whilst E has only two.

On the positive side, we note first that there are eleven elements in E which can be expressed in terms of the generators a and c only, see table opposite; hence, together with e they form a subgroup. Referring to the dicyclic group discussed on pages 59, 130 (for metacyclic representation) and 159, we see that

E has one subgroup isomorphic to Q_3: $\langle a^2c, a \rangle \simeq \langle a, c \rangle$.

Secondly, a^2 commutes with both b and c (Lemma 8.2), and so also with bc, $o(a^2) = 2$ and $o(bc) = 6$. Hence

E has a subgroup isomorphic to $C_6 \times C_2$: $\langle bc, a^2 \rangle$.

No other element of order 2 commutes with an element of order 6, and so E has no further subgroups of this type. The final possibility is D_6. Here again we look for an element x of order 2 and an element y of order 6 which in this case satisfy $xyx = y^5$. We have, noting that $o(a^2c) = 6$ and $o(ab) = 2$,

E has one subgroup isomorphic to D_6: $\langle a^2c, ab \rangle$.

Order 6 The element c and an element of order 2 could form part of a subgroup isomorphic to D_3. After some checking we find that

E has two subgroups isomorphic to D_3: $\langle c, ab \rangle$ and $\langle c, a^3b \rangle$ which form a single conjugacy class.

The cyclic groups of order 6 were listed above.

Order 4 The cyclic case was treated on page 178. As a^2 belongs to the centre by Lemma 8.2 (in fact, $Z(E) = \langle a^2 \rangle$, see below), and there are a further eight elements of order 2, we have after some checking

E has four subgroups isomorphic to the 4-group $T_2 \simeq C_2 \times C_2$: $\langle a^2, b \rangle$, $\langle a^2, ab \rangle$, $\langle a^2, abc \rangle$, and $\langle a^2, abc^2 \rangle$; the first is normal (note that $bab = a^2$), and the remaining three form another conjugacy class.

Having tried all possibilities, the list of subgroups of E is now complete.

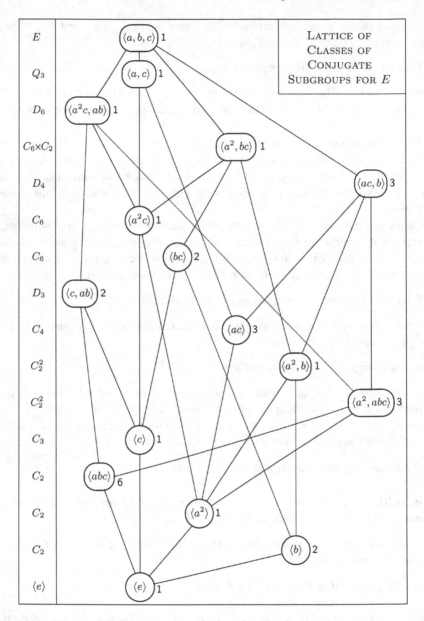

The diagram above illustrates the subgroup structure of the group E. The left-hand column gives the subgroup types; generators, but not relations, are given in the main diagram. For the sake of clarity and simplicity, only one representative of each subgroup conjugacy class is illustrated, the number on the right-hand side of each entry gives the size of the corresponding conjugacy class. If this number is '1', then there is a single subgroup in the class, and it is normal in E (Theorem 2.29).

Finally, we ask which of these thirty subgroups are normal? In fact, nine have this property. A subgroup of index 2 is normal, hence the three subgroups of order 12 just listed are normal, and E and $\langle e \rangle$ are normal by definition. The remaining normal subgroups are obtained by checking the element table on page 178 and noting that a normal subgroup is always a union of conjugacy classes including the neutral class (Theorem 2.29). Hence the remaining normal subgroups are given as follows:

$\langle a^2 \rangle$ containing classes 1 and 2 (order 2),

$\langle c \rangle$ containing classes 1 and 3 (order 3),

$\langle a^2, b \rangle$ containing classes 1, 2 and 3 (order 4), and

$\langle a^2 c \rangle$ containing classes 1, 2, 4, and 6 (order 6).

The reader should check that the other normal subgroups are also unions of conjugacy classes, and no other subgroup has this property.

Centre Previously we noted that $\langle a^2 \rangle \leq Z(E)$ (Lemma 8.2). In fact, equality occurs as we can show now. The centre $Z(E)$ is a normal Abelian subgroup of E that contains $\langle a^2 \rangle$. Referring to the calculations above and the diagram opposite, we see that if $Z(E) \neq \langle a^2 \rangle$ then it can only be $\langle a^2, b \rangle$ or $\langle a^2 c \rangle$. Both of these are impossible because neither b nor c belongs to $Z(E)$ as they do not commute with a^2. Hence $Z(E) = \langle a^2 \rangle$.

Derived Subgroups Using (8.7), we have $[a, b] = a^3 bab = a^2$ and $[a, c]^2 = (a^3 c^2 ac)^2 = c$, hence $a^2, c \in E'$ and so $\langle a^2 c \rangle \leq E'$. On the other hand, $Q_3 \simeq \langle a, c \rangle \lhd E$ and E/Q_3 is Abelian, which shows by Problem 4.6 that $E' \leq Q_3$. But this also applies to both $C_6 \times C_2$ and D_6, and so E' is contained in the intersection of these subgroups, and this is $\langle a^2 c \rangle$. Hence we have equality, that is, $E' = \langle a^2 c \rangle$. As this subgroup is Abelian $E'' = \langle e \rangle$.

Central Series The lower and upper central series both have length 3, they are

Lower series: $\mathcal{D}_1 = E, \mathcal{D}_2 = E' \simeq \langle a^2 c \rangle \simeq C_6$ and $\mathcal{D}_3 = \langle c \rangle \simeq C_3$,

Upper series: $\mathcal{Z}_0 = \langle e \rangle, \mathcal{Z}_1 = Z(E) = \langle a^2 \rangle \simeq C_2$, and $\mathcal{Z}_2 = \langle a^2, b \rangle \simeq C_2 \times C_2$.

Note that $\mathcal{D}_3 = [\mathcal{D}_2, E]$ by definition, and so the first line above is easily checked. Also $\mathcal{Z}_2$ is defined as the preimage of $Z(E/\mathcal{Z}_1) = Z(E/\langle a^2 \rangle)$ under the natural homomorphism from E to $E/\mathcal{Z}_1$, see page 210. We have $Z(E/\langle a^2 \rangle) \simeq \langle a^3 \rangle \simeq C_2$; the reader should now check that the preimage is as stated above. As $\mathcal{D}_n \neq \langle e \rangle$ (all n), E is not nilpotent.

Frattini Subgroup This subgroup is the intersection of the six maximal subgroups of E. It cannot contain a because $a \notin \langle a^2, bc \rangle$, it cannot contain b because $b \notin \langle a, c \rangle$, and it cannot contain c because $c \notin \langle ac, b \rangle$. But each of the six maximal subgroups does contain a^2, and so in this case $\Phi(E) = \langle a^2 \rangle (= Z(E))$.

Fitting Subgroup and Radicals This subgroup is the largest normal nilpotent subgroup of E. Neither Q_3 nor D_6 is nilpotent (they cannot be expressed as direct products of their Sylow subgroups), but $C_6 \times C_2$ is nilpotent—it is Abelian so automatically has this property. Hence the subgroup $\langle a^2, bc \rangle$ is the (unique) Fitting subgroup $F(E)$. Referring to Problem 6.9(vi), we see that the 2-radical $O_2(E) = \langle a^2, b \rangle$, the intersection of the three Sylow 2-subgroups, and the 3-radical $O_3(E) = \langle c \rangle$ as the Sylow 3-subgroup is normal. Note that the product of these radicals equals $F(E) = \langle a^2, bc \rangle$; see Theorem 10.24.

Automorphisms As in the previous example $(SL_2(3))$, the order of the centre of the group of E is two, and so $o(\text{Inn } E) = 12$. In this case, $\text{Inn } E \simeq D_6$ (use Corollary 5.27). Also as in the previous example, there are outer automorphisms and $o(\text{Aut } E) = 24$. One outer automorphism ϕ is defined by

$$a\phi = a, \quad b\phi = b \quad \text{and} \quad c\phi = c^2.$$

Automorphisms map elements of order 3 to elements of order 3, and as E has exactly 2 elements of order 3, the map defined above is an automorphism. After some further checking we see that the full automorphism group $\text{Aut } E$ is isomorphic to $D_6 \times C_2$ generated by the inner automorphisms and ϕ.

Representations As noted at the beginning of this section, the group E has few 'natural' representations. It was defined by a three generator presentation, and it is easily seen that it has the two generator presentation:

$$E \simeq \langle a, d \mid a^4 = d^6 = (ad)^2 = (a^3 d)^2 = e \rangle \tag{8.8}$$

by setting $bc = d$ in the original presentation.

The group E forms a notable subgroup A_7. The alternating group A_7 has 105 involutions each of which consists of pairs of disjoint 2-cycles, and they form a single conjugacy class. (Note S_7 also has 105 involutions each consisting of pairs of disjoint 2-cycles, so use Problem 5.21(iii).) Hence using Theorem 5.19 we see that the centraliser of an A_7 involution has order 24, and for each such involution the corresponding centraliser is isomorphic to E. This can be proved as follows. Consider, for example, the involution $g = (1, 2)(3, 4)$. Clearly, a permutation of the set $\{5, 6, 7\}$ commutes with g, as do the four cycle $(1, 3, 2, 4)$ (note $(1, 3, 2, 4)^2 = g$) and the 2-cycle $\times$ 2-cycle $(1, 3)(2, 4)$. Hence as we know the order of the centraliser, it is straightforward to check that it can be taken in the form

$$\langle (1, 3, 2, 4)(6, 7), (1, 3)(2, 4), (5, 6, 7) \rangle \leq A_7, \tag{8.9}$$

and if we set

$$a \mapsto (1, 3, 2, 4)(6, 7), \quad b \mapsto (1, 3)(2, 4) \quad \text{and} \quad c \mapsto (5, 6, 7)$$

it is also easily checked that the relations in our original definition of E are satisfied by the permutations a, b and c listed above. Therefore, (8.9) provides a permutation representation for the group E. The vital connection between involution centralisers and simple groups is discussed in Chapter 12.

Coxeter and Moser (1984, page 110) give a representation of E as the symmetry group of a polygonal figure, but it cannot be realised in three dimensions. The group is also isomorphic to a subgroup of $SL_2(\mathbb{C})$ generated by the matrices

$$\begin{pmatrix} 0 & 1 \\ -1 & 0 \end{pmatrix}, \quad \begin{pmatrix} \omega & 0 \\ 0 & -\omega^2 \end{pmatrix},$$

where $\omega^3 = 1$ and $\omega \neq 1$; see the presentation (8.8). A second matrix representation is given by the following 7×7 matrices A and D defined over the 2-element field $\mathbb{F}_2$ where the dots stand for zeros, see (8.8),

$$A = \begin{pmatrix} \cdot & \cdot & 1 & \cdot & \cdot & \cdot & \cdot \\ \cdot & \cdot & 1 & \cdot & \cdot & \cdot & \cdot \\ \cdot & 1 & \cdot & \cdot & \cdot & \cdot & \cdot \\ 1 & \cdot & \cdot & \cdot & \cdot & \cdot & \cdot \\ \cdot & \cdot & \cdot & \cdot & 1 & \cdot & \cdot \\ \cdot & \cdot & \cdot & \cdot & \cdot & 1 & \cdot \\ \cdot & \cdot & \cdot & \cdot & \cdot & 1 & \cdot \end{pmatrix} \quad \text{and} \quad D = \begin{pmatrix} \cdot & 1 & \cdot & \cdot & \cdot & \cdot & \cdot \\ 1 & \cdot & \cdot & \cdot & \cdot & \cdot & \cdot \\ \cdot & \cdot & \cdot & 1 & \cdot & \cdot & \cdot \\ \cdot & \cdot & 1 & \cdot & \cdot & \cdot & \cdot \\ \cdot & \cdot & \cdot & \cdot & \cdot & \cdot & 1 \\ \cdot & \cdot & \cdot & \cdot & \cdot & 1 & \cdot \\ \cdot & \cdot & \cdot & \cdot & 1 & \cdot & \cdot \end{pmatrix}.$$

A representation distinct from the above, and also using 4×4 matrices defined over $\mathbb{F}_2$ is given by Problem 8.7, it also uses (8.8).

The group E was defined by a semi-direct product. In fact, it has at least four representations as a semi-direct product as follows:

$E \simeq D_4 \rtimes C_3$ by definition;

$E \simeq C_2 \rtimes Q_3$ if we take $\langle b \rangle$ for C_2 and $\langle a, c \rangle$ for Q_3;

$E \simeq C_2 \rtimes D_6$ if we take $\langle b \rangle$ for C_2 as above, and $\langle a^2 c, ab \rangle$ for D_6;

$E \simeq D_3 \rtimes T_2$ if we take $\langle c, ab \rangle$ for D_3 and $\langle a^2, b \rangle$ for T_2.

Finally, note that E shares many properties with D_{12} and to a lesser extent with $D_4 \times C_3$ and $D_6 \times C_2$, but it is the only group of order 24 which

(a) is indecomposable (it cannot be written as a direct product of smaller but non-neutral subgroups) and has three non-isomorphic subgroups of order 12,
(b) contains six elements of order 4 and six elements of order 6.

8.4 Problems

Problem 8.1 For each of the groups discussed in this chapter, find (a) their largest factor groups, (b) the centralisers of each of their elements, (c) the normalisers of each of their subgroups.

Problem ◆ 8.2 Prove that a group of order 24 possesses a normal subgroup of order 4 or 8. (Hint. Consider Sylow 2-subgroups and their normalisers, and then apply Theorem 5.8.)

Problem 8.3 The group $F_{3,8}$ is defined by

$$F_{3,8} = \langle a, b \mid a^3 = b^8 = e, b^{-1}ab = a^2 \rangle.$$

Establish the following facts/properties concerning this group.

(i) $o(F_{3,8}) = 24$.
(ii) All proper subgroups are cyclic, in this case the group is called *Frobenius* or *metacyclic*; see Theorem 6.18 and Web Section 6.5.
(iii) All of its subgroups except the Sylow 2-subgroups are normal.
(iv) $F_{3,8} \simeq C_8 \rtimes C_3$, describe the implied homomorphism (Section 7.3), and find the centre and the derived subgroup.
(v)   Determine the Frattini and Fitting subgroups.

Problem 8.4 Let G be given by

$$G = \langle c, d \mid c^2 = d^2 = (cd)^3 \rangle.$$

(i)   Show that $c^8 = e$.
(ii) List the elements of G.
(iii) Using the previous problem, prove that $G \simeq F_{3,8}$.

Problem 8.5 (i) Show that $\operatorname{Aut} S_4 \simeq S_4$. Begin by showing that an automorphism preserves 2-cycles if and only if it is inner.

(ii) Using the substitutions $a_1 = b^2, b_1 = abab^2, c_1 = ab$ and $a = c_1 a_1, b = a_1^2$ show that the second presentation of the symmetry group S_4 given on page 170 is valid.

Problem 8.6 (i) Using the permutation representation for $SL_2(3)$ given on page 176 show that the second presentation for this group also given on page 176 is valid. (Hint. Begin with the substitution $a_1 = a^2, b_1 = b^4$.)

(ii) Show that $SL_2(3)$ is not isomorphic to a subgroup of S_n if $n < 8$. One method is as follows. Consider the Sylow 2-subgroups of the groups involved, and note that a Sylow 2-subgroup of S_6 is isomorphic to $D_4 \times C_2$.

Problem 8.7 We have seen that E can be treated as a subgroup of $SL_7(2)$, but it is also isomorphic to a subgroup of $SL_4(2)$, why? Note that $GL_4(2) \simeq SL_4(2) \simeq L_4(2)$.

Problem 8.8 (i) Investigate the dicyclic group $Q_6 = \langle a, b \mid a^6 = b^2 = (ab)^2 \rangle$, see Appendix C. You should determine its subgroups, indicate which are normal and which are maximal, draw a subgroup lattice diagram, find the centre, *et cetera*, and look for possible representations, this is not easy!! See also Problem 3.22.

(ii)* In Appendix C, we state that Q_6 has a 'maximum Cayley count', that is (via Cayley's Theorem (Theorem 4.7)) Q_6 having order 24 is isomorphic to a subgroup of S_{24}, but it is not isomorphic to a subgroup of S_n if $n < 24$. Complete the following sketch proof of this fact. For Q_n to act transitively on m points, m must be a divisor of $4n$. Secondly, note that S_6 contains no dicyclic subgroups. If $m = 8$, the point stabiliser would have an order divisible by 3 but the only subgroup of Q_6 of order 3 is normal, and thus lies in the kernel of the action. Finally, note that a similar argument applies when $m = 12$, hence the only possibility is $m = 24$. (This proof was suggested to B. Fairbairn.)

Problem 8.9 Investigate the group $A_4 \times C_2$, see Appendix C. As in the previous problem, you are asked to find its subgroup structure, and look for representations. Note that it is an example of a group which can be represented as both a direct product and a semi-direct product which is not direct.

Problem 8.10* In this and the next problem, you are asked to show that there are exactly 15 (isomorphism classes of) groups of order 24. By the Sylow theory, each group with this order has one or four Sylow 3-subgroups. In this problem, consider groups G with a unique Sylow 3-subgroup. Using the direct and semi-direct product theory, show that there are 12 types. (Hint. $\mathrm{Aut}(C_3) \simeq C_2$. In Section 6.1, we listed the groups of order 8, so you will need to consider homomorphisms mapping these groups to C_2, that is, you will need to consider their subgroups of order 4.)

Problem 8.11* Continuing the work in the previous problem, now suppose G has four Sylow 3-subgroups. Using the Orbit–Stabiliser Theorem (Theorem 5.7) show that the normaliser of each Sylow 3-subgroup of a group of this type has order 6, and the intersection of all of these normalisers is a subgroup H of order at most 2. Now use Theorem 5.15. Show that if $o(H) = 1$ then $G \simeq S_4$, and if $o(H) = 2$ then G is isomorphic to a semi-direct product $C_3 \rtimes K$ where $K \simeq C_2 \times C_2 \times C_2$ or Q_2 as these last two groups are the only ones of order 8 whose automorphism groups have elements of order 3, a fact that you may take for granted; see Section 4.4. You could also consider what is needed to prove this last statement.

Problem 8.12 (Project—Groups of Order 16) Investigate the class of groups of order 16. You should find nine of 'standard' type, that is direct products, dihedral, *et cetera*, and five of 'exceptional' or semi-direct type; see Problems 3.20, 3.23, 6.4, 6.23, and 7.20. In each case, list the elements and their orders, the subgroups and their orders, find the centres and the derived, Frattini and Fitting subgroups, and describe any unusual features. They can be characterised by considering their 'Abelianisation'. For each group G, this means consider G/G'; see Problem 4.6, Moody (1994) pages 80 to 84, and Appendix C.

Chapter 9
Series, Jordan–Hölder Theorem and the Extension Problem

Continuing our study of group properties, we take a new approach and consider their normal subgroup structure. This will introduce two major classes of groups—nilpotent and soluble—they play a central role in the theory; nilpotent groups will be discussed in Chapter 10 and soluble groups in Chapter 11. But first, in this chapter we prove the Jordan–Hölder Theorem which provides details about a particular series called the composition series. Roughly speaking, this theorem says that we can 'build' all finite (and some infinite) groups from simple groups using extensions, and so it demonstrates the importance of simplicity in our quest to discover as complete picture as possible of all groups. We give a brief introduction to extension theory in the second section of this chapter, it extends the work on semi-direct products discussed in Chapter 7. As we shall see, although this theory has developed over many years it does not yet provide a complete account, and further progress here would greatly benefit the whole of group theory. One major result in this area is the Schur–Zassenhaus Theorem, first proved in full generality in 1937. It states:

> If K is a normal subgroup of a finite group G and $(o(K), o(G/K)) = 1$, then G is isomorphic to a semi-direct product of K by a group A isomorphic to G/K; that is, G can be treated as a particularly simple kind of extension (a 'split' extension) of K by A.

More complex extensions arise when $o(K)$ and $o(A) = o(G/K)$ have common factors which, as noted above, will be discussed in Section 9.2. We shall give a proof of this important result in Web Section 9.4, it introduces some elementary aspects of *cohomology theory* which are needed in the case when the subgroup A is Abelian.

Each of these topics uses the notion of a *series*, that is, a finite sequence of subgroups of the group in question. Properties of their *factors* will provide information about these groups. The nilpotent and soluble groups mentioned above have series whose factors have special properties.

H.E. Rose, *A Course on Finite Groups,*
Universitext,
DOI 10.1007/978-1-84882-889-6_9, © Springer-Verlag London Limited 2009

9.1 Composition Series and the Jordan–Hölder Theorem

We begin with the basic definitions of a series. Given a group G and a subgroup J of G, a *series from J to G* is a finite sequence

$$J = H_0 \leq H_1 \leq \cdots \leq H_n = G \tag{9.1}$$

of subgroups of G where each H_i is a subgroup of its successor. If as is often the case $J = \langle e \rangle$, we say that (9.1) is a *series for G*.

The subgroups H_i in this series are called *terms*.

The *length* of the series is the number of terms excluding G itself, that is, the number of steps in the series, n for (9.1).

A series is called *proper* if no two of the terms are equal, that is, $H_i < H_{i+1}$ for $i = 0, \ldots, n - 1$.

If $H_i \lhd H_{i+1}$, the factor group H_{i+1}/H_i is called a *factor* of the series.

A second series $J = K_0 \leq \cdots \leq K_m = G$ is called a *refinement* of the first series if each term H_i in (9.1) also occurs in the second series.

The refinement $K_0, \ldots, K_m$ given above is called *proper* if there is at least one new term, that is, for all n there exists n' such that $H_n = K_{n'}$, and for some r, $K_r \neq H_i$, for $i = 0, \ldots, n$.

We study three special kinds of series given by

Definition 9.1 (i) A series is called *subnormal* if each term H_i in the series is a normal subgroup of its successor; in (9.1) $H_{i-1} \lhd H_i$, for $i = 1, \ldots, n$.

(ii) A subnormal series is called *normal* if each term in the series is also normal in G; in (9.1) $H_i \lhd G$, for $i = 0, \ldots, n$.

(iii) A proper subnormal series for G is called a *composition series* for G if it has no proper subnormal refinement. The factors of a composition series are called *composition factors*.

Notes (a) Normality is not transitive, and so the property given in (ii) is stronger than that given in (i). (b) A refinement of a normal series need not be normal; see Example (a) below. (c) There is no consistent terminology in the literature, some authors call our subnormal series 'normal', and our normal series 'invariant'; and some older books are different again.

Examples (a) Let $G = SL_2(3)$ (Section 8.2). The diagram on page 175 shows that

$$\langle e \rangle \lhd \langle a^3 \rangle \simeq C_2 \lhd \langle ab, ba \rangle \simeq Q_2 \lhd SL_2(3)$$

is a normal series for $SL_2(3)$. (Reader, what do you notice about the factors of this series?) It has the refinement

$$\langle e \rangle \lhd \langle a^3 \rangle \lhd \langle ab \rangle \simeq C_4 \lhd \langle ab, ba \rangle \lhd SL_2(3).$$

This new series is subnormal and has no further refinements, and so it is a composition series for $SL_2(3)$. It is not a normal series because the subgroup $\langle ab \rangle$ is not normal in $SL_2(3)$.

(b) The group S_5 has the series

$$\langle e \rangle \lhd A_5 \lhd S_5.$$

This is a normal series having no proper refinement because A_5 is simple, and so it is a composition series for S_5. The factors are isomorphic to A_5 and $C_2(\simeq S_5/A_5)$, that is, the factors are simple groups.

(c) One normal series for the group $C_{105} = \langle a \rangle$ (it is Abelian, so all subgroups are normal) is as follows:

$$\langle e \rangle \lhd C_5 \simeq \langle a^{21} \rangle \lhd C_{105}$$

where the factors are isomorphic to C_5 and C_{21}, one of which is not simple. But the series has the proper refinement (which cannot be further refined):

$$\langle e \rangle \lhd C_5 \lhd C_{15} \simeq \langle a^7 \rangle \lhd C_{105}.$$

This is a composition series with simple factors isomorphic to C_5, C_3 and C_7. Further examples can be found in the (subgroup lattice) diagrams given in Chapter 8.

We begin our development of these ideas by proving some elementary facts about composition series. We say that "K is a *maximal normal subgroup* of a group G" if K is a proper normal subgroup of G and there is no proper normal subgroup of G which properly contains K, that is, $K \lhd G$, and if $K \leq J \lhd G$ then $J = K$ or $J = G$. Note that by Definition 2.12(iii) a maximal subgroup is always proper. We have

Lemma 9.2 (i) *Suppose $K \lhd G$. The factor group G/K is simple if and only if K is a maximal normal subgroup of G.*
(ii) *If K_1 and K_2 are maximal normal subgroups of G and $K_1 \neq K_2$, then $K_1 \cap K_2$ is a maximal normal subgroup of both K_1 and K_2.*

Proof (i) This follows directly from the Correspondence Theorem (Theorem 4.16)—if G/K has no proper non-neutral normal subgroup, then G has no proper normal subgroup lying strictly above K, and vice versa.

(ii) We use Theorem 4.15, and write $H = K_1$ and $J = K_2$. We have $H \lhd HJ \lhd G$ by Lemma 4.14, and so as H is maximal, either $HJ = H$ or $HJ = G$. If $HJ = H$ then $J < H$, as H and J are distinct and $J \leq HJ$. This contradicts the maximality of J, and so $HJ = G$. Hence

$$G/J = HJ/J \simeq H/H \cap J,$$

by Theorem 4.15. By (i), G/J is simple, so by (i) again, $H \cap J$ is a maximal normal subgroup of H. We argue similarly for J. $\qquad \square$

Theorem 9.3 (i) *Every finite group G has a composition series.*
(ii) *All composition factors are simple.*

Proof By induction on the order of G. If G is simple, then $\langle e \rangle \lhd G$ is a composition series with a single simple factor; and so both (i) and (ii) follow. Otherwise G has at least one proper non-neutral normal subgroup. Let K be a maximal normal subgroup of G; and so, by Lemma 9.2(i), G/K is simple. By the inductive hypothesis, K has a composition series all of whose factors are simple. If we add G to the top end of this series for K we obtain a composition series for G which again has all of its composition factors simple. Both parts of the theorem follow. $\square$

Result (i) above is not true in general for infinite groups. For example, the group $\mathbb{Z}$ does not have a composition series, for by Theorem 4.19 a non-neutral subgroup of $\mathbb{Z}$ has the form $k\mathbb{Z}$, for some $k \in \mathbb{Z}$, and this subgroup always has proper non-neutral subgroups, for example, $2k\mathbb{Z}$, and all subgroups are normal as $\mathbb{Z}$ is Abelian.

Jordan–Hölder Theorem

We come now to the Jordan–Hölder Theorem, one of the most important in the finite theory. In Example (c) on page 189, we gave a composition series for the group C_{105}, it is not unique. For instance,

$$\langle e \rangle \lhd C_3 \simeq \langle a^{35} \rangle \lhd C_{21} \simeq \langle a^5 \rangle \lhd C_{105},$$

is another composition series for C_{105}. In fact, this group has six composition series, but they all have one property in common:

the set of factors, that is, $\{C_3, C_5, C_7\}$, is the same in each case,

only the order in which they occur changes. It is a remarkable fact that this property holds for *all* groups with composition series, and so in particular for all finite groups. This example relies on the prime factorisation $105 = 3 \cdot 5 \cdot 7$, that is, it relies on the Unique Factorisation Theorem for $\mathbb{Z}$ but this latter result can be treated as a corollary of the Jordan–Hölder Theorem.

To each group G with a composition series we associate a *unique* set of simple groups, its composition factors, which can be treated as the 'building blocks' for G. This is the Jordan–Hölder Theorem, and explains the emphasis given to simple groups in the theory as a whole although it should be noted that non-isomorphic groups can have the same set of composition factors (Problem 9.2). We shall discuss the Extension Problem—ways in which groups can be built up from their simple composition factors—in the next section. Historically, the French mathematician Jordan showed first that the *orders* of the factors are fixed, and later the German mathematician O. Hölder (1859–1937) showed that the factors themselves are fixed. In the example above, the factors C_3, C_5 and C_7 are fixed.

The example above suggests that we make the following

Definition 9.4 Two subnormal series for a group G,

$$\langle e \rangle = H_0 \lhd H_1 \lhd \cdots \lhd H_n = G \quad \text{and} \quad \langle e \rangle = J_0 \lhd J_1 \lhd \cdots \lhd J_m = G,$$

are called *equivalent* if $n = m$, and the set of factors of the first series is the same as the set of factors of the second series ignoring the order but counting multiplicities in each series.

See note on isomorphism classes on page 17. It is easily checked (Problem 9.1) that this defines an equivalence relation on the collection of subnormal series for the group G.

The Jordan–Hölder Theorem states that two composition series for a group are equivalent. We shall give two proofs, the first is fairly short but applies only to finite groups although it can be adapted to apply to all groups with composition series. This is not a major problem because many of the applications are in the finite case. Our second proof, valid for both finite and infinite groups, requires two preliminary results (Lemma 9.7 and Theorem 9.8) which are of interest in their own right.

Theorem 9.5 (Jordan–Hölder Theorem, Finite Case) *All composition series for a fixed finite group are equivalent.*

Proof We use induction on the order of the group G. By hypothesis, we may assume that the result holds for all groups G_1 where $o(G_1) < o(G)$ as the result clearly holds for the neutral group. Let

$$\langle e \rangle = H_0 \lhd \cdots \lhd H_{n-1} \lhd H_n = G \quad \text{and} \quad \langle e \rangle = J_0 \lhd \cdots \lhd J_{m-1} \lhd J_m = G$$

be composition series for G. If $H_{n-1} \simeq J_{m-1}$, then the result holds by the inductive hypothesis because $o(H_{n-1}) < o(G)$. Hence we may assume that $H_{n-1} \not\simeq J_{m-1}$. By Lemma 9.2(ii), we have

$$H_{n-1} \cap J_{m-1} \lhd H_{n-1} \lhd G \quad \text{and} \quad H_{n-1} \cap J_{m-1} \lhd J_{m-1} \lhd G,$$

and both of these are composition series *from* $H_{n-1} \cap J_{m-1}$ *to* G. Also $H_{n-1}J_{m-1} = G$; this follows using the same method as in the proof of Lemma 9.2(ii) above. Hence by assumption and the Second Isomorphism Theorem (Theorem 4.15), we have

$$G/H_{n-1} \simeq J_{m-1}/H_{n-1} \cap J_{m-1} \quad \text{and} \quad G/J_{m-1} \simeq H_{n-1}/H_{n-1} \cap J_{m-1}.$$
$$\tag{9.2}$$

Now if $\langle e \rangle = K_0 \lhd \cdots \lhd K_{r-1} \lhd K_r = H_{n-1} \cap J_{m-1}$ is a composition series for $H_{n-1} \cap J_{m-1}$ (Theorem 9.3), then we have the following four composition series for G, see diagram overleaf:

$$S_1 : \langle e \rangle \lhd H_1 \lhd \cdots \lhd H_{n-1} \lhd G,$$

$$S_2 : \langle e \rangle \lhd K_1 \lhd \cdots \lhd K_{r-1} \lhd H_{n-1} \cap J_{m-1} \lhd H_{n-1} \lhd G,$$

$$S_3 : \langle e \rangle \lhd K_1 \lhd \cdots \lhd K_{r-1} \lhd H_{n-1} \cap J_{m-1} \lhd J_{m-1} \lhd G,$$

$$S_4 : \langle e \rangle \lhd J_1 \lhd \cdots \lhd J_{m-1} \lhd G.$$

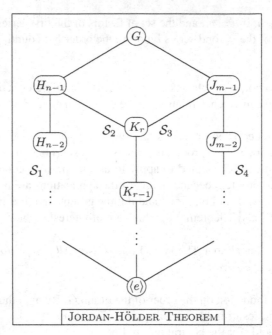

JORDAN-HÖLDER THEOREM

The diagram illustrates the four series $S_1, \ldots, S_4$ used in the first proof of the Jordan–Hölder Theorem.

The series S_1 and S_2 with their last terms (G) removed are composition series for H_{n-1}, a group with order smaller than $o(G)$. Hence by the inductive hypothesis they are equivalent, and so the series S_1 and S_2 themselves are equivalent as the final factor is the same in each case. A similar argument shows that S_3 and S_4 are equivalent. Further, the series S_2 and S_3 are identical except for their penultimate terms, and they have the same length. Also, by (9.2) they have the same set of factors, the only difference being that the last two factors of S_2 are interchanged in S_3. Therefore, S_2 and S_3 are equivalent, which shows finally that S_1 and S_4 are equivalent by Problem 9.1.   □

We give two applications here, but note that the theorem's main significance is the central role it imparts to simple groups via the extension problem. First, let us return to Example (b) on page 189 where we showed that

$$\langle e \rangle \lhd A_5 \lhd S_5$$

is a composition series for S_5 with factors C_2 and A_5. The Jordan–Hölder Theorem implies that this series is unique because C_2 is not isomorphic to a normal subgroup of S_5. Also, S_5 cannot have another normal subgroup (other than A_5). For if $H \lhd S_5$, then $\langle e \rangle \lhd H \lhd S_5$ would be a subnormal series for S_5 which could be refined to a composition series for S_5. But the factors of this new series must be C_2 and A_5 by Theorem 9.5, which is impossible unless H is one of these factors.

For our second application, we characterise completely the Abelian simple groups. In Theorem 2.34, we showed that if the order of a group is prime then it is simple and cyclic. Using the Jordan–Hölder Theorem we have the converse; simpler proofs exist but this one is given to illustrate our main theorem 'in action'.

Theorem 9.6 *A finite simple Abelian group $G \neq \langle e \rangle$ is cyclic with prime order.*

Proof As G is simple and $G \neq \langle e \rangle$, the series

$$\langle e \rangle \lhd G \tag{9.3}$$

is a subnormal series for G with no refinements, and so is a composition series for G. If $o(G) = mn$, $m > 1$, $n > 1$ and $(m, n) = 1$, then G has a subgroup H of order m (Problem 4.15) and $H \lhd G$ because G is Abelian. Hence $\langle e \rangle \lhd H \lhd G$ is a normal series for G which can be refined to a composition series for G one of whose factors is H, or a non-neutral subgroup of H. But this is impossible for by the Jordan–Hölder Theorem all composition series for G are equivalent to the series (9.3). Also, if $o(G) = p^r$ and $r > 0$ then, by Theorem 6.5, G has a (proper non-neutral) normal subgroup. Hence $o(G)$ is prime and so, by Theorem 2.34, G is cyclic of prime order. $\qquad \Box$

We end this section by giving a second proof of our main result, one that does not require the group to be finite. It uses the Schreier Refinement Theorem (Theorem 9.8) which is proved using an extension of the Second Isomorphism Theorem known as Zassenhaus's Lemma (or sometimes the Fourth Isomorphism Theorem). It is also sometimes called the 'Butterfly Lemma' due to the shape of the diagram on the following page giving the subgroup structure. Zassenhaus introduced his lemma to provide a clearer and more straight-forward proof of Schreier's Theorem.

Lemma 9.7 (Zassenhaus's Lemma) *Suppose A', A, B' and B are subgroups of a group G where $A' \lhd A$ and $B' \lhd B$.*

(i) $A'(A \cap B') \lhd A'(A \cap B)$.

(ii) $B'(A' \cap B) \lhd B'(A \cap B)$.

(iii) $\dfrac{A'(A \cap B)}{A'(A \cap B')} \simeq \dfrac{B'(A \cap B)}{B'(A' \cap B)}$.

Proof (i) As $B' \lhd B$ we have, by Problem 2.14(iv), $A \cap B' \lhd A \cap B$, and as $A' \lhd A$ this gives

$$A'(A \cap B') \lhd A'(A \cap B)$$

by Lemma 4.14(iv). This proves (i), and (ii) follows by interchanging A' and A with B' and B; see diagram overleaf.

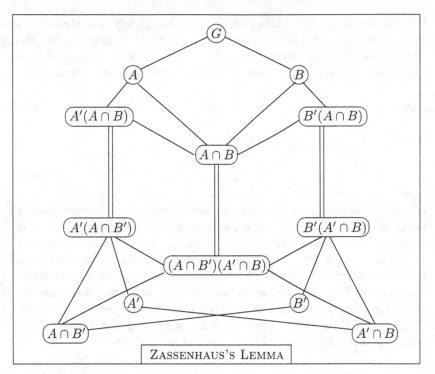

ZASSENHAUS'S LEMMA

The diagram illustrates the subgroup structure used in the proof of Zassenhaus's result, the double lines indicate normality.

Proof of (iii)   We use the Second Isomorphism Theorem. We have $A \cap B \leq A'(A \cap B)$ and $A'(A \cap B') \lhd A'(A \cap B)$ by (i). Also

$$A'(A \cap B) = A'(A \cap B')(A \cap B)$$

because $A \cap B' \leq A \cap B$, and

$$(A \cap B) \cap A'(A \cap B') = (A' \cap B)(A \cap B')$$

by Problem 2.18(i). Using these equations and applying the Second Isomorphism Theorem 4.15 (with $G = A'(A \cap B)$, $H = A \cap B$ and $K = A'(A \cap B')$), we obtain

$$\frac{A'(A \cap B)}{A'(A \cap B')} = \frac{A'(A \cap B')(A \cap B)}{A'(A \cap B')}$$

$$\simeq \frac{A \cap B}{(A \cap B) \cap A'(A \cap B')} = \frac{A \cap B}{(A \cap B')(A' \cap B)},$$

see the left-hand side of the diagram above. Applying the same argument we can, using (ii), prove an exactly similar equation with the pair A', A replaced by B', B, and vice versa throughout, see the right-hand side of the diagram. The last terms of these equations are identical, and so (iii) follows. $\square$

Using this lemma we can now prove

Theorem 9.8 (Schreier Refinement Theorem) *Suppose*

$$\mathcal{S}_1 : \langle e \rangle = H_0 \lhd \cdots \lhd H_n = G, \quad and$$
$$\mathcal{S}_2 : \langle e \rangle = J_0 \lhd \cdots \lhd J_m = G$$

are subnormal series for G. They have equivalent refinements.

Proof We construct a new subnormal series $\mathcal{S}_{1,2}$ by 'inserting a copy' of $\mathcal{S}_2$ between each term in $\mathcal{S}_1$, and a second new series $\mathcal{S}_{2,1}$ by 'inserting a copy' of $\mathcal{S}_1$ between each term in $\mathcal{S}_2$; we then show that $\mathcal{S}_{1,2}$ is equivalent to $\mathcal{S}_{2,1}$; note that repetitions are allowed.

For $0 \leq r \leq n$ and $0 \leq s \leq m$, let the subgroups $K_{r,s}$ be given by

$$K_{r,s} = H_r \left(H_{r+1} \cap J_s \right),$$

where we assume that $H_{n+1} = H_n$ and $J_{m+1} = J_m$. We have

$$K_{r,0} = H_r \left(H_{r+1} \cap \langle e \rangle \right) = H_r \quad and$$
$$K_{r,m} = H_r (H_{r+1} \cap G) = H_r H_{r+1} = H_{r+1},$$

as $H_r \leq H_{r+1}$. Further, for $0 < s < m$,

$$K_{r,s} = H_r(H_{r+1} \cap J_s) \lhd H_r(H_{r+1} \cap J_{s+1}) = K_{r,s+1},$$

using part (i) of Zassenhaus's Lemma (Lemma 9.7) with $H_r = A'$, $H_{r+1} = A$, $J_s = B'$ and $J_{s+1} = B$. The subnormal series $\mathcal{S}_{1,2}$ can now be defined by

$$\mathcal{S}_{1,2} : \langle e \rangle = K_{0,0} \lhd K_{0,1} \lhd \cdots \lhd K_{0,m} \lhd K_{1,0}$$
$$\lhd \cdots \lhd K_{1,m} \lhd K_{2,0} \lhd \cdots \lhd K_{n,m} = G.$$

Interchanging H and J, we define

$$L_{s,r} = J_s \left(J_{s+1} \cap H_r \right),$$

and the subnormal series $\mathcal{S}_{2,1}$ is given by (using (ii) in Lemma 9.7)

$$\mathcal{S}_{2,1} : \langle e \rangle = L_{0,0} \lhd \cdots \lhd L_{0,n} \lhd L_{1,0} \lhd \cdots \lhd L_{m,n} = G.$$

The series $\mathcal{S}_{1,2}$ and $\mathcal{S}_{2,1}$ have the same length (that is, $mn + m + n$), and using the third part of Zassenhaus's Lemma (Lemma 9.7), we obtain

$$\frac{K_{r,s+1}}{K_{r,s}} = \frac{H_r(H_{r+1} \cap J_{s+1})}{H_r(H_{r+1} \cap J_s)} \simeq \frac{J_s(J_{s+1} \cap H_{s+1})}{J_s(J_{s+1} \cap H_r)} = \frac{L_{s+1,r}}{L_{s,r}}.$$

This shows that the sets of factors of $\mathcal{S}_{1,2}$ and $\mathcal{S}_{2,1}$ are identical, and so the result follows. $\square$

We can now reprove the Jordan–Hölder Theorem, the new proof being valid for all groups with composition series.

Second proof of Theorem 9.5   Suppose $\mathcal{S}_1$ and $\mathcal{S}_2$ are composition series for G, they are subnormal and so, by Schreier's Refinement Theorem, they have equivalent refinements, say $\mathcal{S}_1^*$ and $\mathcal{S}_2^*$. But $\mathcal{S}_1$ and $\mathcal{S}_2$ are composition series and so have maximum length, hence $\mathcal{S}_1^*$ ($\mathcal{S}_2^*$) is $\mathcal{S}_1$ ($\mathcal{S}_2$, respectively) with several repeated terms whose corresponding factors are isomorphic to $\langle e \rangle$. Therefore, $\mathcal{S}_1$ and $\mathcal{S}_2$ are equivalent composition series.   □

9.2 Extension Problem

The Jordan–Hölder Theorem states: For groups with composition series, if we can

(a) list all simple groups, and
(b) solve the extension problem: Given groups K and $A \simeq G/K$, construct all possible non-isomorphic groups G,

then we can describe all groups, which is one of the main aims of the theory. Part (a) has been achieved in the finite case, see Chapter 12; hence we need to consider part (b). An answer to (b) has been given by Hölder and Schreier but it has some considerable disadvantages. Given K and A, there may be many solutions G satisfying $K \lhd G$ and $A \simeq G/K$, and the Hölder–Schreier theory provides a characterisation of all possible solutions G. But in general it is not possible to determine whether these solution groups G are isomorphic to one another or not. For example, the theory provides p extension groups G of C_p by C_p, but by Theorem 5.22 we know that there are only two non-isomorphic groups of order p^2. Nevertheless, the theory has a number of applications and so we shall introduce the basic ideas and consider one case: *Cyclic extensions*. In principle, this case is sufficient for all soluble groups, see Chapter 11.

(An alternative approach has been suggested by Kurosh 1955, Volume 2, pages 202 to 210, which can be sketched as follows. A finite group G has a unique maximal normal soluble subgroup J (Chapter 11), which is characteristic and has the property: G/J has no Abelian normal subgroups. Groups of this type are called *semi-simple*, their unique maximal normal subgroups are direct products of non-Abelian simple groups. The factor group G/J can be constructed using permutations and automorphisms of simple groups, and the subgroup J has to be constructed using soluble group methods. One advantage of this procedure is that J is unique.)

We begin by repeating the definition of an extension.

Definition 9.9   A group G is an *extension of K by A* if (i) $K \lhd G$, and (ii) $A \simeq G/K$.

For example, if $G = A \rtimes K$ is a semidirect product of K by A, then $K \lhd G$ and, using the Second Isomorphism Theorem (Theorem 4.15), we have

$$G/K = AK/K \simeq A/A \cap K \simeq A,$$

that is, G is an extension of K by A. Note that for an extension in general there is no condition exactly replacing the property $K \cap A = \langle e \rangle$ as in the semi-direct product case; but see Problem 9.9. The Schur–Zassenhaus Theorem, see the introduction to this chapter, treats the case when the orders of K and A are co-prime. It can be shown that the collection of all extensions of K by A can be treated as a group (the *Second Cohomology Group*), and the neutral element of this new group corresponds to the (collection of) semidirect products. For further details, see one or more of the references quoted at the end of this section and Web Section 9.4.

Examples (a) S_4 is an extension of A_4 by C_2, and $SL_2(3)$ is an extension of C_2 by A_4; for both see Chapter 8.

(b) The group $C_4 = \langle a \rangle$ has a unique normal subgroup $J = \langle a^2 \rangle$, and $G/J \simeq C_2$. Hence C_4 can be treated as an extension of C_2 by C_2 which is not a semi-direct product because the subgroup J is the unique proper non-neutral subgroup of C_4.

(c) The matrix group $GL_2(5)$ is an extension of $SL_2(5)$ by C_4, we shall use this example throughout the section, and so for this chapter we call it the *Standard Example*, that is, we let $G = GL_2(5)$, $K = SL_2(5)$ and $A \simeq C_4$ where we take $(\mathbb{Z}/5\mathbb{Z})^* = \{1, 2, 3, 4\}$ with multiplication modulo 5 to be the representation of C_4 in this example.

Factor Pairs

Our work on the extension problem begins with the following collection of definitions and lemmas.

Definition 9.10 Given an extension G of K by A and an isomorphism $\phi : A \to G/K$, a *section of G through A* is a set $\{s_a : a \in A, s_a \in G\}$ with the properties

(i) $s_e = e$,

(ii) s_a is a representative of the left coset $s_a K$ where $a\phi = s_a K$.

In the Standard Example, there are several choices for the elements of a section. One is: $s_1 = I_2, s_2 = \begin{pmatrix} 2 & 1 \\ 0 & 1 \end{pmatrix}, s_3 = \begin{pmatrix} 3 & 1 \\ 0 & 1 \end{pmatrix}$, and $s_4 = \begin{pmatrix} 4 & 1 \\ 0 & 1 \end{pmatrix}$; note that $\det(s_n) = n$ where $n = 1, 2, 3$ or 4 in this example, that is, n runs through the elements of A.

Much of the theory depends on the next two definitions. As ϕ is an isomorphism, if $a, a' \in A$, then $s_a s_{a'}$ lies in the same coset as $s_{aa'}$, and so they differ by an element of K ($(a, a')\xi$ in the definition below). Note also that to start with we are working with an existing extension investigating its construction and properties.

Definition 9.11 (Part 1) Given G, an extension of K by A, and a section $\{s_a : a \in A\}$ of G through A, the function $\xi : A \times A \to K$ is defined by

$$(a, a')\xi = s_{aa'}^{-1} s_a s_{a'} \quad \text{for } a, a' \in A.$$

We have

$$s_a s_{a'} = s_{aa'}(a, a')\xi, \tag{9.4}$$

$$(a, e)\xi = e = (e, a)\xi. \tag{9.5}$$

Equation (9.4) was given above, and (9.5) follows directly from (i) in Definition 9.10.

Definition 9.11 (Part 2) (i) Given the function ξ of the extension G with section $\{s_a : a \in A\}$ as in Part 1, for $a \in A$ the function $\vartheta_a : K \to K$ is defined by

$$k\vartheta_a = s_a^{-1} k s_a \quad \text{for } k \in K,$$

that is, $k\vartheta_a$ is the conjugate of k by the section element s_a.

(ii) The pair $\xi, \{\vartheta_a : a \in A\}$ is called the *factor pair* of the extension G of K by A with section $\{s_a : a \in A\}$.

Note that as ϑ_a is defined by conjugation, it is an automorphism of K.

In the Standard Example, we have $(m, n)\xi = I_2$, if m or n equals 1 (as $s_1 = I_2$), and $(m, n)\xi = \left(\begin{smallmatrix} 1 & 1/n \\ 0 & 1 \end{smallmatrix}\right)$, if $m \neq 1 \neq n$. Similarly, ϑ_1 is the identity map on $SL_2(5)$ and

$$\vartheta_n : \begin{pmatrix} r & s \\ t & u \end{pmatrix} \mapsto s_a^{-1} \begin{pmatrix} r & s \\ t & u \end{pmatrix} s_a = \begin{pmatrix} r - t & (r + s - t - u)/n \\ nt & t + u \end{pmatrix} \tag{$*$}$$

if $n \neq 1$; the reader should check these facts.

The following two lemmas are basic. To aid clarity in long expressions we have reintroduced the 'dot' notation $\odot$ for the group operation in K. This is not strictly necessary and so can be ignored, but it should help in the understanding of the following proofs.

Lemma 9.12 *Using the notation set out above, where $\xi, \{\vartheta_a : a \in A\}$ is a factor pair, and $a, b, c \in A$, we have*

$$(a, bc)\xi \odot (b, c)\xi = (ab, c)\xi \odot ((a, b)\xi)\vartheta_c.$$

This equation is called the *cocycle identity*.

Proof We use associativity. We have, applying (9.4) twice,

$$s_a(s_b s_c) = s_a\big(s_{bc}(b,c)\xi\big) = s_{abc}\big((a,bc)\xi \odot (b,c)\xi\big),$$

and, introducing the neutral term $s_c s_c^{-1}$ and using (9.4) again,

$$(s_a s_b)s_c = \big(s_{ab}(a,b)\xi\big)s_c = s_{ab}s_c\big(s_c^{-1}(a,b)\xi s_c\big)$$

$$= s_{ab}s_c\big((a,b)\xi\big)\vartheta_c = s_{abc}\big((ab,c)\xi \odot \big((a,b)\xi\big)\vartheta_c\big)$$

using Definition 9.11, part 2 for the penultimate equation. The lemma follows by associativity and cancellation. □

Referring to the Standard Example, suppose, for instance, $a = 2$, $b = 3$ and $c = 4$. (Remember that in this example $A \simeq (\mathbb{Z}/5\mathbb{Z})^* = \{1,2,3,4\}$ with multiplication modulo 5, the neutral element is 1, and $1/2 = 3$, *et cetera*.) We have $ab = 1$, $bc = 2$, $(a,bc)\xi = (2,2)\xi = \left(\begin{smallmatrix} 1 & 3 \\ 0 & 1 \end{smallmatrix}\right)$, $(b,c)\xi = (3,4)\xi = \left(\begin{smallmatrix} 1 & 4 \\ 0 & 1 \end{smallmatrix}\right)$, $(ab,c)\xi = (1,4)\xi = I_2$, $(a,b)\xi = (2,3)\xi = \left(\begin{smallmatrix} 1 & 2 \\ 0 & 1 \end{smallmatrix}\right)$, and using the identity $(*)$ on page 198

$$\big((a,b)\xi\big)\vartheta_4 = \begin{pmatrix} 1 & 2 \\ 0 & 1 \end{pmatrix}\vartheta_4 = \begin{pmatrix} 1 & 3/4 \\ 0 & 1 \end{pmatrix} = \begin{pmatrix} 1 & 2 \\ 0 & 1 \end{pmatrix}.$$

In this example, the left-hand side of the cocycle identity equals $(a,bc)\xi(b,c)\xi = (2,2)\xi(3,4)\xi = \left(\begin{smallmatrix} 1 & 3 \\ 0 & 1 \end{smallmatrix}\right)\left(\begin{smallmatrix} 1 & 4 \\ 0 & 1 \end{smallmatrix}\right) = \left(\begin{smallmatrix} 1 & 2 \\ 0 & 1 \end{smallmatrix}\right)$, and this equals the right-hand side using the displayed equations above because $(ab,c)\xi = (1,4)\xi = I_2$.

The second basic result is

Lemma 9.13 *If $a, a' \in A$ and $k \in K$, then*

$$k\vartheta_a\vartheta_{a'} = \big((a,a')\xi\big)^{-1} \odot k\vartheta_{aa'} \odot (a,a')\xi.$$

Note this result says that the composition of ϑ_a and $\vartheta_{a'}$ is defined by conjugation using the first term of the factor pair.

Proof Applying Definition 9.11 twice and (9.4), we have, for all $k \in K$,

$$(k\vartheta_a)\vartheta_{a'} = s_{a'}^{-1} \odot k\vartheta_a \odot s_{a'} = (s_a s_{a'})^{-1} \odot k \odot s_a s_{a'}$$

$$= \big(s_{aa'}(a,a')\xi\big)^{-1} \odot k \odot s_{aa'}(a,a')\xi$$

$$= \big((a,a')\xi\big)^{-1} \odot s_{aa'}^{-1} k s_{aa'} \odot (a,a')\xi$$

$$= \big((a,a')\xi\big)^{-1} \odot k\vartheta_{aa'} \odot (a,a')\xi. □$$

Returning to the Standard Example, the expression in the last line of the equation above (call it T) is a product of three 2×2 matrices.

For instance, suppose $a = 2$ and $a' = 4$, then $aa' = 3$ and, using the calculation below Definition 9.11, $(a, a')\xi = (2, 4)\xi = \begin{pmatrix} 1 & 4 \\ 0 & 1 \end{pmatrix}$ and $((a, a')\xi)^{-1} = \begin{pmatrix} 1 & 1 \\ 0 & 1 \end{pmatrix}$. If $k = \begin{pmatrix} r & s \\ t & u \end{pmatrix}$, then $k\vartheta_{aa'}$ is given by the right-hand matrix in $(*)$ on page 198 with $n = 3$. Multiplying these three matrices together gives

$$T = \begin{pmatrix} r + 2t & (3r + s + t + 2u)/3 \\ 3t & 3t + u \end{pmatrix}.$$

The reader should check that this equals $k(\vartheta_2 \vartheta_4) = (k\vartheta_2)\vartheta_4$ by applying $(*)$ on page 198 again, first with $n = 2$, and then on the result with $n = 4$. You should also repeat the whole calculation with different a and a'.

Construction of a Group Extension

We come now to the converse of the work described above, that is, given groups K and A, construct G. In this case, we do not have a section to build upon, so we make the following definition as suggested by the lemmas given above. Then using this definition we shall be able to construct G and the corresponding section. As in the semi-direct product case given in Section 7.3, we form a group using the underlying set $A \times K$, and we introduce a new operation by

$$(a_1, k_1)(a_2, k_2) = \left(a_1 a_2, (a_1, a_2)\xi \odot k_1 \vartheta_{a_2} \odot k_2\right), \tag{9.6}$$

where the components of the factor pair ξ and $\{\vartheta_a : a \in A\}$ are given in Definition 9.14 below. This is similar to the semi-direct operation, see page 153, except that we now have the new term $(a_1, a_2)\xi$. Or to put this another way, our extension will be semi-direct if $(a_1, a_2)\xi = e$ for all $a_1, a_2 \in A$.

Definition 9.14 Given groups A and K, the maps $\vartheta_a : K \to K$ and $\xi : A \times A \to K$ satisfy the following three properties.

(i)   The map ϑ_e is the identity map on K, and the map ϑ_a is an automorphism of K for all $a \in A$.

(ii)   The map ξ satisfies (a) $(e, a)\xi = e = (a, e)\xi$ for $a \in A$, and (b) the cocycle equality, that is, for $a_1, a_2, a_3 \in A$,

$$(a_1, a_2 a_3)\xi \odot (a_2, a_3)\xi = (a_1 a_2, a_3)\xi \odot \left((a_1, a_2)\xi\right)\vartheta_{a_3}.$$

(iii)   For all $k \in K$ and $a_1, a_2 \in A$, we have

$$k\vartheta_{a_1}\vartheta_{a_2} = \left((a_1, a_2)\xi\right)^{-1} \odot k\vartheta_{a_1 a_2} \odot (a_1, a_2)\xi.$$

Using this definition we can construct the required extension, it depends on the particular automorphisms ϑ_a chosen.

Theorem 9.15 *Suppose A and K are groups, and ϑ_a, for $a \in A$, and ξ satisfy the properties given in Definition 9.14 above. The set $A \times K$ with the operation (9.6) forms a group G which is an extension of a group K' isomorphic to K by a group A' isomorphic to A, and for which the set $\{(a, e) : a \in A\}$ is a section of G through A'.*

The following proof is straight-forward but somewhat 'involved', the reader should write out a version for him(her)self.

Proof The operation is closed by definition. Next we consider associativity. First, we have using (9.6) twice

$$\big((a_1, k_1)(a_2, k_2)\big)(a_3, k_3)$$
$$= \big(a_1 a_2, (a_1, a_2)\xi \odot k_1 \vartheta_{a_2} \odot k_2\big)(a_3, k_3)$$
$$= \big(a_1 a_2 a_3, (a_1 a_2, a_3)\xi \odot \big((a_1, a_2)\xi \odot k_1 \vartheta_{a_2} \odot k_2\big)\vartheta_{a_3} \odot k_3\big). \quad (9.7)$$

We shall refer below to the argument of the function ϑ_{a_3} in this expression, it is a product of three elements of K. We also have

$$(a_1, k_1)\big((a_2, k_2)(a_3, k_3)\big)$$
$$= (a_1, k_1)\big(a_2 a_3, (a_2, a_3)\xi \odot k_2 \vartheta_{a_3} \odot k_3\big)$$
$$= \big(a_1 a_2 a_3, (a_1, a_2 a_3)\xi \odot k_1 \vartheta_{a_2 a_3} \odot (a_2, a_3)\xi \odot k_2 \vartheta_{a_3} \odot k_3\big). \quad (9.8)$$

Rewriting the second argument of the pair given in the last expression in (9.8) by introducing the neutral term $(a_2, a_3)\xi \odot ((a_2, a_3)\xi)^{-1}$ gives the expression

$$\Big((a_1, a_2 a_3)\xi \odot (a_2, a_3)\xi\Big) \odot \Big(((a_2, a_3)\xi)^{-1} \odot k_1 \vartheta_{a_2 a_3} \odot (a_2, a_3)\xi\Big)$$
$$\odot k_2 \vartheta_{a_3} \odot k_3. \quad (9.9)$$

Using the cocycle identity (Definition 9.14(ii)) the subexpression between the first set of large round brackets in (9.9) satisfies

$$(a_1, a_2 a_3)\xi \odot (a_2, a_3)\xi = (a_1 a_2, a_3)\xi \odot ((a_1, a_2)\xi)\vartheta_{a_3}. \quad (9.10)$$

Also applying (iii) in Definition 9.14 to the subexpression between the second set of large round brackets in (9.9) we obtain

$$((a_2, a_3)\xi)^{-1} \odot k_1 \vartheta_{a_2 a_3} \odot (a_2, a_3)\xi = k_1 \vartheta_{a_2} \vartheta_{a_3}. \quad (9.11)$$

Hence, applying (9.10) and (9.11), and using the fact that ϑ_{a_3} is an isomorphism, Expression (9.9) satisfies

$$(a_1 a_2, a_3)\xi \odot ((a_1, a_2)\xi)\vartheta_{a_3} \odot (k_1 \vartheta_{a_2})\vartheta_{a_3} \odot k_2 \vartheta_{a_3} \odot k_3$$
$$= (a_1 a_2, a_3)\xi \odot ((a_1, a_2)\xi \odot k_1 \vartheta_{a_2} \odot k_2)\vartheta_{a_3} \odot k_3$$

as ϑ_{a_3} is an isomorphism—note its argument, see the comment below (9.7). This expression equals the second argument in the last expression in (9.7); that is, we have established associativity.

The neutral element is (e, e), for by (9.6) and as ϑ_e is the identity map on K

$$(a, k)(e, e) = \big(a, (a, e)\xi \odot k\vartheta_e \odot e\big) = (a, k).$$

Further, the inverse operation is given by

$$(a, k)^{-1} = \big(a^{-1}, \big(((a^{-1}, a)\xi\big)^{-1} \odot k^{-1}\big)\vartheta_a^{-1}\big)$$

as the following calculation shows; note that as ϑ_a is an isomorphism, so is ϑ_a^{-1}. We have using (9.6) and the equation above

$$
\begin{aligned}
(a, k)^{-1}(a, k) &= \big(a^{-1}, \big(((a^{-1}, a)\xi)^{-1} \odot k^{-1}\big)\vartheta_a^{-1}\big)(a, k) \\
&= \big(e, (a^{-1}, a)\xi \odot \big(\big(((a^{-1}, a)\xi)^{-1} \odot k^{-1}\big)\vartheta_a^{-1}\big)\vartheta_a \odot k\big) \\
&= \big(e, (a^{-1}, a)\xi \odot \big((a^{-1}, a)\xi\big)^{-1} \odot k^{-1} \odot k\big) \\
&= (e, e).
\end{aligned}
$$

These equations show that the set $A \times K$ with operation given in (9.6) forms a group G. The reader should show directly (that is, without using Theorem 2.5 and similar facts) that $(e, e)(a, k) = (a, k)$ and $(a, k)(a, k)^{-1} = (e, e)$, see Problem 9.13 (for the second equation apply ϑ).

Next we prove that G contains a normal subgroup K' which is isomorphic to K. Let $K' = \{(e, k) : k \in K\}$. Using (9.6) we obtain for $k, k' \in K$ (note that ϑ_e is the identity map on K)

$$(e, k)\big(e, k'\big) = \big(e, (e, e)\xi \odot k\vartheta_e \odot k'\big) = \big(e, kk'\big) \in K',$$

and $(e, k)^{-1} = (e, (((e, e)\xi)^{-1} \odot k^{-1})\vartheta_e^{-1}) = (e, k^{-1}) \in K'$. For normality

$$
\begin{aligned}
(a, k)^{-1}&\big(e, k'\big)(a, k) \\
&= \big(a^{-1}, \big(((a^{-1}, a)\xi)^{-1} \odot k^{-1}\big)\vartheta_a^{-1}\big)\big(a, (e, a)\xi \odot k'\vartheta_e \odot k\big) \\
&= \big(e, (a^{-1}, a)\xi \odot \big(\big(((a^{-1}, a)\xi)^{-1} \odot k^{-1}\big)\vartheta_a^{-1}\big)\vartheta_a \odot k'\vartheta_a \odot k\big) \\
&= \big(e, k^{-1} \odot k'\vartheta_a \odot k\big) \in K'.
\end{aligned}
$$

An easy exercise shows that $K \simeq K'$. Similarly, we define $A' = \{(a, e)K' : a \in A\}$ and as in the K' case it is clear that $A' \simeq G/K'$. Lastly, we define a map $\psi : A \to G/K'$ by

$$a\psi = (a, e)K' \quad \text{for } a \in A.$$

This map is a homomorphism, for if $a, a' \in A$

$$a\psi a'\psi = (a, e)K'(a', e)K' = (a, e)(a', e)K' = (aa', (a, a')\xi)K'$$
$$= (aa', e)(e, (a, a')\xi)K' = (aa', e)K' = aa'\psi.$$

We leave as an exercise for the reader to show that ψ is an isomorphism and that $\{(a, e) : a \in A\}$ is a section of G through A'. ☐

Using this theorem we can now obtain a representation for our standard example $GL_2(5)$ as an extension of $SL_2(5)$ by C_4 using the operation described on page 197; the reader should check this.

Cyclic Extensions

As an application of the theorem proved above we consider in more detail extensions of a group by a cyclic group, we noted at the beginning of this section that in principle all soluble groups (Chapter 11) can be generated in this way.

Definition 9.16 An extension of a group K by a group A is called *cyclic* if A is cyclic.

Note that the Standard Example is, in fact, a cyclic extension.
    These extensions are characterised by

Theorem 9.17 (i) *Suppose G is an extension of K by a cyclic group $A \simeq G/K$ where $A = \langle a \rangle$ and $o(a) = m$. Choose $s_a \in G$ so that $s_a K$ is a generator of G/K, and suppose $s_a^m = k_0 \in K$. There exists an automorphism ψ of K with the properties*

(a) *$k_0\psi = k_0$ and*
(b) *$k\psi^m = k_0^{-1}kk_0$ for $k \in K$.*

(ii) *Conversely, given $k_0 \in K$ and an automorphism ψ of K satisfying* (a) *and* (b) *above, the set $\{(a^r, k) : 0 \leq r < m$ and $k \in K\}$ with the operation*

$$(a^r, k_1)(a^s, k_2) = (a^{r+s}, (a^r, a^s)\xi \odot k_1\psi^s \odot k_2)$$

where $0 \leq r, s < m$, $k_1, k_2 \in K$, and ξ is defined by

$$(a^r, a^s)\xi = \begin{cases} e & \text{if } 0 \leq r + s < m, \\ k_0 & \text{if } m \leq r + s < 2m, \end{cases} \tag{9.12}$$

forms a group which is a cyclic extension of a group isomorphic to K by a group isomorphic to A.

Proof (i) Define ψ by

$$k\psi = s_a^{-1}ks_a \quad \text{for } k \in K.$$

Then $k_0\psi = s_a^{-1}s_a^m s_a = s_a^m = k_0$, and $k\psi^m = s_a^{-m}ks_a^m = k_0^{-1}kk_0$.

(ii) For the converse, note first that in this case ξ only takes the values e and k_0, and so ξ^r only depends on values of r. Hence we can put $\vartheta_{a^r} = \psi^r$, and we need to show that ξ, $\vartheta_{a^r} (= \xi, \psi^r, 0 \le r < m)$ is a factor pair.

We have $(e, a^r)\xi = (a^r, e)\xi = e$ as $0 \le r < m$ as A is a cyclic group of order m. We also have

$$\left(a^{r+s}, a^t\right)\xi \cdot \left(a^r, a^s\right)\xi = \left(a^r, a^{s+t}\right)\xi \cdot \left(a^s, a^t\right)\xi, \tag{9.13}$$

by the discussion above this is the cocycle identity in this case. Equation (9.13) is derived as follows. First, we show that

$$\left(a^{r+s}, a^t\right)\xi \cdot \left(a^r, a^s\right)\xi = \begin{cases} e & \text{if } 0 \le r+s+t < m, \\ k_0 & \text{if } m \le r+s+t < 2m, \\ k_0^2 & \text{if } 2m \le r+s+t < 3m. \end{cases} \tag{9.14}$$

In the first line of (9.14), we have $0 \le r, s, t, r+s, r+s+t < m$, and so both terms in the product on the left-hand side equal e.

There are two cases to consider for the second line; the first is $0 \le r+s < m$. Here $(a^r, a^s)\xi = e$ and so $(a^{r+s}, a^t)\xi = k_0$ as $r+s+t \ge m$. The second case is $r+s \ge m$, and so $(a^r, a^s)\xi = k_0$ by (9.12). We also have $a^{r+s} = a^{r+s-m}$ and $0 \le r+s-m < m$ (as $o(a) = m$), hence $(a^{r+s}, a^t)\xi = (a^{r+s-m}, a^t)\xi = e$. Therefore, in both of these cases, we obtain the same value, that is, k_0, for the expression on the left-hand side of (9.14).

In the last line, we have $2m > r+s \ge m$ (as $t < m$) and so $(a^r, a^s)\xi = k_0$. Also $m \le r+s-m+t < 2m$ and $r+s-m \ge 0$, and so $(a^{r+s}, a^t)\xi = k_0$; the third line follows.

The conditions in (9.14) are symmetrical in r, s and t, and so these properties also apply to the right-hand side of (9.13). Hence the whole identity (9.13) follows.

Having established (9.13), we need to show that

$$k\psi^r\psi^s = \left((a^r, a^s)\xi\right)^{-1} \odot k\psi^{r+s} \odot (a^r, a^s)\xi, \tag{9.15}$$

where we have defined ϑ_{a^s} to equal ψ^s, $k \in K$ and $0 \le r, s < m$. If $r+s < m$, then $(a^r, a^s)\xi = e$, and so $\psi^r\psi^s = \psi^{r+s}$, and if $r+s \ge m$, we have $r+s = m+t$ for some t where $t \ge 0$, $(a^r, a^s)\xi = k_0$, and $a^ra^s = a^t$. Hence using these properties we have, for $k \in K$,

$$k\vartheta_{a^r}\vartheta_{a^s} = k\psi^r\psi^s = k\psi^{m+t}$$

$$= k_0^{-1} \odot k\psi^t \odot k_0 = \left((a^r, a^s)\xi\right)^{-1} \odot k\vartheta_{a^{r+s}} \odot (a^r, a^s)\xi.$$

Therefore, ξ, ϑ_{a^s}, $0 \le s < m$, form a factor pair, and we can apply Theorem 9.15.

Finally, note that using the hypotheses of the theorem we have $(a, e)^m = (a, e)(a, e)^{m-1} = (a, e)(a^{m-1}, e) = (a^m, (a, a^{m-1})\xi \odot e\vartheta_{a^{m-1}} \odot e) = (e, k_0)$ (because $1 + (m-1) \geq m$) as required. □

An immediate consequence of this result is

Corollary 9.18 *If K is a group, $g \in Z(K)$ and n is a positive integer, then a group G exists satisfying*

$$G = \langle K, c \rangle, \quad K \lhd G, \quad G/K \simeq C_n, \quad c \in C_G(K) \quad and \quad c^n = g.$$

Proof In Theorem 9.17, let $A = C_n$ and ψ be the identity automorphism on K. □

For example, suppose $K = C_3$ and $n = 8$, then this corollary shows that the group $F_{3,8}$ with the presentation given in Problem 8.3 does, in fact, exist. Secondly, suppose K is Abelian, $g \in K$, $o(g) = 2$ and ψ is the automorphism given by $a\psi = a^{-1}$ for $a \in K$, then the corollary asserts the existence of the group G with the properties:

$$G = \langle K, c \rangle, \quad [G : K] = 2, \quad c^2 = a, \quad c^{-1}gc = g^{-1} \quad \text{for } g \in K.$$

This group is called *generalised dicyclic*, and dicyclic if K is cyclic. If in this latter case $o(K) = 4$, then $G \simeq Q_2$ the quaternion group. Reader, referring to Problem 8.12, what group do you obtain if $o(K) = 8$?

We have only been able to give a very brief account of the topics in this section, in particular due to space considerations we have not been able to discuss the basic cohomology theory needed to develop the work. The elementary ideas and results of cohomology theory follow fairly directly from the work of Chapter 2, and Sections 4.1 and 4.2. The reader requiring more details should consult one or more of the following texts: Scott (1964), Chapter 9; Rotman (1994), Chapter 7; Suzuki (1982), Chapter 2; or for a slightly more advanced approach Robinson (1982), Chapter 11, or Benson (1991).

9.3 Problems

Problem 9.1 Show that the relation given in Definition 9.4 is an equivalence relation.

Problem 9.2 (i) Give composition series for the groups: (a) S_4, (b) A_4, (c) $A_4 \times C_2$, (d) $SL_2(3)$, and (e) E (Section 8.3). Which groups have unique series?

(ii) Give an example of a pair of non-isomorphic groups whose composition series have the same factors.

Problem 9.3 Show that if G has a composition series and $K \lhd G$, then G has a composition series one of whose terms is K.

Problem 9.4 (i) Discuss the proposition: An Abelian group has a composition series if, and only if, it is finite.

(ii) Show that the cyclic group C_{p^n} has exactly one composition series.

(iii) Show that $GL_n(F)$ has a composition series if and only if the field F is finite.

(iv) If G has a composition series and $H \leq G$, is it necessary for H to have a composition series? (Hint. Refer to Web Sections 3.6 and 7.5.)

Problem ♦ 9.5 Prove the following statements.

(i) All composition factors of a finite Abelian group have prime order.
(ii) All composition factors of a finite p-group have order p.

Problem 9.6 (i) Suppose G has a composition series with exactly three terms (length 2). Show that G has a unique composition series or it is isomorphic to a direct product of two simple groups.

(ii) Show that groups exist with arbitrarily long composition series.

Problem 9.7 Consider the following two series for $\mathbb{Z}$ for primes p and q:

$$\langle e \rangle \lhd p\mathbb{Z} \lhd \mathbb{Z} \quad \text{and} \quad \langle e \rangle \lhd q\mathbb{Z} \lhd \mathbb{Z}, \quad \text{where } p \neq q.$$

Show that these series have equivalent refinements. Schreier's Theorem (Theorem 9.8) asserts the existence of these refinements, but note that it does not require the group in question to have a composition series.

Problem ♦ 9.8 A subgroup H of a group G is called *subnormal* in G if and only if there exists a subnormal series from H to G, and in this case we write $H \lhd\lhd G$. Note that unlike normality the subnormal relation is transitive, that is, if $J \lhd\lhd H$ and $H \lhd\lhd G$, then $J \lhd\lhd G$. Prove the following three properties.

(i) If $H \lhd\lhd G$ and $J \leq G$, then $H \cap J \lhd\lhd J$.
(ii) If $H \lhd\lhd G$ and $J \lhd\lhd G$, then $H \cap J \lhd\lhd G$.
(iii) If $\theta : G \to G_1$ is a surjective homomorphism, then $H \lhd\lhd G$ if and only if $H\theta \lhd\lhd G_1$.

Problem 9.9 (i) Show that if G is an extension of K by A and both K and A are finite, then $o(G) = o(K)o(A)$.

(ii) Using (i) show that if G is finite and has the subnormal series $\langle e \rangle = H_0 \lhd \cdots \lhd H_m = G$, where $o(H_{i+1}/H_i) = r_i$ for $i = 0, \ldots, m-1$, then $o(G) = \prod_{i=0}^{m-1} r_i$.

Problem 9.10 (i) Construct all possible extensions of C_3 by $C_2 \times C_2$.

(ii) Using Problem 6.5 show that every non-Abelian group with order p^3 is an extension of C_p by $C_p \times C_p$.

Problem 9.11 (i) Show that if G and H are Abelian and $(o(G), o(H)) = 1$, then up to isomorphism there is only one Abelian extension of G by H.

(ii) Construct all extensions of the infinite group $\mathbb{Z}$ by C_2.

Problem 9.12 Show that the quaternion group Q_2 can be written in three different ways as an extension of C_4 by C_2, and one way as an extension of C_2 by $C_2 \times C_2$. Do any of these extensions reduce to a semi-direct product?

Problem 9.13 In the proof of Theorem 9.15, using only Definition 9.14 prove that

(a) $(e, e)(a, k) = (a, k)$ and
(b) $(a, k)(a, k)^{-1} = (e, e)$.

Problem 9.14 Use extensions to construct all groups of order pq where p and q are prime and $p < q$, see Problems 3.6, 6.14 and 7.21, and Web Section 14.3.

Problem 9.15 (Project—Chief Series) Before attempting this project, the reader should read the subsection on minimal normal subgroups given at the end of Section 11.1. A *chief series* for a group G is a proper normal series

$$\langle e \rangle \lhd H_1 \lhd \cdots \lhd H_m = G,$$

with the additional property that if $K \lhd G$ and, for some i, we have $H_i \leq K \leq H_{i+1}$, then either $K = H_i$ or $K = H_{i+1}$. This series is proper, so $H_i < H_{i+1}$ for all i and, whereas a composition series is subnormal with no proper subnormal refinement, a chief series is normal with no proper normal refinement. Also, we define a *chief factor* for G to be a factor of some chief series for G. Now let G be a finite group.

(i)   State and prove the Jordan–Hölder Theorem for chief series of G.
(ii) Prove that finite groups have chief series.
(iii) Show that if G has a chief series, then every chief factor is a minimal normal subgroup of a factor group of G.
(iv) Prove that a minimal normal subgroup of G is a direct product of simple groups H_i where $H_i \simeq H_j$ for all i and j. Use induction on $o(G)$, and begin with a minimal normal subgroup H and its set of conjugates in G. Deduce that a chief factor of G is a direct product, with one or more terms, of mutually isomorphic simple groups.
(v)   Finally, calculate the chief series for the groups in Problem 9.2(i).

Chapter 10
Nilpotency

The material in this chapter is largely disjoint from that given in Chapter 11. Hence with a few minor exceptions these chapters can be read independently.

Nilpotent groups lie between the classes of Abelian and soluble groups, this latter class will be discussed in Chapter 11. Finite nilpotent groups also have a number of similarities with p-groups—for example, both have a plentiful supply of normal subgroups; in fact, some authors describe nilpotent groups as 'generalised p-groups'. In the finite case (which we introduced first in Chapter 6), these groups have many independent definitions showing that they possess many useful properties in common; see Theorem 10.9. Some of these new definitions are 'series based' and apply in both the finite and infinite cases. They also provide a 'measure'—the *nilpotency class*—of how far away from Abelian a particular nilpotent group is, and this measure can itself be used to establish new results. An Abelian group G has nilpotent class 1, and we have

$$[a, b] = e \quad \text{for all } a, b \in G.$$

For nilpotent groups H of 'class n' we have

$$[\ldots [[a_0, a_1], a_2] \ldots, a_n] = e \quad \text{for all } a_0, \ldots, a_n \in H,$$

see Definition 10.5 and Problem 10.2. One origin of the word 'nilpotent' is given in the footnote on page 213.

We begin this chapter by stating the series-based definitions of nilpotency mentioned above, in most cases several series are involved and all need to be considered. This builds on the work given at the beginning of Section 9.1. We shall also establish the equivalence of many of the definitions, some of which at first seem unrelated. For instance, one concerns properties of maximal subgroups whilst another states that elements of coprime order commute. In Section 10.2, we discuss the *Frattini* and *Fitting subgroups* of a group, in the finite case they are both normal and nilpotent and have some remarkable properties and applications. The first introduces a connection between maximal subgroups and the so-called *non-generators* of a group, whilst generalisations of the second have proved useful in the CFSG.

H.E. Rose, *A Course on Finite Groups,*
Universitext,
DOI 10.1007/978-1-84882-889-6_10, © Springer-Verlag London Limited 2009

10.1 Nilpotent Groups

We begin by introducing a new type of series which can be used to define nilpotency in both the finite and infinite cases. The basic series definitions were given at the beginning of Section 9.1.

Central Series

We use the notation $[G, H]$ for the group generated by all commutators of the form $[g, h]$ where $g \in G$ and $h \in H$; see Problem 2.16. A normal series for G

$$\langle e \rangle = H_0 \lhd H_1 \lhd \cdots \lhd H_n = G,$$

see Definition 9.1(ii), is called a *central series* for G if, for each i in the range $0 \le i \le n - 1$, we have

$$H_{i+1}/H_i \le Z(G/H_i),$$

or equivalently, see Problem 4.16(vii),

$$[H_{i+1}, G] \le H_i.$$

Also G is *nilpotent* if it has a central series; see Definition 10.5 below. Clearly all Abelian groups are nilpotent, and so nilpotency can be treated as a generalisation of Abelianness. A group can have several different central series, see the example on page 213, and so we begin by studying these series.

Previously we have defined the derived (commutator) subgroup G' (Problem 2.16) and the centre $Z(G)$ (Definition 2.32) of a group G. Here we extend these notions to define the *higher commutator subgroups* $\mathcal{D}_i(G)$ and the *higher centres* $\mathcal{Z}_i(G)$, and using these, the *lower* and *upper central series* for G. Note the distinction between the higher commutator subgroups given here and the derived subgroups defined on page 234.

Definition 10.1 The ith *higher commutator subgroup* $\mathcal{D}_i(G)$ of G is defined inductively by

$$\mathcal{D}_1(G) = G, \quad \mathcal{D}_{i+1}(G) = [\mathcal{D}_i(G), G].$$

Note that $\mathcal{D}_2(G) = G'$, the derived subgroup of G, and $\mathcal{D}_i(G) = G$ for all i if $G = G'$, that is, if G is *perfect*; see Problem 4.8.

Definition 10.2 The ith *higher centre* $\mathcal{Z}_i(G)$ of G is defined inductively as follows. Set $\mathcal{Z}_0(G) = \langle e \rangle$. If $\mathcal{Z}_i(G)$ (a normal subgroup of G) has been defined, let η_i be the natural homomorphism (Definition 4.13) mapping G to $G/\mathcal{Z}_i(G)$, then $\mathcal{Z}_{i+1}(G)$ is the preimage of $Z(G/\mathcal{Z}_i(G))$ under the homomorphism η_i.

Note that $\mathcal{Z}_1(G) = Z(G)$ because η_0 is the identity map on G and, for $i = 0, 1, \ldots$, the term $\mathcal{Z}_i(G)$ is the unique subgroup of G which satisfies

$$\mathcal{Z}_0(G) = \langle e \rangle, \quad \mathcal{Z}_{i+1}(G)/\mathcal{Z}_i(G) = Z\big(G/\mathcal{Z}_i(G)\big). \tag{10.1}$$

Also $\mathcal{Z}_i(G) = \langle e \rangle$ for all i when G is centreless. If there exists a finite integer k such that $\mathcal{Z}_j(G) < \mathcal{Z}_k(G)$ for all $j < k$, and $\mathcal{Z}_k(G) = \mathcal{Z}_l(G)$ for all $l \geq k$, then the term $\mathcal{Z}_k(G)$ is called the *hypercentre* of G; see Problem 10.2.

The lower and upper central series are now given by

Definition 10.3 The *lower central series* for a group G is the series (*written in reverse order*) from some subgroup of G to G:

$$G = \mathcal{D}_1(G) \rhd \mathcal{D}_2(G) \rhd \cdots,$$

and the *upper central series* for G is the series from $\langle e \rangle$ to some subgroup of G:

$$\langle e \rangle = \mathcal{Z}_0(G) \lhd \mathcal{Z}_1(G) \lhd \cdots.$$

As observed above, for the lower central series note the similarity of its definition with that for the derived series given on page 234. Problem 10.1 provides a slightly different but equivalent definition of the upper series.

When it is clear which group G is involved, we write $\mathcal{D}_r$ for $\mathcal{D}_r(G)$ and $\mathcal{Z}_s$ for $\mathcal{Z}_s(G)$. Both of these series are normal (Definition 9.1); this follows from Problem 4.16 and the Correspondence Theorem (Theorem 4.16). Also neither of these is necessarily a series for G; in the 'worst' case $\mathcal{D}_n(G) = G$ and $\mathcal{Z}_n(G) = \langle e \rangle$ for all $n \geq 1$, for example, when $G = S_3$.

The next result shows that there is a close relationship between the lower and upper central series, and in Problem 10.6 we prove that all other central series lie somewhere 'between' these two.

Theorem 10.4 *If* $\mathcal{Z}_s(G) = \mathcal{Z}_s = G$, *for some integer* s, *then* $\mathcal{D}_{s+1} = \langle e \rangle$ *and*

$$\mathcal{D}_{r+1} \leq \mathcal{Z}_{s-r} \quad \text{for } r = 0, \ldots, s. \tag{10.2}$$

Conversely, if $\mathcal{D}_{s+1}(G) = \mathcal{D}_{s+1} = \langle e \rangle$ *for some integer* s, *then* $\mathcal{Z}_s = G$ *and* (10.2) *again holds.*

Proof For the first part, we use induction on r to prove (10.2). If $\mathcal{Z}_s = G$, then by definition of $\mathcal{D}_1$, $\mathcal{D}_1 = G = \mathcal{Z}_s$ and (10.2) holds when $r = 0$. If $\mathcal{D}_{r+1} \leq \mathcal{Z}_{s-r}$ then, using Definition 10.1, we have

$$\mathcal{D}_{r+2} = [\mathcal{D}_{r+1}, G] \leq [\mathcal{Z}_{s-r}, G] \leq \mathcal{Z}_{s-(r+1)},$$

where the last inequality follows by Problem 4.16(vii) (with G for G, $\mathcal{Z}_{s-r}$ for J and $\mathcal{Z}_{s-(r+1)}$ for K). Proposition (10.2) now follows and, if we put $r = s$, we obtain $\mathcal{D}_{s+1} \leq \mathcal{Z}_0 = \langle e \rangle$.

For the converse, suppose $\mathcal{D}_{s+1} = \langle e \rangle$. As above we apply induction, this time on $t = s - r$ to prove $\mathcal{D}_{(s+1)-t} \leq \mathcal{Z}_t$. By assumption, we have $\mathcal{D}_{s+1} = \langle e \rangle = \mathcal{Z}_0$. Suppose $\mathcal{D}_{(s+1)-t} \leq \mathcal{Z}_t$. We use the enlargement of cosets map ξ given in Problem 4.14 with $J = \mathcal{D}_{(s+1)-t}$, $K = \mathcal{Z}_t$ and $\xi = \xi_t$. This provides the surjective homomorphism

$$\xi_t : G/\mathcal{D}_{(s+1)-t} \to G/\mathcal{Z}_t, \tag{10.3}$$

and shows that, as $\mathcal{D}_{(s+1)-t} \lhd \mathcal{D}_{s-t} \lhd G$,

$$\left(\mathcal{D}_{s-t}/\mathcal{D}_{(s+1)-t} \right)\xi_t = \mathcal{D}_{s-t}\mathcal{Z}_t/\mathcal{Z}_t. \tag{10.4}$$

For if $a \in G$, then $a(\mathcal{D}_{(s+1)-t})\xi_t = a\mathcal{Z}_t$ by definition, see (10.3) and Problem 4.14, hence if $a \in \mathcal{D}_{s-t}$, the cosets on the right-hand side of (10.4) have the form $b\mathcal{Z}_t$ where $b \in \mathcal{D}_{s-t}\mathcal{Z}_t$. Also $\mathcal{D}_{(s+1)-t} = [G, \mathcal{D}_{s-t}]$ by Definition 10.1, and so by Problem 4.16(vii) again

$$\mathcal{D}_{s-t}/\mathcal{D}_{(s+1)-t} \leq Z\left(G/\mathcal{D}_{(s+1)-t}\right),$$

which in turn, by (10.3) and Problem 4.16(vi) (with $\theta = \xi_t$, $G = G/\mathcal{D}_{(s+1)-t}$, $H = \mathcal{D}_{s-t}/\mathcal{D}_{(s+1)-t}$ and $J = G/\mathcal{Z}_t$), gives

$$\left(\mathcal{D}_{s-t}/\mathcal{D}_{(s+1)-t} \right)\xi_t \leq Z(G/\mathcal{Z}_t).$$

Combining this with (10.4), we obtain

$$\mathcal{D}_{s-t}\mathcal{Z}_t/\mathcal{Z}_t = \left(\mathcal{D}_{s-t}/\mathcal{D}_{(s+1)-t} \right)\xi_t \leq Z(G/\mathcal{Z}_t) = \mathcal{Z}_{t+1}/\mathcal{Z}_t,$$

using (10.1) for the last equation. From this we have by the Correspondence Theorem (Theorem 4.16)

$$\mathcal{D}_{s-t} \leq \mathcal{D}_{s-t}\mathcal{Z}_t \leq \mathcal{Z}_{t+1},$$

and the inductive argument is complete. As in the first part, if we put $t = s$, we obtain $\mathcal{D}_1 \leq \mathcal{Z}_s$ and so $\mathcal{Z}_s = G$. $\square$

Nilpotent Groups

Previously we have discussed finite nilpotent groups, see Sections 6.3 and 7.3. Here we begin by giving a new definition which is valid in both the finite and infinite cases. Using Theorem 10.4, we have

Definition 10.5 A group G is called *nilpotent with class* r if r is the least positive integer satisfying $\mathcal{D}_{r+1}(G) = \langle e \rangle$, and G is called *nilpotent* if it is nilpotent with class r for some positive integer r.

By Theorem 10.4, we can replace the equation $\mathcal{D}_{r+1}(G) = \langle e \rangle$ in the above definition by $\mathcal{Z}_r(G) = G$. This theorem also shows that, if G is nilpotent, then the lower and upper central series for G have the same length, see Problem 10.6.

Example 1 Abelian groups are nilpotent with class 1. Nilpotent groups with class 2 satisfy $\mathcal{D}_2 = G' \le Z(G) = \mathcal{Z}_1$ (as $\mathcal{D}_2 \subseteq Z(G)$, we have $\mathcal{D}_3 = \langle e \rangle$). As an example consider the group

$$G_1 = \langle a, b \mid a^8 = b^2 = e, bab = a^5 \rangle$$

discussed in Problem 6.4. It has order 16, so it is a 2-group and hence nilpotent by Theorem 10.6 (or by Theorem 7.9). We also have, using Problem 6.4,

$$\mathcal{D}_1 = G_1, \quad \mathcal{D}_2 = G_1' = \langle a^4 \rangle \quad \text{and} \quad \mathcal{D}_3 = \langle e \rangle,$$

as $\langle a^4 \rangle$ is Abelian and $a^4 \in Z(G)$, and

$$\mathcal{Z}_0 = \langle e \rangle, \quad \mathcal{Z}_1 = Z(G_1) = \langle a^2 \rangle \quad \text{and} \quad \mathcal{Z}_2 = G_1.$$

This last equation follows because $G_1/\mathcal{Z}_1$ is Abelian, and so equals $Z(G_1/\mathcal{Z}_1)$ which in turn equals $\mathcal{Z}_2/\mathcal{Z}_1$ by definition. Therefore, $o(\mathcal{Z}_2) = 4 \cdot o(\mathcal{Z}_1)$, that is, 16; see Problem 6.4 again. These show that G_1 is nilpotent with class 2. This also provides an example of a group with distinct lower and upper central series.

Example 2 Consider the groups $IT_n(\mathbb{Q})$ and $IZT_{n,r}(\mathbb{Q})$; see Problem 3.15. If $A \in IT_n(\mathbb{Q})$ then A is an $n \times n$ upper triangular matrix with 1 at each main diagonal entry, and if $B \in IZT_{n,r}(\mathbb{Q})$ then B has the same form as A except that the first r superdiagonals consist entirely of zeros.[1] Using the quoted problem, we have

$$\langle e \rangle \lhd IZT_{n,n-1}(\mathbb{Q}) \lhd \cdots \lhd IZT_{n,1}(\mathbb{Q}) \lhd IT_n(\mathbb{Q})$$

is a central series for $IT_n(\mathbb{Q})$. Therefore, $IT_n(\mathbb{Q})$ is nilpotent with class at most $n - 1$. In fact, the class is exactly $n - 1$, see Problem 10.2. This example also shows that there exist finite nilpotent groups with arbitrarily high nilpotency class.

The following results give the basic nilpotency properties, they are derived using straightforward group theoretical techniques and so only outline proofs will be given; the reader should fill in the details.

Theorem 10.6 (i) *All Abelian groups are nilpotent* (*with class* 1).
(ii) *All finite p-groups are nilpotent* (*with various classes*).
(iii) *A subgroup of a nilpotent group is nilpotent.*
(iv) *A factor group of a nilpotent group is nilpotent.*
(v) *If G and H are nilpotent, then so is G × H.*

---

[1] In ring theory, an element a is called 'nilpotent' if it is non-zero *and* some positive power a^n of it is zero. In the ring of *all* $n \times n$ matrices defined over a field F, if $B \in IZT_{n,r}(F)$ then $B - I_n$ is nilpotent in the ring theory sense provided $r < n$. This is one of the reasons for the use of the word 'nilpotent' in group theory. There is also an association with 'nilpotent Lie algebras'.

The finiteness condition is necessary in (ii) because there exist infinite non-nilpotent p-groups, see the note at the end of this section and Problem 10.3.

Proof (i) If G is Abelian we have $\mathcal{Z}_1 = G$.

(ii) By induction. Let H be a p-group, if $o(H) = p$ use (i). By Lemma 5.21, $Z(H) \neq \langle e \rangle$, so by the inductive hypothesis $H/Z(H)$ has the series

$$H/Z(H) \triangleright H_1/Z(H) \triangleright \cdots \triangleright H_s/Z(H) = \langle e \rangle,$$

and then $H \triangleright H_1 \triangleright \cdots \triangleright H_s = Z(H) \triangleright \langle e \rangle$ is a lower series for H.

(iii) By induction, we have $H \leq G$ implies $\mathcal{D}_i(H) \leq \mathcal{D}_i(G)$ for all i. Hence $\mathcal{D}_{r+1}(G) = \langle e \rangle$ implies $\mathcal{D}_{r+1}(H) = \langle e \rangle$.

(iv) Let $\theta : G \to G/H$ be the natural homomorphism. By Problem 4.16(vi), $\mathcal{D}_i(G/H) \leq \mathcal{D}_i(G)\theta$, and so if $\mathcal{D}_{r+1}(G) = \langle e \rangle$ then we also have $\mathcal{D}_{r+1}(G/H) = \langle e \rangle$.

(v) Again use induction. $\mathcal{D}_1(G \times H) = G \times H = \mathcal{D}_1(G) \times \mathcal{D}_1(H)$, and $\mathcal{D}_{i+1}(G \times H) = [\mathcal{D}_i(G \times H), G \times H] \leq [\mathcal{D}_1(G) \times \mathcal{D}_i(H), G \times H] \leq [\mathcal{D}_i(G), G] \times [\mathcal{D}_i(H), H]$ by Problem 7.2(i). $\square$

An example of a non-nilpotent group is S_3 (as $Z(S_3) = \langle e \rangle$). This example also shows that the converse of (iii) and (iv) in Theorem 10.6 fails; that is, if $H \triangleleft G$, and H and G/H are nilpotent, it does not follow that G is nilpotent. In this example, both $H = C_3$ and $G/H = C_2$ are Abelian and so nilpotent. See Theorem 10.16 and Problem 10.7 for partial converses.

We have mentioned previously that there are several equivalent definitions for nilpotency in the finite case, we give some more now. The first concerns normalisers and is as follows.

Theorem 10.7 *Let G is a finite group. The following conditions are equivalent*:

(i)   *G is nilpotent*;
(ii) *if $H < G$ then $H < N_G(H)$; and*
(iii) *all maximal subgroups of G are normal in G.*

See also Problem 5.13(iv).

Proof We show that (i) implies (ii) implies (iii) implies (i). Suppose (i) holds, we use the equivalent definition given in Problem 10.1. Let n be the largest integer such that $\mathcal{Z}_n \leq H$. As $H < G$ there exists $j \in \mathcal{Z}_{n+1} \backslash H$, and if $h \in H$, then $j^{-1}hj = hh^{-1}j^{-1}hj = h[h, j] \in H\mathcal{Z}_n$ by Problem 10.1. This in turn shows that $j \in N_G(H)$, and (ii) follows as $j \notin H$.

If (ii) holds and H is a maximal subgroup of G, then $N_G(H) = G$ which gives $H \triangleleft G$ by Lemma 5.25(ii), and (iii) follows.

Lastly, if (iii) holds, then by Theorem 7.9, G is isomorphic to a direct product of its Sylow subgroups, that is, isomorphic to a direct product of p-groups where p ranges over the prime divisors of $o(G)$. (i) now follows by Theorem 10.6(ii) and (v). $\square$

Example 3 Let $G_1 = Q_2 \times C_3$. Both Q_2 and C_3 are normal Sylow subgroups of G_1 by Lemma 7.3, hence by Theorem 7.9 this group is nilpotent. It has the presentation

$$G_1 = \langle a, b, c \mid a^2 = b^2 = (ab)^2, c^3 = e, ac = ca, bc = cb \rangle.$$

Consider $H = \langle a^2, c \rangle$, a subgroup of order 6. By Theorem 10.7, we have $N_{G_1}(H) > H$. This is not surprising for G_1 is *Hamiltonian*, that is, all of its proper subgroups are normal even though it is not Abelian; see Problem 7.13. Hence in this case $N_{G_1}(H) = G_1$. This problem also shows that all maximal subgroups of G_1 are normal, which confirms (iii) in Theorem 10.7 for this group.

Our next equivalent nilpotency condition concerns commuting elements whose orders have no common factors.

Theorem 10.8 *A finite group G is nilpotent if and only if whenever $a, b \in G$ and $(o(a), o(b)) = 1$, then a and b commute.*

Proof There is nothing to prove if G is a p-group. Hence suppose the primes dividing $o(G)$ are $p_1, \ldots, p_s$ where $s > 1$. If G is nilpotent, then by Theorem 7.9, it is a direct product of its Sylow p_i-subgroups P_i, $1 \leq i \leq s$, say. Re-ordering these factors if necessary, let the prime factors of $o(a)$ be $p_1, \ldots, p_t$, and so by hypothesis the prime factors of $o(b)$ are included in the set $\{p_{t+1}, \ldots, p_s\}$. Now if $R_1 = P_1 \times \cdots \times P_t$ and $R_2 = P_{t+1} \times \cdots \times P_s$, then $G \simeq R_1 \times R_2$, $a \in R_1$ and $b \in R_2$, and so a and b commute by Theorem 7.4. For the converse, we can reverse this argument. If for all $a \in R_1$ and $b \in R_2$, a and b commute, then as in the proof of Theorem 7.4, R_1 and R_2 are normal subgroups of G, $R_1 \cap R_2 = \langle e \rangle$ and $R_1 R_2 = G$ (by Theorem 5.8), that is $G \simeq R_1 \times R_2$; the theorem follows. $\square$

Example 4 Let G_2, a subgroup of S_7, be generated by the permutations $(1, 2, 3, 4)$, $(1, 3)$ and $(5, 6, 7)$. The orders of the elements of G_2 are $1, 2, 3, 4, 6$ and 12, and an easy calculation shows that all of the elements of order 2 or 4 commute with the elements of order 3, hence by Theorem 10.8 it is nilpotent. In fact, it is isomorphic to $D_4 \times C_3$, see Appendix C. Now $J = \langle (1, 3), (5, 6, 7) \rangle \leq G_2$ with order 6. Therefore, Theorem 10.7 gives $N_{G_2}(J) > J$, as the reader can check directly by showing that $N_{G_2}(J) = \langle (1, 3), (2, 4), (5, 6, 7) \rangle$, a subgroup of G_2 of order 12, thus showing that J is a non-normal subgroup of G_2.

The statement on the next page collects together the ten main properties of finite nilpotent groups, it is important to note that many of these properties fail in the infinite case, see comments on the next page.

Theorem 10.9 *For a finite group G, propositions* (i) *to* (x) *are equivalent.*

(i) G *is nilpotent.*

(ii)   *For some positive integer r, $\mathcal{D}_r(G) = \langle e \rangle$.*

(iii)   *For some positive integer s, $\mathcal{Z}_s(G) = G$.*

(iv)   *If $H < G$ then $H < N_G(H)$.*

(v)   *All subgroups of G are 'subnormal' in G.*

(vi)   *All maximal subgroups of G are normal in G.*

(vii)   *All Sylow subgroups of G are normal in G.*

(viii) *G is isomorphic to a direct product of its Sylow subgroups.*

(ix)   *If $a, b \in G$ and $(o(a), o(b)) = 1$, then a and b commute in G.*

(x)   *The derived subgroup G' of G is a subgroup of the 'Frattini' subgroup $\Phi(G)$ of G, see Section* 10.2.

Proof Property (v) is discussed in Problem 10.4 and property (x) will be considered in the next section where the Frattini subgroup is defined. For the remaining equivalences, use Theorems 7.9, 10.4, and 10.6 to 10.8.   □

Properties (iv) to (x) do not apply in the infinite case, they are all too weak; for a discussion of this point see Robinson (1982), page 126. Also there exist infinite non-nilpotent p-groups, an example is given in Problem 10.3. The list of equivalences given in Theorem 10.9 is not complete, four more of slightly less importance are as follows.

(xi) Every factor of a chief series $\{H_i\}$ for G is central, that is, if the factor is H_1/H_2 then $H_2 \lhd G$ and $H_1/H_2 \leq Z(G/H_2)$.

Chief series were discussed in Problem 9.15, and a proof of this equivalence is given in Rose (1978), page 144. This is an example of an equivalence which is valid in both the finite and infinite cases. The remaining three properties are:

(xii) The factor group G/J is nilpotent when $J \leq Z(G)$; see Problem 4.16.

(xiii) For all proper normal subgroups K of G, $Z(G/K) > \langle e \rangle$.

(xiv) For all non-neutral subgroups K of G, $[K, G] < K$.

Their equivalence to nilpotency is established in Problem 10.9.

Lastly, we ask: Is there a condition on n such that all groups of order n are nilpotent? One is as follows. Suppose the prime factorisation of n is given by $n = p_1^{r_1} \cdots p_k^{r_k}$, then the condition is:

$$p_i \nmid p_j^{s_j} - 1 \quad \text{for all } i, j, s_j \quad \text{satisfying} \quad 1 \leq i, j \leq k \quad \text{and} \quad 1 \leq s_j \leq r_j.$$

For example, all groups of order 891 are nilpotent; see Problem 7.12. For a proof of this result, see Scott (1964), page 217.

10.2 Frattini and Fitting Subgroups

Most groups have a number of subgroups with special properties; the centre is often the most important. Here we introduce two further examples: The *Frattini* and *Fitting subgroups*. All finite groups possess these subgroups, but they may be degenerate in the sense that they may equal either the whole group or the neutral subgroup. For instance, if G is Abelian then $Z(G)$ equals G as does the Fitting subgroup of G. Both the Frattini and the Fitting subgroups have a number of useful properties mainly related to the topics discussed earlier in this chapter, we shall introduce them now.

First, we consider the Frattini subgroup.

Definition 10.10 Given a group G, the intersection of all maximal subgroups of G is called the *Frattini subgroup* of G, it is denoted by $\Phi(G)$. If G has no maximal subgroups we set $\Phi(G) = G$.

Clearly, $\Phi(G) \leq G$ (Theorem 2.15), in fact, it is both normal and (for finite groups) nilpotent as we shall see below (Lemma 10.13 and Theorem 10.15). This subgroup was first defined by G. Frattini (1852–1925) in 1885.

Examples We have (i) $\Phi(A_5) = \langle e \rangle$, (ii) $\Phi(C_4 \rtimes C_4) \simeq C_2 \times C_2$, (iii) $\Phi(C_{32}) = C_{16}$, and (iv) $\Phi(\mathbb{Q}) = \mathbb{Q}$; see Problems 7.20 and 10.13.

There is a second independent definition which involves 'non-generators', the reader should revisit Problem 2.13(ii).

Definition 10.11 An element a of a group G is called a *non-generator* of G if whenever the set X generates G, then the set $X \backslash \{a\}$ also generates G.

For example, the neutral element is a non-generator of all groups whose orders are larger than 1.

The first result shows that the Frattini subgroup equals the set of non-generators. As with many of the results in this section its proof relies mainly on simple logical arguments involving maximality.

Theorem 10.12 *For all finite groups G, the set of non-generators of G equals the Frattini subgroup of G.*

Proof Suppose a is a non-generator of G and H is a maximal subgroup. If $a \notin H$, then $\langle H, a \rangle > H$, and so by maximality $\langle H, a \rangle = G$. But a is a non-generator, and hence $H = G$ which contradicts the maximality of H (note maximal subgroups are always proper). Therefore, $a \in H$, and as this holds for all maximal subgroups H, we also have $a \in \Phi(G)$.

Conversely, suppose $a \in \Phi(G)$, so a belongs to all maximal subgroups of G; we show that a is a non-generator. Assume that, for some set X, $a \notin X$ and $\{a\} \cup X$ generates G. If $\langle X \rangle \neq G$, then there exists a maximal subgroup J of G which contains $\langle X \rangle$, possibly $\langle X \rangle$ itself, that is,

$$\langle X \rangle \leq J < G.$$

By supposition, $a \in J$, and so $\langle a, X \rangle \leq J$. But $\langle a, X \rangle = G$, hence $G \leq J$ which is impossible. Hence $\langle x \rangle = G$ and a is a non-generator. $\qquad\square$

As noted above, Frattini subgroups have many novel properties, we shall derive some of the main ones now. Four of these apply to all finite groups whilst the remaining three provide some more information about p-groups. First, we have

Lemma 10.13 $\Phi(G) \lhd G$.

Proof If G has no maximal subgroups, then $\Phi(G) = G$ and the result holds trivially. An automorphism maps a maximal subgroup to a maximal subgroup, this is a consequence of the Correspondence Theorem. Hence this holds for all inner automorphisms, and normality follows. $\qquad\square$

The stronger property: $\Phi(G)$ char G also holds, see Problem 4.22.

The next lemma is needed in the proof of Theorem 10.15, it was first proved by Frattini in 1885; *cf.* Theorem 10.12.

Lemma 10.14 *If G is finite, $H \leq G$ and $G = H\Phi(G)$, then $H = G$.*

Proof If $H \neq G$, then there exists a maximal subgroup J of G which contains H possibly H itself. Now $\Phi(G) \leq J$ by definition, therefore $G = H\Phi(G) \leq J$, which is impossible. The result follows. $\qquad\square$

Theorem 10.15 *If G is finite then $\Phi(G)$ is nilpotent.*

Proof We show that all Sylow subgroups of $\Phi(G)$ are normal in $\Phi(G)$; there is nothing to prove if $\Phi(G) = \langle e \rangle$. Suppose $\Phi(G) > \langle e \rangle$ and let P be a non-neutral Sylow subgroup of $\Phi(G)$. Using the Frattini Argument (Lemma 6.14), we have $G = N_G(P)\Phi(G)$, as $\Phi(G) \lhd G$ by Lemma 10.13. Now applying Lemma 10.14 with $H = N_G(P)$, we obtain $N_G(P) = G$, which gives $P \lhd \Phi(G)$. The result follows as this holds for all Sylow subgroups P of $\Phi(G)$. $\square$

The next two results provide further conditions for nilpotency; note that in general if $H \lhd G$, and H and G/H are nilpotent, it does not follow that G is nilpotent; S_3 is an example as we noted above.

Theorem 10.16 *If G is finite, then G is nilpotent if and only if $G/\Phi(G)$ is nilpotent.*

Proof If G is nilpotent, then all of its factor groups are also nilpotent by Theorem 10.6(iv). For the converse, we prove that all Sylow subgroups P of G are normal in G as in the proof above. By Problem 6.10(iii) and Lemma 10.13, $P\Phi(G)/\Phi(G)$ is a Sylow subgroup of $G/\Phi(G)$, and as $G/\Phi(G)$ is nilpotent by hypothesis, this shows that

$$P\Phi(G)/\Phi(G) \lhd G/\Phi(G)$$

because all Sylow subgroups are normal in nilpotent groups. The Correspondence Theorem now gives $P\Phi(G) \lhd G$. As P is a Sylow p-subgroup of $P\Phi(G)$, applying Frattini argument (Lemma 6.14) we obtain

$$G = N_G(P)P\Phi(G) = N_G(P)\Phi(G),$$

as $P \leq N_G(P)$. Hence we have $G = N_G(P)$ by Lemma 10.14, that is, $P \lhd G$. As this holds for all Sylow subgroups P of G, the result follows. $\square$

We also have: G is soluble if and only if $G/\Phi(G)$ is soluble, see Section 11.2. The second new nilpotency criterion completes the proof of Theorem 10.9.

Theorem 10.17 *If G is a finite group, then G is nilpotent if and only if $G' \leq \Phi(G)$.*

Proof By Theorem 7.9, G is nilpotent if and only if all of its maximal subgroups are normal. If H is maximal, then by hypothesis, $H \lhd G$ and $H \neq G$, so by Problem 4.13 we see that G/H is cyclic (Abelian). Hence, $G' \leq H$ by Problem 4.6. This holds for all maximal H and so $G' \leq \Phi(G)$.

Conversely, if $G' \leq \Phi(G)$ and H is maximal, then $G' \leq H$ by definition of $\Phi(G)$, and so $G' \lhd H$ by Problem 2.14(i). It follows that $H/G' \leq G/G'$, but as G/G' is Abelian (Problem 2.16), we also have $H/G' \lhd G/G'$ and so, by the Correspondence Theorem, $H \lhd G$. This holds for all maximal H, so G is nilpotent by Theorem 7.9 again. $\square$

A weaker result involving the centre holds for all finite groups as the following theorem due to W. Gaschütz shows.

Theorem 10.18 *If G is finite, then $G' \cap Z(G) \leq \Phi(G)$.*

Proof Suppose $G' \cap Z(G) \nleq \Phi(G)$. Then there exists a maximal subgroup H of G which satisfies $G' \cap \Phi(G) \nleq H$. Therefore,

$$G = (G' \cap Z(G))H, \tag{10.5}$$

as H is maximal, and so the set

$$(G' \cap Z(G))\backslash H \quad \text{is not empty.} \tag{10.6}$$

Now if $g \in G$, then $g = ah$ where $a \in G' \cap Z(G)$ and $h \in H$ (by (10.5)). As a belongs to $Z(G)$ it commutes with all elements in H, and so

$$g^{-1}Hg = h^{-1}a^{-1}Hah = h^{-1}Hh = H,$$

which shows that $H \lhd G$. Hence by Problem 4.13, G/H is cyclic and so Abelian, therefore $G' \leq H$ by Problem 2.16. But $G' \cap Z(G) \leq G' \leq H$, which contradicts (10.6). Hence $G' \cap Z(G) \leq H$ for all maximal H, and so finally $G' \cap Z(G) \leq \Phi(G)$. $\qquad\qquad\qquad\qquad\qquad\qquad\qquad\qquad\qquad\qquad\qquad\qquad$ $\square$

Example 5 We illustrate these last two results using the groups discussed in Chapter 8. Firstly for S_4, both the centre and the Frattini subgroup have order 1 and the derived subgroup is A_4, S_4 is not nilpotent, and we have equality in Gaschütz's result. Secondly for $Q_2 \times C_3$, the derived and Frattini subgroups are equal, and isomorphic to C_2, whilst the centre is isomorphic to C_6 and contains the first two subgroups, $Q_2 \times C_3$ is nilpotent, and again we have equality in Gaschütz's result. Finally for $F_{3,8}$, see Problem 8.3, the centre and the Frattini subgroup are both isomorphic to C_4, whilst the derived subgroup is isomorphic to C_3, $F_{3,8}$ is not nilpotent, and we have a strict inequality in Gaschütz's result.

For p-groups the Frattini subgroup has some useful properties as we show now. See Problem 4.18 for the definition of elementary Abelian.

Lemma 10.19 *If G is an elementary Abelian group, then $\Phi(G) = \langle e \rangle$.*

Proof In an elementary Abelian group, every non-neutral element can act as a generator, and so the result follows from Theorem 10.12. $\qquad\qquad\qquad$ $\square$

Lemma 10.20 *If G is a p-group, then $G/\Phi(G)$ is an elementary Abelian p-group.*

Proof Suppose K is a maximal subgroup of G, then $K \lhd G$ and $[G : K] = p$ (Theorem 6.6), hence G/K is elementary Abelian. We use Problem 4.12 to complete the proof. In this problem, let $K_1, \ldots, K_n$ be the maximal subgroups of G (and so G/K_i is elementary Abelian for $i = 1, \ldots, n$). It follows that

$$G \Big/ \bigcap_{i=1}^{n} K_i$$

is also elementary Abelian (a subgroup of an elementary Abelian group is elementary Abelian). The result follows using the definition of $\Phi(G)$. $\qquad$ $\square$

The last result, which is one of many due to Burnside, has a number of applications, for example, it provides information about the automorphism groups of p-groups; see, for instance, Scott (1964), pages 161 and 162.

Theorem 10.21 *If G is a finite p-group and $o(G/\Phi(G)) = p^n$, then we can find $a_1, \ldots, a_n \in G$ to generate G.*

Proof By Lemma 10.20 and Problem 4.18, $G/\Phi(G)$ is isomorphic to a direct product of n copies of the cyclic group C_p. Let the generators of these cyclic groups be $a_1 \Phi(G), \ldots, a_n \Phi(G)$, where $a_i \in G$ for $i = 1, \ldots, n$. Then

$$G \simeq \langle a_1, \ldots, a_n \rangle \Phi(G) = \langle a_1, \ldots, a_n \rangle$$

using Lemma 10.14 for the last equality.   □

Example 6 In Problem 6.4, we gave an example of a 2-group G_1 with order 16. It has three maximal subgroups: $\langle a \rangle$, $\langle ab \rangle$ and $\langle a^2, b \rangle$, and so $\Phi(G_1) = \langle a^2 \rangle$ with order 4. Hence Burnside's result confirms the fact that G_1 has a 2-element generating set. But if we take the elementary Abelian 2-group of order 16, its Frattini subgroup has order 1 and so the integer n in Burnside's result is 4, and this elementary group has a 4-element generating set; see Problem 4.18 for details.

Fitting Subgroup

We come now to the Fitting subgroup, it has fewer elementary properties, but generalisations have proved useful in recent work on CFSG. The following result was proved by H. Fitting (1906–1938) in 1938.

Theorem 10.22 (Fitting's Theorem) *If J and K are nilpotent normal subgroups of a group G, with classes r and s, respectively, then JK is also a nilpotent normal subgroup of G with class at most $r + s$.*

Proof The normality of JK follows immediately from Lemma 4.14. For nilpotency we argue as follows. First, note that if $L_i \lhd G$ for $i = 1, 2, 3$, then

$$[L_1 L_2, L_3] = [L_1, L_3][L_2, L_3] \quad \text{and} \quad [L_1, L_2 L_3] = [L_1, L_2][L_1, L_3].$$

These identities follow from Problem 2.17 using normality (the reader should verify them). Hence by Definition 10.1, we have, using the second identity above with $L_1 = \mathcal{D}_i(JK)$, $L_2 = J$ and $L_3 = K$,

$$\mathcal{D}_{i+1}(JK) = \left[\mathcal{D}_i(JK), JK\right] = \left[\mathcal{D}_i(JK), J\right]\left[\mathcal{D}_i(JK), K\right]. \qquad (10.7)$$

This suggests the following: For subgroups $H_i, i = 1, 2, \ldots$ of G, let $[H_1] = H_1$ and $[H_1, \ldots, H_i, H_{i+1}] = [[H_1, \ldots, H_i], H_{i+1}]$ where $[G, H]$ is as usual the subgroup generated by the commutators of the elements of G and H. Then we have by induction, and using (10.7),

$\mathcal{D}_i(JK)$ equals the product of all terms of the form $[H_1, \ldots, H_i]$, where

$H_j = J$ or $H_j = K$ for $j = 1, \ldots, i$.

Now set $i = r + s + 1$. In the term $L = [H_1, \ldots, H_{r+s+1}]$, J will occur at least $r + 1$ times, or K will occur at least $s + 1$ times. Therefore, L is contained in either $\mathcal{D}_{r+1}(J)$ or $\mathcal{D}_{s+1}(K)$ (Problems 2.16(v) and 10.2(iv)) both of which equal $\langle e \rangle$ by hypothesis. This shows that $\mathcal{D}_{r+s+1}(JK) = \langle e \rangle$, and the result follows. $\square$

This result leads us to make the following

Definition 10.23 For a finite group G, the product of all nilpotent normal subgroups of G is called the *Fitting subgroup* of G, and it is denoted by $\mathsf{F}(G)$.

There is no concensus amongst authors on the notation for the Fitting subgroup, some use $F(G)$, Fit(G) or Fitt(G). By Theorem 10.22, $\mathsf{F}(G)$ is the unique largest nilpotent normal subgroup of G; see Appendix C for some examples. Note that by Lemma 10.13 and Theorem 10.15,

$$\Phi(G) \leq \mathsf{F}(G).$$

Suppose G is a finite group, p is prime and $P_1, \ldots, P_k$ is a list of its Sylow p-subgroups. Then we define

$$O_p(G) = \bigcap_{i=1}^{k} P_i.$$

The (normal) subgroup $O_p(G)$ is called the *p-radical* of G. If G has only one Sylow p-subgroup P_1, then $O_p(G) = P_1 \lhd G$ by Sylow 3. Also, if G has several Sylow p-subgroups (and so not normal), we still have $O_p(G) \lhd G$ by Problem 6.9(vi). This leads to (for an example, see page 182)

Theorem 10.24 *If* $p_1, \ldots, p_r$ *are the (distinct) primes dividing* $o(G)$, *then*

$$\mathsf{F}(G) = O_{p_1}(G) \cdots O_{p_r}(G).$$

Proof The Sylow p-subgroups of G form a class of conjugate subgroups in G (Theorem 6.9), and so $O_p(G)$ is the core (Problem 2.24) of each member of this class, and hence is normal in G. Therefore, as p-groups are nilpotent, $O_p(G) \leq \mathsf{F}(G)$ for each p dividing $o(G)$. Now by Theorem 10.22, this shows that

$$\prod_{i=1}^{r} O_{p_i}(G) \subseteq \mathsf{F}(G).$$

Conversely, if Q is a Sylow p-subgroup of $\mathsf{F}(G)$, then it is contained in some Sylow p-subgroup P of G (Theorem 6.9). By Problem 6.20(i), we have $Q \lhd G$, hence $Q \subseteq \text{core}(P) = O_p(G)$, and so $\mathsf{F}(G) \subseteq \prod_{i=1}^{r} O_p(G)$. The result follows. $\square$

It is not difficult to show that

$$C_G(F(G)) \leq F(G),$$

provided G is finite and soluble; see Chapter 11 and Kurzweil and Stellmacher (2004), page 123. The solubility condition can be removed if we replace $F(G)$ by the *Generalised Fitting Subgroup* $F^*(G)$ which is defined as the product of $F(G)$ and the subgroup of G generated by the *components* of G. A component H of G is a subgroup of G which is perfect and subnormal in G (Problem 10.4), and also satisfies the condition $H/Z(H)$ is simple. For example, A_5 is a component of $G = A_5$ wr C_2; see page 156. The subgroup A_5 is perfect, and in G no copy of A_5 is normal but $A_5 \triangleleft A_5 \times A_5 \triangleleft G$. Generalised Fitting subgroups play a important role in CFSG. For example, in a series of papers published in the Pacific Journal of Mathematics during the 1970s John Thompson showed that if G is a non-soluble group and the normalisers of all of its non-neutral p-subgroups are soluble, then $F^*(G)$ is isomorphic to one of the following groups most of which will be considered in Chapter 12:

$$L_2(q), \quad Sz(2^{2n-1}), \quad A_7, \quad M_{11}, \quad L_3(3), \quad U_3(3), \quad {}^2F_4(2)',$$

where $q > 3$ and $n > 1$. This means that in most cases (that is, except those listed above whose properties are well known) a non-soluble group will have a non-soluble normaliser attached to at least one of its p-subgroups. For further details, the reader should consult the references quoted above; the Suzuki groups $Sz(2^{2n-1})$ are discussed in Web Section 14.3, and the last group listed is known as the *Tits group*, it has order 17971200 and some authors (ATLAS 1985, for example) treat it as 'nearly' sporadic.

10.3 Problems

Problem ◆ 10.1 Show that the following statement gives an equivalent definition for the upper central series of a group G, see page 211: $Z_0(G) = \langle e \rangle$ and

$$Z_{n+1}(G) = \{g : g \in G \text{ and } [a, g] \in Z_n(G) \text{ for all } a \in G\}.$$

Problem 10.2 (i) Find the hypercentres of the three groups discussed in Chapter 8.

(ii) Show that the dihedral group D_n is nilpotent if and only if n is a power of 2.

(iii) Using Problem 3.15, show that the nilpotency class of the group $IT_n(\mathbb{Q})$ is exactly $n - 1$.

(iv) Establish the nilpotency condition given in the introduction to this chapter (page 209).

Problem 10.3 For $n = 1, 2, \ldots$, let H_n be a finite p-group with nilpotency class n. Further, let G be the direct product of the H_i, $i = 1, 2, \ldots$. Show that G is an infinite p-group which is not nilpotent. The definition of infinite direct products is given in Web Section 7.5.

Problem ♦ 10.4 (i) Show that if H is a subgroup of a nilpotent group G and $HG' = G$, then $H = G$. (Hint. Use induction on the nilpotency class number and Problem 2.17.)

(ii) A subgroup J of G is called *subnormal* in G if there exists a sequence of subgroups $\{J_i\}$ with the property

$$J = J_0 \lhd J_1 \lhd \cdots \lhd J_m = G.$$

This is expressed by $J \lhd \lhd G$. Note that this relation is transitive: if $H \lhd \lhd J$ and $J \lhd \lhd G$, then $H \lhd \lhd G$; see Problem 9.8. Prove that G is nilpotent if and only if all subgroups of G are subnormal in G. This establishes the fifth equivalence in Theorem 10.9.

Problem ♦ 10.5 (i) Show that if G is nilpotent with class 2 and $g \in G$, then the function $\theta : G \to G$ defined by $a\theta = [g, a]$, for $a \in G$, is a homomorphism, and use it to show that $C_G(g) \lhd G$.

(ii) Prove that if G is nilpotent with class r, then $G/Z(G)$ is nilpotent with class $r - 1$.

(iii) What is the nilpotency class of the dihedral group D_{2^n}; see Problem 10.2(ii).

Problem 10.6 Suppose G is nilpotent, and $\langle e \rangle = H_0, \ldots, H_n = G$ is a central series for G; see page 210. Show that $\mathcal{D}_{r+1}(G) \leq H_{n-r} \leq \mathcal{Z}_{n-r}(G)$ for $0 \leq r \leq n$.

Problem ♦ 10.7 Suppose G is a group and $J \leq Z(G)$. Show that G is nilpotent if and only if G/J is nilpotent. (Hint. Use Theorem 10.9(v).)

Hall has shown that if $K \lhd G$, and both K and G/K' are nilpotent where K' denotes the derived subgroup of K, then G is also nilpotent. See Problem 10.11*, or for a different proof Robinson (1982), page 129.

Problem ♦ 10.8 Suppose G is nilpotent, $o(G) = n$, and $m \mid n$. Show that G has a subgroup of order m; it is reverse Lagrange. (Compare with Problem 7.7).

Problem 10.9 Prove that the statements (xii), (xiii) and (xiv)* given on page 216 are, in fact, equivalent to nilpotency. (Hints. For (xii) use Lemma 5.21, and for (xiv) use induction on $o(G)$ and Problem 10.7.)

Problem 10.10 This problem uses the Hall–Witt Identity, see Problem 2.17(iv).
Let $H_i \leq G$ for $i = 1, 2, 3$, and let $[H_1, H_2, H_3] = \langle [[h_1, h_2], h_3] : h_i \in H_i, i = 1, 2, 3 \rangle$.

(i) Show that if $K \lhd G$, $[H_2, H_3, H_1] \leq K$ and $[H_3, H_1, H_2] \leq K$, then

$$[H_1, H_2, H_3] \leq K.$$

(ii) Prove that if $[H_2, H_3, H_1] = \langle e \rangle$ and $[H_3, H_1, H_2] = \langle e \rangle$, then $[H_1, H_2, H_3] = \langle e \rangle$. This result is known as the *Three Subgroup Lemma*.

(iii) Using (i) show that if $H_i \lhd G$ for $i = 1, 2, 3$, then

$$[H_1, H_2, H_3] \leq [H_2, H_3, H_1][H_3, H_1, H_2].$$

(iv) Prove that if G is perfect (Problem 4.8), then $G/Z(G)$ is centreless by applying the Three Subgroup Lemma to two copies of G and one of $Z_2(G)$.

(v) Lastly, prove that $[\mathcal{D}_i(G), \mathcal{Z}_j(G)] \leq \mathcal{Z}_{j-i}(G)$ if $i \leq j$.

Problem 10.11* (i) Show that if G has nilpotency class j, then $\mathcal{D}_{[j/2]+1}(G)$ is Abelian. (Hint. Use Problem 10.10(v).)

(ii) Suppose $K \lhd G$. Define K_i and L_n by

$$K_0 = K, \quad K_{i+1} = [K_i, G], \quad \text{and} \quad L_n = \prod_{i=0}^{n}[K_{n-i}, K_i].$$

Prove that $[G, L_n] \leq L_{n+1}$.

(iii) Using the notation set up in (ii) suppose K has nilpotency class k and G/K' has nilpotency class l, show that the nilpotency class of G is at most $2^{k-1}(2l - 1)$.

Problem 10.12 Suppose G is a nilpotent group.

(i) Show that $K \cap Z(G) \neq \langle e \rangle$ if K is a non-neutral normal subgroup of the group G.

(ii) Deduce $H \leq Z(G)$, if H is a minimal normal subgroup of G; for the definition of minimal normal see page 235.

(iii) Prove that $J = C_G(J)$, if J is a maximal Abelian normal subgroup of G. See also Problem 5.11.

Problem 10.13 Calculate the Frattini and Fitting subgroups of A_5, C_{32}, D_{12} and S_n.

Problem ◆ 10.14 Let $K \lhd G$. Show that $K \leq \Phi(G)$ if and only if G has no proper subgroup H which satisfies $HK = G$.

Problem 10.15 (i) If θ is a homomorphism of a group G to itself (that is an endomorphism) show that $\Phi(G)\theta \leq \Phi(G\theta)$.

(ii) Show by an example that the inequality given in (i) can be strict.

Problem ◆ 10.16 Prove that if J and K are normal subgroups of G, $K \leq \Phi(G)$ and J/K is nilpotent, then J is nilpotent. (Hint. One method begins by considering Sylow subgroups of J, and using the Frattini Argument (Lemma 6.14) and Lemma 10.14.)

Problem 10.17 Suppose $H_1, \ldots, H_m$ are groups. Using Problem 7.3(ii) show that

$$\Phi(H_1 \times \cdots \times H_m) \leq \Phi(H_1) \times \cdots \times \Phi(H_m).$$

For a discussion of the reverse inequality, see Scott (1964), page 163.

Problem 10.18 Suppose G is a finite group and $K \lhd G$. Prove the following statements.

(i)   If $H \leq G$, it does not follow that $\Phi(H) \leq \Phi(G)$. (Hint. Try $G = S_4$.)
(ii) If $H \leq G$ and $K \leq \Phi(H)$, then $K \leq \Phi(G)$. (Hint. Assume the contrary and use Problem 2.18 and Theorem 10.12.)
(iii) $\Phi(K) \leq \Phi(G)$.

Problem 10.19 (i)* Let G be a finite group with an Abelian normal subgroup K which satisfies the property: $K \cap \Phi(G) = \langle e \rangle$. Show how to find a subgroup J of G such that $G = JK$ and $J \cap K = \langle e \rangle$. (Hint. Begin by considering a subgroup J which is minimal subject to the first of these conditions, then use Problem 10.18(ii).)
   (ii) Give an example from Chapter 8 to illustrate (i).

Problem 10.20 (i) Suppose $K \lhd G$ and $K \leq \Phi(G)$. Using Problems 10.15 and 10.18, show that $\mathrm{Inn}(K) \leq \Phi(\mathrm{Aut}(K))$.
   (ii) Prove that if G is finite, then $\mathrm{Inn}(\Phi(G)) \leq \Phi(\mathrm{Aut}(\Phi(G)))$.
   (iii) Now applying this result to the dihedral group D_4, prove that $o(\Phi(D_4)) = 2$ and $o(\mathrm{Inn}(D_4)) = 4$, and so deduce D_4 cannot be the Frattini subgroup of a finite group. This shows it is *not* true that every group is isomorphic to a Frattini subgroup of some other group.

Problem 10.21 Suppose G is a finite p-group.

(i) Prove that $\Phi(G) = G'G^p$ where as usual G^p denotes the subgroup of G generated by its pth powers. (Hint. Use Theorem 6.6 and Problem 4.18.)
(ii) Use (i) (but not Lemma 10.20) to show that the factor group $G/\Phi(G)$ can be treated as a vector space over the field $\mathbb{F}_p$.

Problem 10.22 A finite p-group G is called *extra-special* if $Z(G)$ is cyclic, and $Z(G) = G' = \Phi(G)$.

(i) Prove that if G is extra-special, then $G/Z(G)$ is elementary Abelian.
(ii) Show that if $o(G) = p^3$, then G is extra-special; see Problem 6.5.

Further facts concerning these groups can be found in Aschbacher (1994), Chapter 1.

Problem 10.23 (i) If G is a 2-group, show that $\Phi(G) = \langle a^2 \mid a \in G \rangle$.
   (ii) Show that the proposition given in (i) can be false if 2 is replaced by a larger prime. (Hint. See Problem 6.5.)

Problem 10.24 (i) For a finite group G show that $\mathsf{F}(G/\Phi(G)) = \mathsf{F}(G)/\Phi(G)$. (Hint. Use Problem 10.16.)
   (ii) Verify directly the result given in (i) for the three groups discussed in Chapter 8.

Problem 10.25 Suppose G is a non-neutral finite soluble group, see Chapter 11.

(i) Show that if $\langle e \rangle < K \lhd G$, then (a) K contains a non-neutral normal Abelian subgroup of G, and (b) $K \cap F(G) > \langle e \rangle$. (Hint. One proof uses the derived series of G as defined on page 234.)

(ii)* Deduce $C_G(F(G)) = Z(F(G))$. (Hint. Use terms in the lower central series for a suitably chosen subgroup of G.)

(iii) Show that $F(G) > \langle e \rangle$.

(iv) Deduce $\Phi(G) < F(G)$.

Note (ii) shows that the Fitting subgroup of a soluble group plays a similar role to the centre in a nilpotent group, see Problem 10.11*.

Problem 10.26 (Project—Supersoluble Groups) A group is called *supersoluble* if and only if it has a normal series with cyclic factors; for further details, see Scott (1964), Chapter 7.

(i) Give an example of a supersoluble group, and another of a group that is not supersoluble.

(ii) Using Theorem 10.6 show that we may suppose all cyclic factors of a normal series of a supersoluble group have prime order.

(iii) Prove that subgroups and factor groups of a supersoluble group are themselves supersoluble, but the converse can be false. One example is given using S_4.

(iv) Show that a direct product of supersoluble groups is supersoluble.

(v) A finite nilpotent group is supersoluble, and a supersoluble group is soluble; see Chapter 11. Prove these statements and give examples to show that the reverse inclusions are false.

(vi) Prove that if G is supersoluble, then G' is nilpotent. (Hint. Use the fact that the automorphism group of a cyclic group is Abelian.)

(vii) If K is a cyclic normal subgroup of a finite group G and G/K is supersoluble, prove that G is also supersoluble.

(viii) The set of odd-order elements of a finite supersoluble group forms a characteristic subgroup. Investigate what is needed to prove this. See Robinson (1982), page 146.

(ix)* Show that every subgroup of a finite group G is reverse Lagrange if and only if G is supersoluble; see Rose (1978), page 292.

(x) Is it true that all Abelian groups are supersoluble? (Hint. Consider what happens in the uncountable case.)

Chapter 11
Solubility

See note at the head of Chapter 10.

Évariste Galois, who died following a duel at the age of 20 in 1832 (he was in-volved in French anti-monarchist politics), was one of the main early group theory pioneers. He introduced normal subgroups, soluble groups and some of the results concerning symmetric groups in order to solve a long-standing problem: Can a general polynomial equation be solved by radicals (that is using square-roots, cube-roots, *et cetera*). He did rely on the work of some earlier mathematicians. Ruffini was the first to claim that some quintic equations are unsoluble, and Abel proved this in the 1820s without making the connection with the permutation group of the roots. Nevertheless, Galois's contributions, which also included much of the theory of finite fields (Section 12.2), were of the highest originality and importance, and they have had a profound influence on the development of group theory and algebra in general ever since. A brief account of his life and work can be found in van der Waerden (1985) or Stewart (1989).

It had been known since the sixteenth century that quadratic, cubic and quartic polynomial equations are soluble by radicals, see the references quoted above and Problem 11.1, so why not quintics, sextics, *et cetera*? The answer depends on the fact to be proved in this chapter that certain groups related to the polynomials in question, S_5 amongst them, are not 'soluble', whereas others are soluble including S_n for $n \leq 4$. Briefly, these groups can be defined as follows; for a more detailed explanation, see any of the standard texts on Galois theory, for example, Artin (1948) or Stewart (1989). Let $f(x)$ be an irreducible polynomial defined over a field F, and let $F^\diamond$ be an extension field formed by adding one or more of the roots of the polynomial equation $f(x) = 0$ to F. Now consider the set $\mathcal{F}$ of automorphisms θ of $F^\diamond$ which fix F. (If $\theta \in \mathcal{F}$, then it is a bijection of $F^\diamond$ to itself, $z\theta = z$ for all $z \in F$, and $(x + y)\theta = x\theta + y\theta$ and $(xy)\theta = x\theta y\theta$ for all $x, y \in F^\diamond$.) It is a simple matter to show that $\mathcal{F}$ forms a group under composition of maps, and this group is called the *Galois group* $\mathrm{Gal}_{F^\diamond/F}(f)$ of f for the extension $F^\diamond$ over F. When the extension field $F^\diamond$ includes all of the roots of the polynomial equation $f(x) = 0$, the so-called *splitting field* for f, the Galois group is denoted by $\mathrm{Gal}(f)$. For example, working over the rational field $\mathbb{Q}$ we have, see Cohen (1993), page 327 (where several more examples are given),

H.E. Rose, *A Course on Finite Groups*,   229
Universitext,
DOI 10.1007/978-1-84882-889-6_11, © Springer-Verlag London Limited 2009

(a) if $f(x) = x^5 - x + 1$, then $\mathrm{Gal}(f) \simeq S_5$,

(b) if $g(x) = x^5 - 5x + 12$, then $\mathrm{Gal}(g) \simeq D_5$, and

(c) if $h(x) = x^2 + x + 1$, then $\mathrm{Gal}(h) \simeq C_2$.

In 1829, Abel showed that a quintic polynomial with a commutative (Abelian) 'Galois group' is soluble using radicals (that is, using terms some positive power of which belong to the base field F; see van der Waerden 1985). Galois greatly generalised Abel's result by showing that in a sequence of extensions with Galois groups $G_1, G_2, \ldots$, if each factor group G_{i+1}/G_i is Abelian, then the corresponding polynomial is soluble by radicals—in Example (c) above, the roots of the polynomial equation $x^2 + x + 1 = 0$ are defined using square roots and the corresponding Galois group is cyclic (Abelian). Repeated applications of these ideas leads naturally to the definition of solubility, see Definition 11.1 below; in fact, the use of the word 'soluble' in group theory arises from this result.

We begin this chapter by giving the basic facts about solubility. The second section discusses some of the many equivalent definitions including an important one due to Philip Hall, this generalises the Sylow theory but it only applies to soluble groups. In Chapter 10, we studied nilpotent groups, they form an important subclass of the class of soluble groups. For finite groups, we have the (strict) inclusions

$$\text{Abelian} \subset \text{Nilpotent} \subset \text{Soluble} \subset \text{Generic group.}$$

The class of 'supersoluble groups' lies strictly between nilpotent and soluble, see Problem 10.26. Some of these inclusions do not hold for infinite groups. The modern development of solubility theory is covered extensively in Doerk and Hawkes (1992).

11.1 Soluble Groups

We begin with the basic definition, the reader should refer to the first part of Chapter 9 for the elementary facts about series.

Definition 11.1 A group is called *soluble*[1] if it has a subnormal series all of whose factors are Abelian, when this holds the series is called a *soluble series*.

In Galois's work on the solution of polynomial equations over fields, the several extensions needed to form the splitting field are mirrored by the steps in the corresponding soluble series.

Clearly, all finite Abelian groups are soluble and all non-Abelian simple groups are not soluble. The group S_5 is an example which is neither simple nor soluble, see Example (b) on page 189. A wide range of finite groups are soluble including

---

[1] Some, mainly American, authors use the word 'solvable' for soluble.

- all Abelian, nilpotent and p-groups (Theorem 11.6);
- all groups whose orders have at most three prime factors (Theorem 11.7), and if we exclude A_5 then 'three' can be replaced by 'four'. This count can be further increased to 'five' if a handful of well-known cases are excluded;
- all groups which have square-free order, these groups are *metacyclic*, see Theorem 6.18, Problem 6.20 and Web Section 6.5;
- all groups of order $p^r q^s$ where p and q are prime—this is Burnside's $p^r q^s$-theorem which we shall prove in Web Section 14.2;
- all groups of odd order—this famous result, which appeared in 1963 and was first conjectured by Burnside in the early 1900s, is due to Walter Feit and John Thompson. It is remarkable for its fundamental importance to the theory especially to CFSG, and for its very long proof (about 270 pages)!

We begin our development by showing that subgroups and factor groups of soluble groups are soluble, and vice versa. This is an indication of the importance of solubility in the theory, see also page 116. Note that the constructions and proofs presented in this section rely entirely on material given in Chapters 2 and 4.

Theorem 11.2 *A subgroup J of a soluble group G is soluble.*

Proof Suppose $\{H_0, \ldots, H_n\}$ is a soluble series for G. We prove the result by showing that

$$\{H_0 \cap J, \ldots, H_n \cap J\} \tag{11.1}$$

is a soluble series for J. It may have some repetitions but this does not affect the result. Note first that

$$H_0 \cap J = \langle e \rangle, \quad H_n \cap J = J, \quad \text{and} \quad H_i \cap J \lhd H_{i+1} \cap J$$

(Problem 2.14(iv)). Also, for each i we have (by definition) $H_i \lhd H_{i+1}$ and $H_{i+1} \cap J \leq H_{i+1}$, and using the Second Isomorphism and Correspondence Theorems we obtain

$$\frac{H_{i+1} \cap J}{H_i \cap J} \simeq \frac{H_{i+1} \cap J}{H_i \cap (H_{i+1} \cap J)} \quad \text{as } H_i \leq H_{i+1}$$

$$\simeq \frac{H_i (H_{i+1} \cap J)}{H_i} \quad \text{by Theorem 4.15}$$

$$\leq \frac{H_{i+1}}{H_i} \quad \text{by Theorem 4.16.}$$

This shows that the factors of the series (11.1) for J are subgroups of the factors of the original series for G, and are therefore Abelian. $\square$

Theorem 11.3 *If G is soluble and $K \lhd G$, then G/K is also soluble.*

Proof Let $\{H_0, \ldots, H_n\}$ be a soluble series for G. By Lemma 4.14, Problem 4.6, and as $K \lhd G$, the series from K to G

$$K = K H_0 \lhd K H_1 \lhd \cdots \lhd K H_n = G$$

is subnormal. Arguing as in the proof above we have, for $i = 0, \ldots, n-1$,

$$\frac{K H_{i+1}}{K H_i} = \frac{(K H_i) H_{i+1}}{K H_i} \qquad \text{as } H_i \leq H_{i+1}$$

$$\simeq \frac{H_{i+1}}{(K H_i) \cap H_{i+1}} \qquad \text{by Theorem 4.15}$$

$$\simeq \frac{H_{i+1}}{H_i} \Big/ \frac{(K H_i) \cap H_{i+1}}{H_i} \qquad \text{by Theorem 4.17 as } H_i, H_{i+1} \cap J \leq H_{i+1}.$$

This shows that $K H_{i+1} / K H_i$ is isomorphic to a factor group of the Abelian group H_{i+1}/H_i, and so is itself Abelian (Problem 4.6). Hence, as $K H_0 = K$,

$$\langle e \rangle = \frac{K H_0}{K H_0} \lhd \frac{K H_1}{K H_0} \lhd \cdots \lhd \frac{K H_n}{K H_0} = \frac{G}{K}$$

is a subnormal series for G/K with Abelian factors (by Theorem 4.17), that is, G/K is soluble.   □

The converse of these two results is

Theorem 11.4 *If $K \lhd G$, and both K and G/K are soluble, then G is also soluble.*

This implies that the extension of one soluble group by another is itself soluble.

Proof Let $\{H_0, \ldots, H_n\}$ be a soluble series for K, and let $\{J_0, \ldots, J_m\}$ be a soluble series for G/K. Further, let θ be the natural homomorphism from G to G/K (Definition 4.13). By the Correspondence Theorem,

$$\left\{ J_0 \theta^{-1}, J_1 \theta^{-1}, \ldots, J_m \theta^{-1} \right\}$$

is a subnormal series from $K = J_0 \theta^{-1}$ to $G = J_m \theta^{-1}$ with Abelian factors. Hence

$$\left\{ H_0, \ldots, H_{n-1}, J_0 \theta^{-1}, \ldots, J_m \theta^{-1} \right\}$$

is a soluble series for G because $H_n = K = J_0 \theta^{-1}$, the result follows.   □

In the finite case, the next result provides a second definition of solubility.

Theorem 11.5 *If G is a finite group, then G is soluble if and only if it has a composition series all of whose factors are cyclic of prime order.*

Proof Clearly, if G has a composition series with cyclic factors, then it is soluble as all cyclic groups are Abelian.

Conversely, suppose G finite and soluble. If G is Abelian, the result follows from Lemmas 7.5 and 7.13. For the non-Abelian case, we use induction on the order of G. If the order of G is 1 or a prime, there is nothing to prove. Hence we may assume that (a) $o(G)$ is composite, and (b) the result holds for groups whose orders are less than $o(G)$. Using our supposition, the series

$$\langle e \rangle \lhd G$$

cannot be the only subnormal series for G as this series has a non-Abelian factor. Hence there exists a subgroup K satisfying $\langle e \rangle \lhd K \lhd G$ with $1 < o(K) < o(G)$. By the inductive hypothesis, both K and G/K have composition series with factors which are cyclic of prime order. As in the proof of Theorem 11.4, we can combine these series to form a series for G with cyclic factors of prime order. This proves the result.   □

We shall now consider some examples. Clearly, all Abelian groups are soluble, as are the dihedral and dicyclic groups; see Problem 11.2. The next two results provide a further wide range of examples.

Theorem 11.6 *All finite nilpotent groups G are soluble.*

By Theorem 10.6, this shows that all finite p-groups are also soluble.

Proof If G is a p-group, this follows directly from Theorem 6.5. Otherwise, by Theorem 7.9, a finite nilpotent group can be expressed as a direct product of p-groups for various primes p, and the result follows by Problem 11.2. □

Theorem 11.7 *If G is a finite group whose order has at most three (not necessarily distinct) prime factors, then G is soluble.*

Proof Let p, q and r be distinct primes. There are a number of cases. If $o(G) = p$, p^2 or p^3, the result follows from Theorem 11.6. If $o(G) = pq$ and $p < q$, then G has a normal subgroup K of order q (Theorem 6.11) and $\langle e \rangle \lhd K \lhd G$ is a soluble series for G. If $o(G) = p^2 q$ then, by Theorem 6.13, G has a normal subgroup of order p^2 or q. In both cases, these subgroups are soluble, and so the solubility of G follows by Theorem 11.4. Finally, if $o(G) = pqr$, then by Theorem 6.13 again, G has a normal subgroup whose order has one or two prime factors, and the solubility of G again follows from Theorem 11.4 by applying the pq case above.   □

Using Problem 6.15, this shows that the smallest non-soluble group is isomorphic to A_5 with order 60; see Problem 11.3. We noted above that no non-Abelian simple groups is soluble. But groups can be both non-simple and non-soluble as the following result shows.

Theorem 11.8 *If $n > 4$, then S_n is not soluble.*

Proof The series $\langle e \rangle \lhd A_n \lhd S_n$ is a composition series for G one of whose factors is not Abelian (Theorem 3.14). Hence by the Jordan–Hölder Theorem (Theorem 9.5), S_n cannot have a subnormal series with Abelian factors, and so S_n for $n > 4$ is not soluble.   □

Derived Series

The derived subgroup G' of a group G is generated by the set of commutators $[a, b] = a^{-1}b^{-1}ab$ where $a, b \in G$. We have, if $H \lhd G$,

$$G' \lhd G \quad \text{and} \quad G/H \quad \text{is Abelian} \quad \text{if and only if} \quad G' \leq H,$$

see Problems 2.16 and 4.6. (It is not sufficient just to take the set of commutators in the definition of the derived subgroup. Scott (1964), page 455, gives an example of a group containing two commutators whose product is not a commutator.) Therefore, the series $G' \lhd G$ can be used as the final step of a soluble series for G, and an extension of this procedure provides an efficient method for constructing this series. We begin with

Definition 11.9 (i) The nth *derived subgroup* $G^{(n)}$ of G is defined inductively by:

$$G^{(0)} = G, \quad G^{(n+1)} = \left(G^{(n)}\right)'.$$

(ii) The *derived series* for G is the normal series from some subgroup of G to G given by

$$\cdots G^{(k)} \lhd G^{(k-1)} \lhd \cdots \lhd G.$$

The *derived length* of a soluble group G is the least n such that $G^{(n)} = \langle e \rangle$, see below. Groups with arbitrarily long derived lengths have been constructed; see page 156. Also compare this with Definition 10.3 for lower central series.

Theorem 11.10 (i) $G^{(n)} \lhd G$.
(ii) *G is soluble if and only if $G^{(n)} = \langle e \rangle$ for some positive integer n.*

The proof of this result uses 'characteristic subgroups' which were discussed in Problem 4.22. We use this notion because, unlike normality, the characteristic property is transitive, see the quoted problem. Note also G' char G.

Proof (i) We use induction on n. By Problem 2.16, $G' \lhd G$. Problem 4.22 and the definition of $G^{(k)}$ give $G^{(k+1)}$ char $G^{(k)}$. Combining this with the inductive hypothesis (that is, $G^{(k)} \lhd G$), we obtain $G^{(k+1)} \lhd G$. Now use Problem 4.22 again. (We also have $G^{(n)}$ char G.)

(ii) Suppose first G has the soluble series $\langle e \rangle = H_m \lhd \cdots \lhd H_0 = G$ where for this proof we write the suffixes in reverse order. First, we show

$$G^{(i)} \leq H_i \quad \text{for } i = 0, \ldots, m, \tag{11.2}$$

by induction on i. We have $H_0 = G = G^{(0)}$, and by Definition 11.9 and the inductive hypothesis

$$G^{(i+1)} = \left(G^{(i)}\right)' \leq (H_i)'.$$

As H_i/H_{i+1} is Abelian by definition, we also have $(H_i)' \leq H_{i+1}$ by Problem 4.6. Combining these we obtain $G^{(i+1)} \leq H_{i+1}$, and (11.2) follows. Now as $\{H_m, \ldots, H_0\}$ is a soluble series for G, this shows that $G^{(m)} \leq H_m = \langle e \rangle$. Conversely, if $G^{(n)} = \langle e \rangle$ for some $n \leq m$, then by Problem 4.6 again

$$\langle e \rangle = G^{(n)} \lhd G^{(n-1)} \lhd \cdots \lhd G^{(0)} = G$$

is a soluble series for G.   □

Minimal Normal Subgroup

Finally, in this section we consider the smallest (minimal) normal subgroups of a soluble group, these subgroups will be used in the proof of Hall's theorems given in the next section. A subgroup K of a group G is called *minimal normal* if it is a non-neutral normal subgroup of G, and if $J \lhd G$ and $\langle e \rangle \leq J \leq K$, then either $J = \langle e \rangle$ or $J = K$. If G is simple, then G itself is the only minimal normal subgroup of G. For example, referring to Section 8.1, the subgroup V generated by a product of two 2-cycles in S_4 is an example of a minimal normal subgroup (of S_4). In this case, the group S_4 is soluble, and the subgroup V (isomorphic to $C_2 \times C_2$) is elementary Abelian. We show in the next theorem that this is always true for soluble groups. It does not hold in general for non-soluble groups, for instance, consider S_5. On page 189, we showed that the only non-neutral proper normal subgroup of S_5 is A_5, so A_5 is the only minimal normal subgroup of S_5 and it is clearly not elementary Abelian! See also (17) on page 293 for the *sockel*, the product of the minimal normal subgrups of the group in question.

Theorem 11.11 *If G is a finite soluble group and H is a minimal normal subgroup of G, then H is an elementary Abelian p-group for some prime p.*

Proof Let K be a minimal normal subgroup of G. By Problem 4.22, K' char K (where K' denotes the derived subgroup of K), and so $K' \lhd G$. As K is minimal, this implies that either $K' = K$ or $K' = \langle e \rangle$. The first possibility is ruled out by Theorem 11.10 as G is soluble; hence $K' = \langle e \rangle$, that is, K is Abelian. By Theorems 7.9 and 10.6, K is a direct product of its Sylow subgroups, and each of these subgroups is normal in K. Hence by minimality, K can only have one such subgroup, and so K is an Abelian p-group for some prime p.

Further, note that if $K_1 = \{k \in K : k^p = e\}$, then K_1 is a characteristic sub-group of the Abelian p-group K (Problem 7.12); and so by minimality again we have $K_1 = K$; that is, K is an elementary Abelian p-group. $\quad\square$

It can be shown that if H is a minimal normal subgroup of general finite group G, then H is either simple or a direct product of isomorphic simple groups; see, for example, Rotman (1994), page 106. This result follows from the fact that if G has no characteristic subgroups (except $\langle e \rangle$ and itself), then it is simple or a direct product of isomorphic simple groups. Therefore, Theorem 11.11 shows that in the soluble case these simple groups (factors of the direct product) are always cyclic with prime order.

11.2 Hall's Theorems and Solubility Conditions

There are a number of equivalent definitions for solubility (for nilpotency, see Theorem 10.9). Perhaps the most remarkable relates to the extension of the Sylow theory to subgroups with coprime orders. In 1928, Philip Hall (1904–1982) showed that if G is soluble and $o(G) = mn$ where $(m, n) = 1$, then G contains subgroups of orders m and n; and in 1937 he proved the converse, so providing a new characterisation of solubility, one apparently far removed from Galois's original definition. We shall prove these results in this section, discuss some related results and applications, and give some further characterisations. We begin with

Theorem 11.12 (Hall's First Theorem) *If G is a soluble group with order mn where $(m, n) = 1$, then G contains at least one subgroup of order m.*

We give two proofs of this result. The first uses the Schur–Zassenhaus Theorem, see page 187, in fact, the proof essentially points out that Hall's Theorem can be treated as a corollary of the result of Schur and Zassenhaus which appeared in its final form in 1937 nine years after Hall's first paper. The second proof is a version (based on one given in Rose 1978) of Hall's original proof; it is quite long and needs careful study, but it does not use any methods not already discussed.

First Proof We use induction on $o(G)$, hence we may assume that the result holds for all groups G_1 where $o(G_1) < o(G)$. Let K be a minimal normal subgroup of G (Theorem 11.11). For some prime p, K is an (elementary Abelian) p-group of order $s(= p^t)$, say. There are two cases.

Case 1: $p \mid m$.

In this case, the inductive hypothesis gives a factor group J/K of G/K which has order m/s. The Correspondence Theorem (Theorem 4.16) now provides a subgroup J of G with order m completing this case.

Case 2: $p \nmid m$.

As above the inductive hypothesis gives $H/K \leq G/K$ for some H (a subgroup of G) with $o(H/K) = m$. This implies that $o(H) = ms$ and $K \vartriangleleft H$. Now $(o(K), [H : K]) = 1$, and so the Schur–Zassenhaus Theorem states that H is isomorphic to a semi-direct product of the form $H \simeq J \rtimes K$ where J is a complement of K in H. This result also implies that $o(J) = m$ which completes the proof. $\square$

Before giving the second proof we make the following, see also Definition B.5(ii) on page 285.

Definition 11.13 Given a set of primes π, a subgroup H of G is called a *Hall* π-*subgroup* of G if all the prime factors of $o(H)$ belong to π, *and* no prime factor of $[G : H]$ is contained in π.

If $\pi = \{p\}$ for some prime p, then a Hall π-subgroup is simply a Sylow p-subgroup. Using Definition 11.13, we can restate the Hall's Theorem as:

For all sets of primes π, G contains a Hall π-subgroup.

The main case is when π equals the set of prime factors of m.

Second Proof As in the first proof, we use induction on $o(G)$, there is nothing to prove if $o(G)$ is 1 or a prime. Let

$$Q = O_\pi(G).$$

For the elementary properties of the radical $O_\pi(G)$, see Problem 6.9(vi) and Theorem 10.24; the fact it is a normal subgroup can be proved using Lemma 4.14, and Problems 4.11 and 6.9. There are two cases.

Case 1: $Q > \langle e \rangle$.

As $Q \vartriangleleft G$, the factor group G/Q has a subgroup H/Q for some $H \leq G$ where $o(H)$ is a π-number (by the induction hypothesis). Now $o(H) = o(H/Q)o(Q)$ and so $o(H/Q)$ is a π-number (Definition B.5). Hence as $[G : H] = [G/Q : H/Q]$, this shows that $[G : H]$ is a π'-number (it only involves primes outside π), so H is a Hall π-subgroup as required.

Case 2: $Q = \langle e \rangle$.

Let $K = O_{\pi'}(G)$. If $K = G$, then G is a π' group, there is only one π-subgroup—that is, $\langle e \rangle$, and the theorem follows in this case. Hence we may suppose

$$K < G.$$

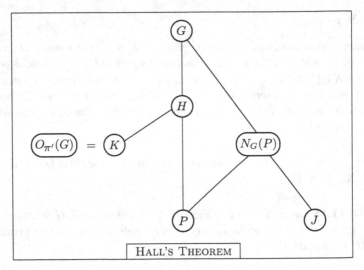

HALL'S THEOREM

The diagram illustrates the subgroup structure used in the last part of the second proof of Hall's First Theorem.

$$* \quad * \quad * \quad * \quad * \quad *$$

We apply Problem 9.15. (Note that using Theorem 11.11 most properties of *chief series* given in this problem are derived in an exactly similar manner to those for composition series given in Section 9.1.) Using this problem we see that G has a chief factor H/K which is an elementary Abelian p-group for some prime p. We have

$$p \in \pi.$$

Suppose not, that is, suppose $p \in \pi'$. As $o(H) = o(H/K)o(K)$, this would imply that H is a normal (the terms in a chief series are normal in G) π'-subgroup of G with $K < H$. But this is not possible by the definition of K and our Case 2 assumption. Therefore, $p \in \pi$.

Let P be a Sylow p-subgroup of H. We have

$$H = PK, \tag{11.3}$$

for both P and K are subgroups of H, P is Sylow, and K is normal. Further, as $P \neq \langle e \rangle$ and $Q = \langle e \rangle$, we have $N_G(P) < G$. (If we had equality, then $P \triangleleft G$, but this would contradict our assumption.) The inductive hypothesis implies that $N_G(P)$ has a Hall π-subgroup J, see diagram above. We complete the proof by showing that J is a Hall π-subgroup of G, that is, we need to show that $[G : J]$ is a π'-number.

By Problem 2.15, we have

$$[G : J] = [G : N_G(P)][N_G(P) : J]. \tag{11.4}$$

Also by the Frattini Argument (Lemma 6.14), (11.3) and the fact that a normaliser of a group always contains that group, we have

$$G = N_G(P)H = N_G(P)PK = N_G(P)K.$$

Using this and the Second Isomorphism Theorem (Theorem 4.15), we have

$$[G : N_G(P)] = [N_G(P)K : N_G(P)] = [K : N_G(P) \cap K],$$

which is a π'-number. Also $[N_G(P) : J]$ is a π'-number by the definition of J. Combining these facts with (11.4) shows that $[G : J]$ is a π'-number, and the theorem follows. □

As in the Sylow theory, it can be shown that two Hall π-subgroups of a finite group G are conjugate in G, we shall not prove this fact (it is also due to Philip Hall); a proof can be found in Rose (1978), page 285.

Hall's First Theorem cannot be extended to all finite groups, for example, consider A_5 which is not soluble. We have $o(A_5) = 4 \cdot 3 \cdot 5$, and A_5 does have a subgroup of order 12 (isomorphic to A_4) and $(12, 5) = 1$. But by the example on page 101, A_5 does not have subgroups of order 15 or 20 even though $(15, 4) = (20, 3) = 1$.

If the prime factorisation of $o(G)$ is $p_1^{r_1} \cdots p_n^{r_n}$ and G is soluble, then both Hall's and Sylow's First Theorems state that G has a subgroup of order $p_i^{r_i}$ for each i. In the soluble case, Hall's Theorem extends this to assert the existence of Hall subgroups with order $p_{i_1}^{r_{i_1}} \cdots p_{i_k}^{r_{i_k}}$ where $I = \{i_1, \ldots, i_k\} \subseteq \{1, \ldots, n\}$. Perhaps the most important case is when $I = \{1, \ldots, j-1, j+1, \ldots, n\}$ for some j between 1 and n; in this case, the corresponding subgroup is called a p_j-*complement* of G, it has order $o(G)/p_j^{r_j}$.

For example, a group of order 30 (soluble by Theorem 11.7) will have a 2-complement, that is, a subgroup of order 15; a 3-complement, a subgroup of order 10; and a 5-complement, a subgroup of order 6. One consequence of Hall's First Theorem is: For all soluble groups G and for all primes p dividing $o(G)$, G has a p-complement. Hall's Second Theorem gives the converse.

Theorem 11.14 (Hall's Second Theorem) *Suppose G is a finite group. If G has a p-complement for each prime p dividing $o(G)$, then it is soluble.*

The proof of this result relies on Burnside's $p^r q^s$-theorem which we shall prove in Web Section 14.2. It states that if $o(G) = p^r q^s$, and p and q are prime, then G is soluble; via the Sylow theory, this is Hall's Second Theorem in the case when $o(G)$ has exactly two distinct prime factors. As with the second derivation of Hall's First Theorem above, the proof given below is quite long, but it only uses methods applied previously; the main one being consideration of the 'smallest counterexample'.

Proof Again we use induction on $o(G)$. Suppose the result is false and G_1 is a counter-example of smallest order. So G_1 has p-complements for all p dividing $o(G_1)$, G_1 is not soluble, and the result holds for all groups whose orders are less than $o(G_1)$.

We show first that G_1 is simple. Suppose not, then G_1 has a proper non-neutral normal subgroup K. If $p \mid o(G_1)$, then by hypothesis G_1 has a p-complement J of order $o(G_1)/p^r$ where p^r is the largest power of p dividing $o(G_1)$. Now $J \cap K$ is a p-complement of K, and JK/K is a p-complement of G_1/K (by Theorem 5.8, $o(JK/K) = o(J/J \cap K) \mid o(J)$). This holds for all p-complements J, that is for all primes p dividing $o(G)$. Now by the inductive hypothesis both K and G_1/K are soluble, and so G is soluble by Theorem 11.4.

We complete the proof by constructing a proper normal subgroup of G_1 which, of course, is impossible because we have just shown that we may assume G_1 is simple. This then shows that our main assumption—that the counter-example G_1 exists—is false and the theorem follows.

Suppose

$$o(G_1) = p_1^{r_1} \cdots p_n^{r_n} \quad \text{where } n > 2 \text{ and } r_i > 0, \ i = 1, \ldots, n.$$

We may assume that $n > 2$ by Theorem 11.6 and Burnside's $p^r q^s$-theorem. For $i = 1, \ldots, n$, let H_i be a p_i-complement of G_1 as given by the hypothesis, then

$$[G_1 : H_i] = p_i^{r_i} \quad \text{and} \quad o(H_i) = o(G_1)/p_i^{r_i}.$$

Let $L = H_3 \cap \cdots \cap H_n$, by Theorem 5.8 this gives

$$[G_1 : L] = p_3^{r_3} \cdots p_n^{r_n} \quad \text{hence} \quad o(L) = p_1^{r_1} p_2^{r_2}.$$

Now Burnside's $p^r q^s$-theorem asserts that L is soluble, and by Theorem 11.11 it has a minimal normal subgroup K which is an (elementary Abelian) p-group for some p dividing $o(L)$. Relabelling if necessary, we may assume that $p = p_1$, and so K is a p_1-group. Applying Theorem 5.8 again,

$$[G : L \cap H_2] = p_2^{r_2} \cdots p_n^{r_n} \quad \text{and so} \quad o(L \cap H_2) = p_1^{r_1},$$

that is, $L \cap H_2$ is a Sylow p_1-subgroup of L. Hence, by Problem 6.10(iii),

$$K \leq L \cap H_2, \quad \text{and so} \quad K \leq H_2. \tag{11.5}$$

Also using the same argument, we have $o(L \cap H_1) = p_2^{r_2}$, and so

$$G = (L \cap H_1)H_2.$$

(Note $o(H_2) = p_1^{r_1} p_3^{r_3} \cdots p_n^{r_n}$.) Therefore, if $g \in G_1$, then $g = lh$ where $l \in L \cap H_1$ and $h \in H_2$. Further, if $k \in K$, we have

$$g^{-1}kg = h^{-1}l^{-1}klh = h^{-1}yh \quad \text{where } y = l^{-1}kl \in K \text{ as } K \triangleleft L.$$

This gives by (11.5)

$$g^{-1}kg = h^{-1}yh \in H_2 \qquad (11.6)$$

for $g \in G_1$ and $k \in K$. Finally, let M be the normal closure of K in G; see Problem 2.25. We have $K \leq M \lhd G_1$ and $M \leq H_2 < G_1$ by (11.6), that is, M is a non-neutral proper normal subgroup of G_1. Hence G_1 is not simple, contrary to our assumption, and the theorem follows. $\qquad \square$

Example Let H be the group given by

$$H = \langle a, b \mid a^5 = b^6 = e, \ b^{-1}ab = a^4 \rangle,$$

a Frobenius group, see Theorem 6.18. The order of this group is 30, and it contains subgroups (p-complements for $p = 2, 3$ and 5): $\langle b \rangle$ of order 6, $\langle a, b^3 \rangle$ of order 10 ($\simeq D_5$), and $\langle ab^2 \rangle$ of order 15 ($\simeq C_{15}$). Hence the conditions of Hall's Second Theorem are satisfied and so H is soluble. This fact, of course, also follows from Theorem 11.7.

Solubility Conditions

We end this section by listing some solubility conditions. By definition, G is soluble if it has a subnormal series with Abelian factors. By Theorems 11.2 and 11.3, it is sufficient for the factors themselves to be soluble. For a finite group, each of the Conditions 1 to 10 below is equivalent to solubility.

Condition 1 For some positive integer k, the kth derived subgroup $G^{(k)}$ satisfies $G^{(k)} = \langle e \rangle$; see Theorem 11.10.

Condition 2 All factors of all composition series are cyclic of prime order; see Theorem 9.3. This is an easy consequence of the definitions.

Condition 3 There exists a normal series all of whose factors are p-groups for various primes p. This is also an easy consequence of the definitions; see Problem 11.5 for this and the previous condition.

Condition 4 Hall subgroups exist for orders m where $(m, o(G)/m) = 1$ and $m \mid o(G)$; see Theorems 11.12 and 11.14 above.

Condition 5 All 'chief factors' are elementary Abelian (Problem 9.15).

Condition 6 (Sylow systems) Given a group G, let $p_1, \ldots, p_k$ be a list of the distinct prime factors of $o(G)$. Suppose for $1 \leq i \leq k$ we choose a single Sylow p_i-subgroup P_i, and this set $\mathcal{P}$, say, has the property

$$P_i P_j = P_j P_i \quad \text{for } 1 \leq i < j \leq k.$$

Then $\mathcal{P}$ is called a *Sylow system* for G. Using his two major theorems, Hall showed that a group G has a Sylow system if and only if it is soluble. Note that some authors give a slightly different, but equivalent, definition using Hall subgroups rather than Sylow subgroups.

Two examples First, consider the group H given in the example on page 241. It has order 30, and is soluble by Theorem 11.7, or see the quoted example. One Sylow system for H is defined by $\{P_1 = \langle b^3 \rangle, P_2 = \langle b^2 \rangle, P_3 = \langle a \rangle\}$, for it is easy to see that $P_i P_j = P_j P_i$ for $1 \leq i, j \leq 3$ as P_2 and P_3 are both normal subgroups of H.

Second, consider A_5 which of course is not soluble. Its Sylow 3-subgroups are cyclic generated by 3-cycles, an example is $B = \langle (1, 2, 3) \rangle$, and its Sylow 5-subgroups are also cyclic generated by 5-cycles, $C = \langle (1, 2, 3, 4, 5) \rangle$ is an example. We have $BC \neq CB$ with a similar result for other choices of the Sylow 3- and 5-subgroups, so confirming Hall's result in this case.

Condition 7 If G is finite, it is *not* possible to find $a, b, c \in G$ satisfying

$$ab = c \quad \text{and} \quad \bigl(o(a), o(b)\bigr) = \bigl(o(b), o(c)\bigr) = \bigl(o(c), o(a)\bigr) = 1, \tag{11.7}$$

when a, b and c are all non-neutral.

This number-theoretic condition was first studied by Hall (1937), the equivalence was proved by Thompson (1968). There is a connection with the Burnside's $p^r q^s$-theorem which was used in the proof of Theorem 11.14, for if the condition applies then the order of the group must involve at least three primes. For example, consider S_5. If $a = (1, 2)(3, 4)$, $b = (2, 3, 5)$ and $c = (1, 3, 4, 5, 2)$, it is easy to see that $a, b, c \in S_5$, $ab = c$, $o(a) = 2$, $o(b) = 3$ and $o(c) = 5$, hence a, b and c satisfy (11.7), and by Theorem 11.8 the group S_5 is not soluble.

Condition 8 If G is finite and every pair of elements of G generate a soluble group.

This equivalence was also established by Thompson in his 1968 paper. For example, we can use this condition to reprove the insolubility of S_5 by noting that its insoluble subgroup A_5 is generated by two of its cycles, for instance, by $(1, 2, 3)$ and $(3, 4, 5)$.

Condition 9 This equivalence is partly due to Galois and is not quite exact. Galois showed that if G is soluble and H is a maximal subgroup, then $[G : H] = p^n$ for some prime p and positive integer n.

Using Hall's theorems, it is easy to show that if G is soluble and $p \mid o(G)$, then there exists a maximal subgroup J which satisfies: $[G : J] = p^m$ for some positive integer m. For a 'converse' we have:

If for all non-neutral subgroups J of G, there exists a subgroup L of J with the property: $[J : L] = p^k$ for some positive integer k, and this applies for all primes p dividing $o(J)$, then G is soluble.

For a further discussion of this condition see Rose (1978), page 279. In Problem 11.9, we show that there exists a non-soluble group G_1 with a subgroup with index a power of p for every prime p dividing $o(G_1)$; so in the condition above we need to consider all subgroups J of G and not just G itself.

Condition 10 The factor group $G/\Phi(G)$ is soluble where $\Phi(G)$ denotes the Frattini subgroup of G; see also Theorem 10.17.

At the beginning of Section 11.1, we listed some number-theoretic conditions on the group order which imply solubility, none is equivalent to solubility. There are also several conditions of a more group-theoretic nature that imply solubility, some of the simpler ones are as follows:

Condition 11 If all maximal subgroups of a finite group G have an index which is a prime or the square of a prime, then G is soluble.

In Problem 12.9, we show that $L_2(7)$ is simple, hence not soluble, and it has a maximal subgroup index 8. This shows that Condition 11 is best possible.

Condition 12 If all proper subgroups of a finite group G are nilpotent, then G is soluble. A similar condition applies with 'nilpotent' replaced by 'supersoluble'; see Problem 10.26. Also Thompson has shown that if G has a single nilpotent subgroup which is maximal and has odd order, then G is soluble. The simple group $L_2(17)$ has a maximal subgroup of order 16, and so the oddness condition above is essential; see Passman (1968), page 141, for further details.

For example, if $o(G) = p^2 q^2$ and $p \nmid q - 1$, then all subgroups of G are nilpotent (in fact, Abelian), and it is an example of a soluble group; see Problem 11.4*.

Condition 13 There are a number of conditions of the following type. Suppose G is finite, $H, J \leq G$ and $G = HJ$, then G is soluble if certain conditions apply to H and J. We give three of the simpler ones here; see Scott (1964), Chapter 13.

(a) H and J are nilpotent and $(o(H), o(J)) = 1$;
(b) H is nilpotent and J has a cyclic subgroup K with $[J : K] = 2$;
(c) H and J are both dihedral or dicyclic.

Some more conditions which imply solubility, but are not equivalent to solubility, are given in the problem section below; see Problems 11.6(i) and (iii), 11.8, 11.10(ii), 11.11, 11.14*, and 11.15*.

11.3 Problems

Problem 11.1 (i) Complete the following sketch of the proof that the three zeros of a cubic polynomial equation with rational coefficients can be defined using radicals, that is, by using square and cube roots. This was first proved by three early sixteenth century Italian mathematicians—S. del Ferro, N. Fontana (Tartaglia) and G. Cardano. The proof was an important step in the development of algebra (and so of science in general) because it forced mathematicians to take complex numbers seriously (R. Feynmann).

(a) Using a suitable substitution show that a general cubic polynomial equation with coefficients in $\mathbb{Q}$ can be written in the form

$$f(x) = x^3 + ax + b \quad \text{where } a, b \in \mathbb{Q}.$$

(b) Suppose z is a root of $f(x) = 0$. Write $z = u + v$, and so show that

$$u^3 + v^3 + (3uv + a)z + b = 0.$$

(c) Set $3uv + a = 0$ to obtain $u^3 - a^3/(3u)^3 + b = 0$. Solve this for u^3, and hence obtain radical expressions for the zeros of the equation $f(x) = 0$.

(ii) In 1545, L. Ferrari showed that quartic polynomial equations also have radical solutions. You are asked to complete the following sketch proof which was given by R. Descartes in 1637.

(a) As above show that a general quartic polynomial equation with coefficients in $\mathbb{Q}$ can be written as

$$g(x) = x^4 + ax^2 + bx + c,$$

where we may assume that $b \neq 0$ (for otherwise we have a quadratic in x^2).

(b) Factorising $g(x)$ as a product of two quadratics

$$x^4 + ax^2 + bx + c = (x^2 + rx + s)(x^2 - rx + t),$$

show that $2s = r^2 + a - b/r$ and $2t = r^2 + a + b/r$. Deduce

$$r^6 + 2ar^4 + (a^2 - 4c)r^2 - b^2 = 0,$$

and so complete the proof using part (i).

Problem 11.2 (i) Show that the following groups are soluble (a) D_n, (b) Q_n, (c) S_4.

(ii) Give two proofs of the result: If both G and H are soluble, then so is $G \times H$.

Problem 11.3 (i) Using Problem 3.19, or otherwise, show that $SL_2(5)$ is not soluble.

(ii) Show that if G is not soluble and $o(G) \leq 200$, then $o(G) = 60, 120, 168$, or 180, and give examples in each case. (Hint. The only simple groups satisfying the inequality above have orders 60 or 168, this will be proved in the problem section of Chapter 12.)

Problem 11.4* Apply the Sylow theory to show that each group G of order $p^n q$ or $p^2 q^2$ where p and q are primes, is soluble using the following method. In the second case, suppose $p < q$. In both cases, assume that G is simple using the minimum counter-example principle, and P_1 and P_2 are distinct Sylow p-subgroups. Again in both cases, if $P_1 \cap P_2 = \langle e \rangle$, use a counting argument to find a unique Sylow q-subgroup, and if this condition is false use the normal closure in the first case, and Theorem 5.15 in the second.

Problem ♦ 11.5 (i) Show that a refinement of a soluble series is also soluble series.

(ii) Must a soluble group with a composition series be finite?

(iii) Prove that a finite group is soluble if and only if all factors of all of its composition series are cyclic of prime order.

(iv) Finally, prove that a finite group is soluble if and only if it has a normal series all of whose factors are p-groups for various primes p.

Problem 11.6 (i) Using character theory (Web Chapter 13), Burnside proved that if a group G has a conjugacy class of order p^r (p prime and $r > 0$), then G is not simple. Use this to prove that if $o(G) = p^r q^s$ where p and q are prime, and r and s are non-negative integers, then G is soluble. (Hint. Use Theorem 5.19.)

(ii) Feit and Thompson (1963) proved that every group of odd order is soluble, show that this statement is equivalent to: 'every non-Abelian finite simple group has even order'.

(iii) A group is called 'metacyclic' (Theorem 6.18) if all of its Sylow subgroups are cyclic. Show that all metacyclic groups are soluble.

Problem ♦ 11.7 Suppose $o(G) > 1$. Show that

(i) if G is soluble it has a non-neutral normal Abelian subgroup, and

(ii) if G is not soluble it has a non-neutral normal perfect subgroup.

Problem 11.8 (i) Give an example of a soluble group G which is not 'reverse Lagrange'. By reverse Lagrange we mean that G has at least one subgroup of order m for all positive divisors m of $o(G)$, see page 101.

(ii) Discuss the proposition: 'if G is not soluble then it is also not reverse Lagrange'.

Problem ♦ 11.9 (i) Suppose G is soluble, $J \lhd G$ and K is minimal subject to the conditions

$$ J < K \quad \text{and} \quad K \lhd G. $$

Show that K/J is elementary Abelian. (Hint. Use Problem 9.15.)

(ii) Prove that the index of a maximal subgroup H of a soluble group G has prime power order—a result due (essentially) to Galois. (Hint. Begin by considering $J = \text{core}(H)$ and K as defined in (i) for this J. Use (i) to show that $KH = G$ and $K \cap H = J$, then use Theorem 4.15.)

Problem ◆ 11.10 (i) Suppose H and J are soluble normal subgroups of a group G. Show that HJ is also a soluble normal subgroup of G, and so deduce that if $HJ = G$ then G is soluble.

(ii) Let $S(G)$ be the product of all soluble normal subgroups of G where $o(G) < \infty$. Show that $S(G)$ is the maximal soluble normal subgroup of G.

(iii) Can a maximal normal subgroup of G be larger than $S(G)$?

Problem 11.11 Find a non-soluble group H which has a subgroup of index p for all p dividing $o(H)$. One example is a direct product of a simple group and a suitably chosen cyclic group.

Problem 11.12* Suppose G is a soluble group, and $G^{(n)}/G^{(n+1)}$ is cyclic for $n = 1, 2, \ldots$. Using the following method, show that $G^{(2)} = G^{(3)} = \langle e \rangle$. First, use Theorem 11.10 to show that $G^{(3)} = \langle e \rangle$ and so $G^{(2)}$ is cyclic, then use the N/C-theorem (Theorem 5.26) to show $G^{(2)} \leq Z(G')$, and apply Problem 4.16(ii).

Problem 11.13 Using the previous problem and Burnside's Normal Complement Theorem (Theorem 6.17) show that if G is finite and all of its Sylow subgroups are cyclic, then both G' and G/G' are also cyclic. This is the starting point for the proof of Theorem 6.18.

Problem 11.14* (i) A theorem of O. Schmidt states: If every maximal subgroup of a finite group G is nilpotent, then it is soluble. Give a proof of this result using the following method.

(a) Let G_1 be a counter-example of smallest order. Show that G_1 is simple.
(b) Prove that there exist maximal subgroups H_1 and H_2 of G_1 with the property $H_1 \cap H_2 > \langle e \rangle$. If not, take conjugates and count elements.
(c) Choose H_1 and H_2 as in (b) so that $J = H_1 \cap H_2$ has the largest possible order, then show that J cannot exist, and so deduce Schmidt's result.

(ii) Use (i) to show that if all proper subgroups of a finite group G are Abelian, then (a) at least one is normal, and (b) $G'' = \langle e \rangle$.

Problem 11.15* If G has Abelian subgroups H and J with the property $G = HJ$, prove that G is soluble. One method uses Problem 2.17 several times to show that G' is Abelian.

Problem 11.16 For the groups A_5, A_6 and $GL_3(2)$, see Chapter 12, determine for which prime divisors p of their orders they do not possess p-complements.

Problem 11.17 Given a integer n and primes p and q which satisfy $p < q \leq n$, show that the symmetric group S_n has a Hall $\{p, q\}$-subgroup if $n = 3, 4, 5, 7$, or 8, $p = 2$ and $q = 3$. Thompson has shown that if H is a proper non-soluble Hall subgroup of S_n, then $n > 5$, n it is prime, and $H \simeq S_{n-1}$.

Problem 11.18 (Properties of Hall Subgroups) Derive the following properties of Hall subgroups, see Definition 11.13.

(i) Suppose G is a finite soluble group and π is a set of prime numbers. If H is a maximal π-subgroup of G (so there is no π-subgroup of G lying strictly between H and G), then H is a Hall π-subgroup of G.

(ii) Let $\pi_1 = \{3, 5\}$ and consider the group S_5. Show that both C_3 and C_5 are isomorphic to maximal π_1-subgroups of S_5. Hence maximal π-subgroups of a non-soluble group need not be isomorphic.

(iii) The symmetric group S_5 has maximal $\{2, 5\}$-subgroups which are not Hall $\{2, 5\}$-subgroups.

(iv) Consider the group $GL_3(2)$; see Sections 3.3 and 12.2. This group has Hall $\{2, 3\}$-subgroups which are not conjugate.

(v) Finally, show that $L_2(11)$ (Problem 12.9) has non-isomorphic Hall $\{2, 3\}$-subgroups. (Hint. These subgroups have order 12.) Gross (1987) has shown that if $2 \notin \pi$, then all Hall π-subgroups are conjugate.

Problem ♦ 11.19 Let H be a Hall π-subgroup of G. Derive the following properties.

(i) If $J \leq G$ and $o(J) = o(H)$, then J is a Hall π-subgroup of G.

(ii) For $g \in G$, the conjugate subgroup $g^{-1} H g$ is a Hall π-subgroup of G.

(iii) If $H \leq J \leq G$, then H is a Hall π-subgroup of J.

(iv) LK/K is a π-subgroup of G/K, if $K \lhd G$ and L is a π-subgroup of G.

Problem 11.20 (i) Give Sylow systems for the groups discussed in Chapter 8.

(ii) Why does the group S_5 not have a Sylow system?

(iii) Show that the solubility of S_4 can be deduced from Condition 6 or Condition 7 both of which are given on page 242.

Problem 11.21 (Project—Maximal Subgroups of Soluble Groups) Suppose G is soluble with maximal subgroup H, see also Suzuki (1986), Section 5. Firstly, revisit Problem 11.9(ii) which is due to Galois.

Secondly, suppose $K = \text{core}(H) = \langle e \rangle$ and let J be a minimal normal subgroup of G. Prove that

(a) $JH = G$ and $J \cap H = \langle e \rangle$. (Hint. Use first part.)

(b) If H_1 maximal, then $J \leq H_1$ or H_1 is conjugate to H in G.

(c) $J = C_G(J)$, and it is the unique minimal normal subgroup of G.

Lastly, suppose H' is another maximal subgroup of G. Show that the following four conditions are equivalent.

(d) H and H' are conjugate.

(e) If $G = H'L$ for $L \leq G$, then $G = HL$.

(f) $G \neq H'H$.

(g) $\text{core}(H) = \text{core}(H')$.

Chapter 12
Simple Groups of Order Less than 10000

The Jordan–Hölder Theorem states that every finite group can be 'constructed' from simple groups using extensions. The nature of these extensions is not fully understood (Section 9.2), but a complete list of all finite simple groups is now known—indeed its discovery and development was one of the greatest achievements of twentieth-century mathematics; see the ATLAS (1985), Gorenstein (1982), and Gorenstein *et al.* (1994). To add to our collections of group examples we shall give an account of the groups in this list whose orders are less than 10000. This may seem a large bound (Appendix D) but, if we exclude Abelian groups, it contains only 16 non-isomorphic groups of three main types (b), (c) and (d) below, and gives a brief 'snapshot' of the total picture.

The collection of all finite simple groups can be put into four broad categories:

(a) Cyclic groups C_p of prime order p, measured by the size of their order 'most' simple groups are of this type, there are 1229 with order less than 10000;
(b) Alternating groups A_n, for $n > 4$, three have order less than 10000;
(c) A number of collections of matrix groups defined over finite fields including the *linear groups*, and a number of other types with similar constructions called the *Classical Groups* and the *Chevalley Groups*; there are 16 with order less than 10000 but some isomorphisms between individual groups occur, see below;
(d) Twenty-six *sporadic groups* having a variety of constructions, only one of which has order less than 10000.

As noted above, there is some overlap between these classes; see Problems 6.16, 12.4 and 12.13, Web Section 12.6, and Conway and Sloane (1993).

Categories (a) and (b) above have been discussed previously in Chapters 3 and 4 where their simplicity properties were established. Each alternating group A_n has a presentation of the form

$$A_n \simeq \langle a_1, \dots, a_{n-2} \mid a_i^3 = (a_i a_j)^2 = e \rangle,$$

provided $n \geq 4$, $1 \leq i, j \leq n$ and $i \neq j$. This can be shown to be correct using a similar method to that given in Problem 3.21* for a presentation of S_n. Also A_5, A_6 and A_8 have easily defined matrix representations, these are given on pages 260 and 295.

In the first section of this chapter, we introduce a new method for constructing groups using so-called *Steiner systems*. This method is not more important than several others, but we give it because it can be defined with a minimum amount of preliminary material, and it illustrates some of the more complex methods used. The second section introduces the linear groups, see Category (c) above, provides a proof of their simplicity (in all but two cases), and discusses some of their other representations. The third section introduces one class of classical groups—*unitary groups*, and discusses an example in this class, $U_3(3)$, in some detail.

The existence of the sporadic groups (Category (d) above) is perhaps the most remarkable single feature of group theory. The first five, the *Mathieu groups*, were discovered between 1861 and 1873 although their simplicity was established by others two decades later. The remaining 21 were discovered between 1965 and 1980, the first J_1 by Z. Janko in 1965. Reasons for this timing include the introduction of electronic computation at this time, and the work of Brauer, Suzuki and others beginning in the 1950s that stimulated mathematicians to take up afresh the challenges of finite group theory. For details on the discovery and early history of the sporadic groups, see Aschbacher (1994) and Solomon (2001). A wide range of methods are used in the construction of simple groups including permutation and matrix theory, certain lattice types (where the groups arise as automorphism groups of these lattices) and game theory, *et cetera*;[1] Steiner systems are one of these.

This chapter is more 'descriptive' than the others in this book, and many proofs are either sketched or omitted altogether. In some cases, the lack of proof is partly offset by the provision of numerical examples. A whole book would be needed to give all the background and proofs. The properties of the outer automorphism (see pages 81 to 84) and the Schur multiplier groups, and their connections with extensions of the simple groups under discussion, will be treated in Web Section 12.6. The reader is strongly urged to refer to the ATLAS (1985) and the other references for further details.

We wish to thank Robert Curtis for advice and helpful suggestions made during the writing of this chapter.

12.1 Steiner Systems

In Chapter 3, we introduced three methods for constructing groups. Although these are the main ones in the theory, a number of other constructions have been investigated, and these need to be considered if all simple groups are to be constructed.

---

[1] For more details, see Conway and Sloane (1993).

Some occur as 'automorphisms' of lattices or 'games', and this holds for the constructions to be studied in this section. Further methods are given in the ATLAS (1985), and Conway and Sloane (1993). As noted above the Steiner system method is not more important than several others, but it only requires a minimal amount of preliminary material. Some authors consider that these systems should be named after Kirkman rather than Steiner. T.P. Kirkman was an English mathematician who lived from 1806 to 1895, and J. Steiner was Swiss and lived from 1796 to 1863.

We begin with the basic

Definition 12.1 Let r, s and t be integers satisfying $1 < r < s < t$, let X be a t-element set, and let $\mathcal{Y}$ denote the collection of all (unordered) r-element subsets Y of X. A *Steiner system* $S(r, s, t)$ for X is a collection $\mathcal{Z}$ of (unordered) subsets Z of X with the properties:

(a) $o(Z) = s$ for each $Z \in \mathcal{Z}$, and
(b) each $Y \in \mathcal{Y}$ is a subset of exactly one $Z \in \mathcal{Z}$.

A set Z in $\mathcal{Z}$ is called a *block*, so each block has s members. Usually X is taken to be a subset of the positive integers because they form an easily recognised set of labels for the elements of X.

It is not known precisely for which r, s and t Steiner systems exist (but see Problem 12.3). There is no system $S(2, 3, 4)$, for instance, but versions of $S(2, 3, 7)$ do exist as the following example shows.

Example 1 Let $X_1 = \{1, 2, 3, 4, 5, 6, 7\}$ (so $t = 7$), $r = 2$, $s = 3$, and let

$$\mathcal{Z}_1 = \big\{\{1, 2, 4\}, \{1, 3, 7\}, \{1, 5, 6\}, \{2, 3, 5\}, \{2, 6, 7\}, \{3, 4, 6\}, \{4, 5, 7\}\big\}. \quad (12.1)$$

It is a straightforward exercise to check that every 2-element subset of X_1 occurs in exactly one member of $\mathcal{Z}_1$, for instance, the pair $\{1, 5\}$ occurs in the third member of $\mathcal{Z}_1$ and no other. This system can be generated as follows:

Working modulo 7, start with the triple $\{1, 2, 4\}$, that is, the non-zero quadratic residues modulo 7, and apply the map $x \mapsto x + 1$ six times.

As an exercise the reader should construct a version of the Steiner system $S(2, 3, 9)$; it has 12 members, see Problem 12.2.

A relatively simple counting argument gives the order of $S(r, s, t)$, provided this system actually exists, as follows.

Theorem 12.2 *Using the notation set up in Definition 12.1, if the Steiner system* $S(r, s, t)$ *exists, then*

$$o(\mathcal{Z}) = \frac{t(t - 1) \cdots (t - (r - 1))}{s(s - 1) \cdots (s - (r - 1))}.$$

Proof Let $\mathcal{V}$ denote the collection of all pairs (Y, Z) where $Y \in \mathcal{Y}$, that is, Y is an r-element subset of X, and Z is a block in $\mathcal{Z}$. Further, let $\mathcal{W}$ denote the subset of $\mathcal{V}$ of those pairs (Y, Z) where $Y \subset Z$. We count $\mathcal{W}$ in two ways. First, by hypothesis, each $Y \in \mathcal{Y}$ occurs in exactly one $Z \in \mathcal{Z}$, so

$$o(\mathcal{W}) = o(\mathcal{Y}) = \binom{t}{r} = t(t-1)\cdots(t-(r-1))/r!,$$

the number of ways of choosing r elements from a t-element set. Second, $o(\mathcal{W})$ is the product of $o(\mathcal{Z})$, the number of blocks, and $\binom{s}{r}$, the number of ways of choosing an r-element subset from an s-element block. Hence

$$o(\mathcal{W}) = o(\mathcal{Z}) \cdot \binom{s}{r} = o(\mathcal{Z}) \cdot s(s-1)\cdots(s-(r-1))/r!.$$

The result follows from these two equations by cancelling the term $r!$. $\square$

An extension of this result is given in Problem 12.3.

A Steiner system for r, s and t is never unique, for instance, in Example 1 above, the set

$$\{\{3,2,4\},\{3,5,1\},\{3,6,7\},\{2,5,6\},\{2,7,1\},\{5,4,7\},\{4,6,1\}\} \qquad (12.2)$$

is another version of the Steiner system for the parameters $r = 2$, $s = 3$ and $t = 7$. Compared with (12.1), the elements of the set $X = \{1, \ldots, 7\}$ have been permuted by the 5-cycle $(1, 3, 5, 6, 7)$. But, for example, if we permute the elements of X in (12.1) by the permutation $\xi_1 = (1, 6, 2)(4, 5, 7)$ we obtain the Steiner system

$$\{\{6,1,5\},\{6,3,4\},\{6,7,2\},\{1,3,7\},\{1,2,4\},\{3,5,2\},\{5,7,4\}\},$$

that is, we obtain the original system (12.1) again, the only difference being that the order of the sets and the order of the elements in these sets have been changed by ξ_1, which is immaterial. We shall see below that the transformation ξ_1 is a member of the automorphism group of the Steiner system $S(2, 3, 7)$. These examples suggest that we should make the following

Definition 12.3 For $i = 1, 2$, suppose $\mathcal{Z}_i$ is a Steiner system for X_i. These systems are said to be *isomorphic* if

(a) there exists a bijection $\xi : X_1 \to X_2$, and
(b) $Z \in \mathcal{Z}_1$ if and only if $Z\xi \in \mathcal{Z}_2$.

The isomorphism ξ is called an *automorphism* if $X_1 = X_2$ *and* $\mathcal{Z}_1 = \mathcal{Z}_2$.

In the examples given above, the system (12.1) is isomorphic to (12.2), and the permutation ξ_1 is an automorphism of the system $\mathcal{Z}_1$. In some cases, if a Steiner system exists for r, s, t, then there is effectively only one such system, that is, they are all isomorphic as in the example above. But in many cases, this is not so, for instance, there are two non-isomorphic systems for the parameters 2, 3, 13 (Problem 12.2) and 80 non-isomorphic systems for the parameters 2, 3, 15.

The automorphisms of a Steiner system form a group as we can easily show now.

Theorem 12.4 *The collection of automorphisms of the Steiner system $S(r, s, t)$ forms a group under composition which is isomorphic to a subgroup of symmetric group S_t.*

Proof Composition of bijections is closed and associative, and the inverse of a bijection is a bijection (Appendix A). In fact, as all sets are finite, this also follows from Theorem 2.7(iii). The identity map ι acts as the neutral element. Note that Condition (b) in Definition 12.1 applies as $\mathcal{Z}_1 = \mathcal{Z}_2$ in all cases. Finally, as ξ is a bijection on X its images are permutations of X, and hence they are elements of S_t. □

The group given by Theorem 12.4 is denoted by $\mathrm{Aut}(S(r, s, t))$.

We shall illustrate this result by returning to Example 1. In this case, $\mathrm{Aut}(S(2, 3, 7)) \simeq GL_3(2)$, see Section 12.2, so $S(2, 3, 7)$ has 168 automorphisms.

The system $S(2, 3, 7)$, and some other Steiner systems, can be treated as a kind of 'finite geometry', that is, as a geometry defined over a finite field. Working over the 2-element field $\mathbb{F}_2$ and in 'dimension 2', the 'plane' of this geometry consists of seven 'points' labelled $A, B, \ldots, G$ and seven 'lines' where each line consists of exactly three points. This is illustrated in the diagram below which is called the *Fano Plane*.

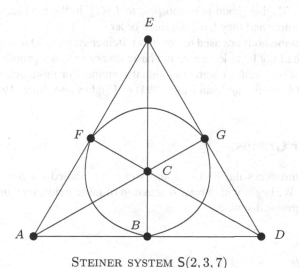

STEINER SYSTEM $S(2, 3, 7)$

In this diagram the strokes, A to B to D, for example, are not part of the geometry, but are included to indicate which sets of three points form lines, so $\{A, B, D\}$ is one of the seven lines. If we refer now to Example 1 and (12.1), we see that the seven members of $S(2, 3, 7)$ correspond to the seven lines in the diagram above if we set $A \mapsto 1, B \mapsto 2, C \mapsto 3, D \mapsto 4, E \mapsto 5, F \mapsto 6$ and $G \mapsto 7$. In this geometry, as in most others, two distinct points determine a line; so in $S(2, 3, 7)$ there is a unique line (triple of points) passing through each pair of points as given by the properties of this Steiner system. To be more precise, the geometry given in this diagram is a kind of 2-dimensional 'projective geometry' defined over $\mathbb{F}_2$. Note also that there is no notion of 'end-point', so it might be better to treat each of the seven strokes as closed loops. We claimed above that

$$\mathrm{Aut}\big(S(2, 3, 7)\big) \simeq GL_3(2) \simeq L_2(7).$$

We will not prove this fact as it would take too long to set up the basic geometric properties needed, the following sketch will suffice; further details can be found in Rotman (1994), Chapter 9, for example. The 'affine' part of our geometries can be treated as the geometry of the underlying space of a vector space V, and the linear maps (automorphisms) correspond to the elements of the general linear group $GL(V)$ represented by the collection of $m \times m$ non-singular matrices where $m = \dim V$, they map points to points, and lines to lines.

This work can easily be extended to 'projective' geometry. Effectively this can be done by adding one extra coordinate, and then the seven points in the diagram on page 253 have the projective coordinates: $A = (0 : 0 : 1)$, $B = (0 : 1 : 1)$, $C = (1 : 1 : 1)$, $D = (0 : 1 : 0)$, $E = (1 : 0 : 0)$, $F = (1 : 0 : 1)$, and $G = (1 : 1 : 0)$ working modulo 2; note that this choice of coordinates is not unique. Hence in our example there is a correspondence between the automorphisms of $S(2, 3, 7)$ and the non-singular 3-dimensional linear maps defined over the 2-element field, that is, $GL_3(2)$. By Problem 12.7*, this group is isomorphic to $L_2(7)$, in the next section we prove that both are simple and they have the same order.

A number of methods are used to construct Steiner systems. Due to space considerations, we shall not be able to give the basic theory or many proofs, but practical procedures will be given for some individual systems. For more details, the reader should consult Cameron and van Lint (1991) or Hughes and Piper (1985).

12.2 Linear Groups

This section introduces the *linear groups* sometimes called the *projective special linear groups*. We begin with a brief discussion of finite fields, they underlie all the matrices and groups discussed.

Finite Fields

The finite field $\mathbb{F}_q$ has q elements where $q = p^m$. These fields were discovered by Galois and as a consequence are sometimes called *Galois fields*. Their basic properties are as follows:

(i) $\mathbb{F}_q$ is a field, that is, the standard operations of addition, subtraction, multiplication and division apply without restriction, except that division by zero is not allowed. Divisors are constructed using the Euclidean Algorithm (Theorem B.2).

(ii) $o(\mathbb{F}_q) = q = p^m$ where p is a prime and m is a positive integer. In fact, $\mathbb{F}_q$ can be treated as a vector space of dimension m over the p-element field $\mathbb{F}_p$ with a vector multiplication. So in particular $\mathbb{F}_q$ has *characteristic* p, that is, we work 'modulo p'.

(iii) For each prime p and positive integer m, there is a unique (up to isomorphism) field with $p^m = q$ elements.

(iv) The multiplicative group of the field $\mathbb{F}_q$ is cyclic; see Problem 7.5(iii).

(v)   In all fields, and so in particular in $\mathbb{F}_q$, a one-variable polynomial equation of degree n cannot have more than n roots, see Theorem B.13 in Appendix B.

For example, consider the field $\mathbb{F}_4$. Its elements can be written in the form 0, 1, c and $c + 1$, where $c^2 = c + 1$ and $c^3 = 1$. We have $c^{-1} = c + 1$, $(c + 1)^{-1} = c$, and c can be taken as the generator of the multiplicative group of the field; see (iv) above. For further details, see Problem 12.1 and a standard text on modern algebra, for example, Herstein (1964).

2-Dimensional Linear Groups

In this subsection and the next, we discuss some simple matrix groups. The first class is defined by

Definition 12.5 Given the underlying field $\mathbb{F}_q$ and the dimension $n \geq 1$, the *linear group* (or sometimes the *projective special linear group*) $L_n(q)$ is formed by factoring the special linear group $SL_n(q)$ (page 53) by its centre.

Some authors write $PSL_n(q)$ for $L_n(q)$. We shall show that these groups are simple, except when $n = 2$, and $q = 2$ or 3. The proofs follow similar lines to those used to show the simplicity of A_n for $n > 4$, and given in Chapter 3. In particular, the role played by the 3-cycles in those proofs is played here by a set of matrices called *transvections*. These are given by

Definition 12.6 The $n \times n$ *transvection* $E_{i,j}(r)$ is the matrix formed from the $n \times n$ identity matrix I_n by replacing 0 with r at the (i, j)th place where $r \neq 0$ and $i \neq j$.

Notes (a) The identity matrix is *not* a transvection, that is, the inequalities in the above definition are essential.

(b) The determinant of a transvection equals 1.

(c) The inverse of a transvection is another transvection.

The reader should check these facts.

We begin by determining the centre of $SL_n(F)$ for an arbitrary field F, this will lead to a formula for the order of $L_n(q)$.

Theorem 12.7 (i) *The centre Z of $SL_n(F)$ is the set of matrices aI_n where I_n is the $n \times n$ identity matrix, $a \in F$, and $a^n = 1$.*
(ii) *The order of $L_n(q)$ equals $q^{n(n-1)/2}(q^n - 1) \cdots (q^2 - 1)/d$ where $d = (n, q-1)$, the GCD of n and $q - 1$.*

Proof (i) Note first that each matrix of the form aI_n, where $a^n = 1$, belongs to Z.

For the converse, let $A = (a_{ij}) \in SL_n(F)$, and consider $E_{1,2}(r)$. As $r \neq 0$ we have

$$E_{1,2}(r)A = \begin{pmatrix} a_{11} + ra_{21} & a_{12} + ra_{22} & \cdots & a_{1n} + ra_{2n} \\ a_{21} & a_{22} & \cdots & a_{2n} \\ \vdots & \vdots & \ddots & \vdots \\ a_{n1} & a_{n2} & \cdots & a_{nn} \end{pmatrix},$$

and

$$AE_{1,2}(r) = \begin{pmatrix} a_{11} & ra_{11} + a_{12} & a_{13} & \cdots & a_{1n} \\ a_{21} & ra_{21} + a_{22} & a_{23} & \cdots & a_{2n} \\ \vdots & \vdots & \vdots & \ddots & \vdots \\ a_{n1} & ra_{n1} + a_{n2} & a_{n3} & \cdots & a_{nn} \end{pmatrix}.$$

If these two matrices are equal then, as $r \neq 0$, $a_{21} = a_{23} = \cdots = a_{2n} = 0$, $a_{21} = a_{31} = \cdots = a_{n1} = 0$, and $a_{12} + ra_{22} = ra_{11} + a_{12}$, which gives $a_{11} = a_{22}$. Applying exactly similar arguments using $E_{1,j}(r)$ for $2 \leq j \leq n$, it follows that if A commutes with all transvections, then all off-diagonal entries in A are zero, and each main diagonal entry equals $a_{11} = a$, say. This proves (i).
    (ii) Use (i), Theorems 3.15 and 4.11, and Problem 7.5(iii) as $d = (n, q-1)$ is the number of solutions of the equation $a^n = 1$ in $\mathbb{F}_q$. $\square$

The matrices aI_n in the proof above are called *scalar matrices*. For $n = 2$, the equation given in the first part of the theorem above, that is, $a^2 = 1$, has solutions 1 and -1 and none other if the characteristic of F is larger than 2, and the single solution $a = 1$ if the characteristic is 2, as $1 = -1$ in a field of this type. For $n = 3$, the equation $a^3 = 1$ has the single solution $a = 1$ if the field characteristic is 3 or is congruent to 2 modulo 3, and it has 3 solutions if the characteristic is congruent to 1 modulo 3; see Problem 12.1.
    We prove the main simplicity result for $L_n(q)$ in two stages: $n = 2$, and $n > 2$. The proofs given below have a similar structure, the first uses a minimum amount of linear algebra whilst the second relies on the theory of echelon and rational canon-

ical forms.[2] For the first stage, there are two preliminary lemmas which determine the crucial role played by the transvections. The first is

Lemma 12.8 *The group $SL_2(q)$ is generated by its transvections.*

This group also has a 2-generator matrix representation, see Problem 12.5.

Proof Let $A = \begin{pmatrix} a & b \\ c & d \end{pmatrix} \in SL_2(q)$ and $c \neq 0$. We have for $r, s, t \in \mathbb{F}_q \setminus \{0\}$,

$$\begin{pmatrix} 1 & r \\ 0 & 1 \end{pmatrix} \begin{pmatrix} a & b \\ c & d \end{pmatrix} \begin{pmatrix} 1 & s \\ 0 & 1 \end{pmatrix} \begin{pmatrix} 1 & 0 \\ t & 1 \end{pmatrix}$$

$$= \begin{pmatrix} a + cr + (as+b)t + (cs+d)rt & as + b + (cs+d)r \\ c + (cs+d)t & cs + d \end{pmatrix}.$$

As $c \neq 0$ we can choose s to satisfy $cs + d = 1$, then this matrix equals

$$\begin{pmatrix} a + (as+b)t + (c+t)r & as + b + r \\ c + t & 1 \end{pmatrix}.$$

If we choose r and t to satisfy $as + b + r = 0$ and $c + t = 0$, respectively, then this matrix equals the 2×2 identity matrix as the determinant is 1. Therefore, with these choices of r, s and t, and as $\begin{pmatrix} 1 & x \\ 0 & 1 \end{pmatrix}^{-1} = \begin{pmatrix} 1 & -x \\ 0 & 1 \end{pmatrix}$, we have

$$A = \begin{pmatrix} 1 & -r \\ 0 & 1 \end{pmatrix} \begin{pmatrix} 1 & 0 \\ -t & 1 \end{pmatrix} \begin{pmatrix} 1 & -s \\ 0 & 1 \end{pmatrix},$$

a product of transvections, s and t are non-zero as c is non-zero. The entry r may be zero in which case A is a product of two transvections. Secondly, if $c = 0$ then $d \neq 0$ as the matrix A is non-singular, and we can argue as follows. Using the argument above, we have $\begin{pmatrix} 0 & 1 \\ -1 & 0 \end{pmatrix} = D$ is a product of transvections (the lower left-hand entry is non-zero), then post-multiplying the given matrix by D will provide a matrix whose lower left-hand entry is non-zero and we can return to the first case.   □

The second preliminary lemma shows the connection between transvections and normal subgroups; note the similarity to Lemma 3.13.

Lemma 12.9 *Suppose $q > 3$. If $K \lhd SL_2(q)$ and K contains a transvection, then $K = SL_2(q)$.*

---

[2]An extension of these results known as 'Iwasawa's Lemma' (see, for example, Cameron's notes at www.maths.qmul.ac.uk/~pjc/class_gps/ or Web Section 5.4) can be used to establish the simplicity of a wide range of matrix groups including $L_n(q)$.

Proof By Lemma 12.8, it suffices to show that K contains all transvections. As K is normal, conjugates of elements of K belong to K. Using this fact, suppose $\left(\begin{smallmatrix} 1 & r \\ 0 & 1 \end{smallmatrix}\right)$ is the given transvection, and $\left(\begin{smallmatrix} a & b \\ c & d \end{smallmatrix}\right) \in SL_2(q)$, then

$$\begin{pmatrix} d & -b \\ -c & a \end{pmatrix} \begin{pmatrix} 1 & r \\ 0 & 1 \end{pmatrix} \begin{pmatrix} a & b \\ c & d \end{pmatrix} = \begin{pmatrix} 1+cdr & d^2r \\ -c^2r & 1-cdr \end{pmatrix} \in K.$$

Hence by setting $c = 0$, and then $d = 0$, we see that

$$\begin{pmatrix} 1 & t^2r \\ 0 & 1 \end{pmatrix}, \begin{pmatrix} 1 & 0 \\ -t^2r & 1 \end{pmatrix} \in K,$$

for all $t \in \mathbb{F}_q$. This gives, as K is also closed under inverses and products,

$$\begin{pmatrix} 1 & (t^2 - u^2)r \\ 0 & 1 \end{pmatrix}, \begin{pmatrix} 1 & 0 \\ (u^2 - t^2)r & 1 \end{pmatrix} \in K,$$

for all $t, u \in \mathbb{F}_q$. There are now two cases to consider: The characteristic of the underlying field is (a) not equal to 2, or (b) exactly 2.

Case (a): Characteristic larger than 2.

We have $a = ((a+1)/2)^2 - ((a-1)/2)^2$ for all a in $\mathbb{F}_q$. This provides suitable values for t and u above, and so all transvections belong to H.

Case (b): Characteristic equals 2.

Suppose $A = \left(\begin{smallmatrix} 1 & r \\ 0 & 1 \end{smallmatrix}\right) \in K$, then as above $\left(\begin{smallmatrix} 1 & 0 \\ -r & 1 \end{smallmatrix}\right)$, $\left(\begin{smallmatrix} 1 & ra^2 \\ 0 & 1 \end{smallmatrix}\right)$ and $\left(\begin{smallmatrix} 1 & 0 \\ -ra^2 & 1 \end{smallmatrix}\right)$ also belong to K for all $a \in \mathbb{F}_q$. Now for $a, c \in \mathbb{F}_q$, we have

$$\begin{pmatrix} 1 & 0 \\ -r & 1 \end{pmatrix} \begin{pmatrix} 1 & rc^2 \\ 0 & 1 \end{pmatrix} \begin{pmatrix} 1 & 0 \\ -ra^2 & 1 \end{pmatrix} = \begin{pmatrix} 1-r^2a^2c^2 & rc^2 \\ -r(1+a^2-r^2a^2c^2) & 1-r^2c^2 \end{pmatrix}$$

is in K. If we choose r to satisfy $a^{-1} = 1 + rc$ (note the map $r \mapsto 1 + rc$ is a bijection), then $1 + a^2 = r^2a^2c^2$ (remember the field characteristic is 2) and the lower left-hand entry in the matrix above is 0. Hence with this value of c, if we construct the commutator of this matrix with $\left(\begin{smallmatrix} 1 & b \\ 0 & 1 \end{smallmatrix}\right)$ we obtain (the square brackets denote the commutator)

$$\left[\begin{pmatrix} 1-r^2a^2c^2 & rc^2 \\ 0 & 1-r^2c^2 \end{pmatrix}, \begin{pmatrix} 1 & b \\ 0 & 1 \end{pmatrix} \right] = \begin{pmatrix} 1 & d \\ 0 & 1 \end{pmatrix} \in K,$$

this is valid as K is normal. Also

$$d = (1 - r^2c^2)r^2c^2(1 - a^2)b = (a^{-4} - 1)b$$

using the value of r chosen above; reader, check. Now as the order of the field is a power of 2, we can choose a to satisfy $a^{-4} \neq 1$ as $q > 3$. Hence, as the above calculation is valid for all $b \in \mathbb{F}_q$, d can take all values in $\mathbb{F}_q$, that is K contains all transvections.   □

To be precise, the condition $q > 3$ in the lemma above is not necessary because none of the normal subgroups of either $SL_2(2)$ or $SL_2(3)$ contain a transvection; the reader should check this using Section 8.2. We can now derive the main result in the 2-dimensional case.

Theorem 12.10 *If $q > 3$, the group $L_2(q)$ is simple.*

Proof We prove this result by showing that if $K \lhd SL_2(q)$ and K properly contains the centre Z, then K contains a transvection. Lemma 12.9 then shows that $K = SL_2(q)$, that is, $SL_2(q)$ contains no proper normal subgroup larger than Z if $q > 3$. The simplicity of $L_2(q)$, for $q > 3$, follows by the Correspondence Theorem (Theorem 4.16). Hence it remains to show that the supposed normal subgroup K contains a transvection if $q > 3$.

Case 1. Suppose first $B = \begin{pmatrix} a & b \\ 0 & d \end{pmatrix} \in K$ and $B \notin Z$.

If $a = 1$, then $d = 1$ as $\det B = 1$, and B is a transvection. If $a = -1$ then $d = -1$, and as $\begin{pmatrix} -1 & 0 \\ 0 & -1 \end{pmatrix} \in Z \leq K$, we have

$$\begin{pmatrix} a & b \\ 0 & d \end{pmatrix} \begin{pmatrix} -1 & 0 \\ 0 & -1 \end{pmatrix} = \begin{pmatrix} 1 & -b \\ 0 & 1 \end{pmatrix} \in K,$$

and again K contains a transvection. In both subcases, $b \neq 0$ because $B \notin Z$. This case does not apply if the field characteristic is 2. Lastly, suppose $a \neq \pm 1$, and so $d \neq \pm 1$. We have, as $K \lhd SL_2(q)$,

$$\begin{pmatrix} 0 & 1 \\ -1 & 0 \end{pmatrix}^{-1} \begin{pmatrix} a & b \\ 0 & d \end{pmatrix}^{-1} \begin{pmatrix} 0 & 1 \\ -1 & 0 \end{pmatrix} = \begin{pmatrix} a & 0 \\ b & d \end{pmatrix} \in K,$$

and

$$\begin{pmatrix} 1 & -1 \\ 1 & 0 \end{pmatrix}^{-1} \begin{pmatrix} a & b \\ 0 & d \end{pmatrix} \begin{pmatrix} 1 & -1 \\ 1 & 0 \end{pmatrix} = \begin{pmatrix} d & 0 \\ d-a-b & a \end{pmatrix} \in K,$$

hence

$$\begin{pmatrix} a & 0 \\ b & d \end{pmatrix} \begin{pmatrix} d & 0 \\ d-a-b & a \end{pmatrix} = \begin{pmatrix} ad & 0 \\ d^2-ad & ad \end{pmatrix} = \begin{pmatrix} 1 & 0 \\ d^2-1 & 1 \end{pmatrix} \in K,$$

because $ad = \det B = 1$. Therefore, as $d \neq \pm 1$, K contains a transvection. Note that this last case does not require b to be non-zero.

Case 2. Secondly, suppose $B = \begin{pmatrix} a & b \\ c & d \end{pmatrix} \in K$ and $c \neq 0$.

We reduce this to Case 1. As K is a normal subgroup we have

$$\begin{pmatrix} 1 & -t \\ 0 & 1 \end{pmatrix}\begin{pmatrix} a & b \\ c & d \end{pmatrix}\begin{pmatrix} 1 & t \\ 0 & 1 \end{pmatrix} = \begin{pmatrix} a - ct & (a - ct)t + b - dt \\ c & ct + d \end{pmatrix} \in K,$$

for all $t \in \mathbb{F}_q$. So, if t satisfies $a = ct$ (note $c \neq 0$), we obtain

$$\begin{pmatrix} 0 & b - dt \\ c & a + d \end{pmatrix} \in K, \tag{12.3}$$

and $c(dt - b) = 1$ as this matrix has determinant 1. Further, if $r, s \in \mathbb{F}_q$ and $rs = 1$, then

$$\begin{pmatrix} r & 0 \\ 0 & s \end{pmatrix}^{-1}\begin{pmatrix} 0 & b - dt \\ c & a + d \end{pmatrix}^{-1}\begin{pmatrix} r & 0 \\ 0 & s \end{pmatrix}\begin{pmatrix} 0 & b - dt \\ c & a + d \end{pmatrix}$$

$$= \begin{pmatrix} s^2 & (a + d)(b - dt)(1 - s^2) \\ 0 & r^2 \end{pmatrix}, \tag{12.4}$$

and this matrix belongs to K. Therefore, we can return to Case 1 provided $r, s \in \mathbb{F}_q \backslash 0$ satisfy $r^2 \neq \pm 1 \neq s^2$, and this is possible for all q except $q = 5$. We can also return to Case 1 if $q = 5$ and

$$(a + d)(b - dt)(1 - s^2) \neq 0.$$

We have $b - dt \neq 0$ by the note above, and we can choose s to satisfy $s^2 \neq 1$. Hence finally we need to consider the case $q = 5$ and $a + d = 0$. By (12.3), $\begin{pmatrix} 0 & u \\ c & 0 \end{pmatrix} \in K$ where $u = b - dt \neq 0$ (as $-uc = 1$), and

$$\begin{pmatrix} 1 & v \\ 0 & 1 \end{pmatrix}\begin{pmatrix} 0 & u \\ c & 0 \end{pmatrix}\begin{pmatrix} 1 & -v \\ 0 & 1 \end{pmatrix}\begin{pmatrix} 0 & -u \\ -c & 0 \end{pmatrix} = \begin{pmatrix} c^2v^2 + 1 & v \\ c^2v & 1 \end{pmatrix} \in K,$$

for all $v \in \mathbb{F}_5$. Choose v to satisfy $cv = 1$, then this matrix equals $\begin{pmatrix} 2 & v \\ c & 1 \end{pmatrix}$ with $c \neq 0 \neq v$. Hence we can return to the first subcase of Case 2. We have constructed a transvection in all cases, so the theorem follows. $\qquad\square$

The cases $q = 2$ and 3 are genuine exceptions for $L_2(2) \simeq GL_2(2) \simeq S_3$ (Problem 4.2) and $L_2(3) \simeq A_4$ (Problem 4.4(iv)). In both cases, the groups contain proper non-neutral normal subgroups. We also have

$$L_2(4) \simeq L_2(5) \simeq A_5, \quad L_2(7) \simeq L_3(2) \quad \text{and} \quad L_2(9) \simeq A_6.$$

The first of these statements follows from Problem 6.16(ii), the second from Problem 12.7*, and for the third see Problem 12.4. At the end of the last section, we showed that $GL_3(2)$ is isomorphic to the automorphism group of the Steiner system $S(2, 3, 7)$, thus providing another representation of this group. Further, some presentations have been constructed for these groups. We have

$$L_2(p) \simeq \langle a, b \mid a^p = \left(a^4 ba^{(p+1)/2}b\right)^2 = e, \ (ab)^3 = b^2 \rangle, \qquad (12.5)$$

provided p is prime and greater than 3; see Problem 12.6. It can also be shown that $o(b) = 2$ in this presentation. Another presentation is given in Conway and Sloane (1993), page 268; in some ways, it is more natural being related to work presented in Problem 12.8, but it only applies in the case $p \equiv 1 \pmod 4$.

Permutation representations for many of these groups have been constructed. Each of the groups $L_2(q)$ is isomorphic to a subgroup of S_{q+1} (and in some cases S_u for suitably chosen $u \leq q$)—a result due to Galois; see Problem 12.8 and Appendix E. We consider some subgroups of one of these groups in Problem 12.9.

Groups $L_n(q)$ with $n > 2$

The general case for $L_n(q)$ where $n > 2$ will be treated now. In the main proof below, we make use of some standard results from linear algebra, in particular the reduced echelon and rational canonical forms of a matrix; see, for example, Halmos (1974a) or Rose (2002). The structure of the proof is similar to that for the case $n = 2$, and so we begin with

Lemma 12.11 *The group $SL_n(q)$ is generated by its transvections.*

Proof Use elementary row operations, the details are left as an exercise for the reader; see Problem 12.4. $\qquad \square$

Next we consider conjugation.

Lemma 12.12 *Two transvections belonging to $SL_n(q)$ are conjugate in $SL_n(q)$, provided $n > 2$.*

Proof Case 1: The transvections are $E_{i,j}(r)$ and $E_{i,j}(s)$ for $r \neq s$, that is, the 'new entries' r and s are in the same position in each transvection.

Let $t = r^{-1}s$ and let D_1 be the $n \times n$ diagonal matrix with t at the (j, j)th place, t^{-1} at the (k, k)th place, and 1 at all other diagonal places where k is least such that $i \neq k \neq j$. Clearly, $D_1 \in SL_n(q)$ and

$$D_1^{-1} E_{i,j}(r) D_1 = E_{i,j}(s).$$

Case 2: The transvections are $E_{i,k}(r)$ and $E_{j,k}(r)$ where $i \neq j$, that is, the 'new entries' have the same value and horizontal position but are in different vertical positions.

Let D_2 be I_n with the following four changes: (i, i)th place has entry 0, (i, j)th place has entry 1, (j, i)th place has entry -1, and (j, j)th place has entry 0. Clearly, $D_2 \in SL_n(q)$ and

$$D_2^{-1} E_{i,k}(r) D_2 = E_{j,k}(r).$$

The remaining cases can be dealt with similarly, we leave them as exercises for the reader.   □

Every non-singular square matrix defined over a field F is similar[3] to a matrix in *rational canonical form* C as follows: C is a diagonal block matrix and each block has the form

$$
\begin{pmatrix}
0 & 1 & 0 & \cdots & 0 & 0 \\
0 & 0 & 1 & \cdots & 0 & 0 \\
\vdots & \vdots & \vdots & \ddots & \vdots & \vdots \\
0 & 0 & 0 & \cdots & 0 & 1 \\
c_1 & c_2 & c_3 & \cdots & c_{t-1} & c_t
\end{pmatrix},
$$

where $c_i \in F$ and $c_1 \neq 0$. The blocks are called *companion matrices*. If $t = 1$ then C is the 1×1 matrix (c_1).

We can now prove our main result in the n-dimensional case; Theorem 12.10 treated the case $n = 2$.

Theorem 12.13 *If $n > 2$, then $L_n(q)$ is a simple group.*

Proof By Lemmas 12.9, 12.11 and 12.12, we need to show that if $K \triangleleft SL_n(q)$ and $K \not\leq Z(SL_n(q)) = Z$, then K contains a transvection.

Let $A \in K$. As noted above, we may assume that A is in rational canonical form (as K is normal); there are three cases to consider:

Case 1. A is diagonal (all companion matrices are 1×1), and the diagonal entries are non-zero and not all equal (for otherwise $A \in Z$).
Case 2. All companion matrices of A are 2×2 or 1×1, and at least one of the first type is present.
Case 3. At least one companion matrix of A is $r \times r$ where $r > 2$.

We shall construct a transvection in each case.
    Case 1. Suppose the diagonal of A has the form

$$(a_1, \ldots, a_1, a_2, \ldots, a_2, \ldots),$$

---

[3]The term 'similar' means 'conjugate in $GL_n(q)$', but using Lemmas 12.11 and 12.12 it can easily be shown that this is equivalent to 'conjugate in $SL_n(q)$' in the cases under discussion.

with r_1 copies of a_1, r_2 copies of $a_2, \ldots$, and $a_i \neq a_j$ if $i \neq j$. We have

$$\left[A, E_{1,r_1+r_2}(1)\right] = E_{1,r_1+r_2}(1 - a_2/a_1) \in K,$$

that is K contains a transvection.

Case 2. A has the block form $\left(\begin{smallmatrix} B & O \\ O & C \end{smallmatrix}\right)$ where B is a 2×2 companion matrix. We may assume that B has the form $\left(\begin{smallmatrix} 0 & 1 \\ -1 & c \end{smallmatrix}\right)$ and $C = I_{n-2}$, the $n - 2 \times n - 2$ identity matrix. To see this, we argue as follows. Firstly, by pre-multiplying by a suitable diagonal matrix we may assume that $\det B = \det C = 1$, and so $B = \left(\begin{smallmatrix} 0 & 1 \\ -1 & a \end{smallmatrix}\right)$ for some $a \in \mathbb{F}_q$. Secondly, by constructing the commutator of A with a matrix of the form $\left(\begin{smallmatrix} D & O \\ O & I_{n-2} \end{smallmatrix}\right)$, we may assume that $C = I_{n-2}$. We let $D = \left(\begin{smallmatrix} 1 & -1/a \\ 0 & 1 \end{smallmatrix}\right)$, this replaces B by the matrix $\left(\begin{smallmatrix} 0 & a \\ -1/a & d \end{smallmatrix}\right)$ where $d = 2 + 1/a^2$. Finally, pre-multiplying this by the diagonal block matrix, with $\left(\begin{smallmatrix} 1/a & 0 \\ 0 & a \end{smallmatrix}\right)$ in the top left-hand corner and I_{n-2} in the bottom right, gives A in the required form with $c = 2a + 1/a$. Now with A in this form we have

$$\left[A, E_{1,3}(-1)\right] = E_{1,3}(c - 1) + E_{2,3}(1) - I_n,$$

and the required transvection is given by

$$\left[E_{1,3}(c - 1) + E_{2,3}(1) - I_n, E_{1,2}(1)\right] = E_{1,3}(-1) \in K.$$

Case 3. Suppose the first companion matrix is $r \times r$ where $r > 2$, and its $(r, 1)$th entry is a_1 where $a_1 \neq 0$ (by definition of companion matrix). As above we have

$$\left[A, E_{r,1}(-1)\right] = E_{1,2}(1/a_1) + E_{r,1}(-1) - I_n \in K,$$

and

$$\left[E_{1,2}(1/a_1) + E_{r,1}(-1) - I_n, E_{r,1}(-1)\right] = E_{r,2}(1/a_1) \in K,$$

another transvection in K. This proves the result. $\square$

For $n > 2$, only two of these groups occur in our chosen range: $L_3(2)$ with order 168 and $L_3(3)$ with order $5616 = 2^4 \cdot 3^3 \cdot 13$. The first is isomorphic to $L_2(7)$, see Problem 12.7* (both are simple and have order 168). The second has a number of distinct representations which we discuss now. First, using a suitable computer package it is a straightforward matter to check that

$$L_3(3) \quad \text{is generated by} \quad \begin{pmatrix} 0 & 2 & 0 \\ 1 & 1 & 0 \\ 0 & 0 & 1 \end{pmatrix} \quad \text{and} \quad \begin{pmatrix} 0 & 1 & 0 \\ 0 & 0 & 1 \\ 1 & 0 & 0 \end{pmatrix}, \qquad (12.6)$$

over $\mathbb{F}_3$ where the first matrix has order 6 and the second has order 3; there are, of course, many other choices. Further, this group has the presentation

$$L_3(3) \simeq \langle a, b \mid a^6 = b^3 = (ab)^4 = \left(a^2 b\right)^4 = \left(a^3 b\right)^3 = \left[a^2, \left(ba^2 b\right)^2\right] = e \rangle. \quad (12.7)$$

If we map a in this presentation to the first matrix in (12.6) and b to the second, then the equations in the presentation follow, and so it is not difficult to show that (12.7) is a correct presentation for $L_3(3)$. Thirdly, we have

$$L_3(3) \simeq \mathrm{Aut}\bigl(\mathrm{S}(2, 4, 13)\bigr).$$

A Steiner system of the form $\mathrm{S}(2, 4, 13)$ can be constructed as follows. Noting that $\{0, 1, 3, 9\}$ is the set of fourth powers modulo 13, apply the map $x \mapsto x + 1 \pmod{13}$ to this set twelve times to obtain the collection of thirteen 4-tuples:

$$\{\{0, 1, 3, 9\}, \{1, 2, 4, 10\}, \{2, 3, 5, 11\}, \{3, 4, 6, 12\},$$

$$\{0, 4, 5, 7\}, \{1, 5, 6, 8\}, \{2, 6, 7, 9\}, \{3, 7, 8, 10\}, \{4, 8, 9, 11\}, \{5, 9, 10, 12\},$$

$$\{0, 6, 10, 11\}, \{1, 7, 11, 12\}, \{0, 2, 8, 12\}\}.$$

It is easily checked that every pair of integers, each of which lie between 0 and 12, occur in exactly one of the 4-tuples above. This can be done by showing that the set of differences of elements of the set $\{0, 1, 3, 9\}$ modulo 13 gives all twelve non-zero integers modulo 13. Hence we have the Steiner system $\mathrm{S}(2, 4, 13)$; note that by Theorem 12.4 this system has 13 members. We can associate the plane projective geometry defined over the 3-element field with this system, the geometry has 13 points and 13 lines, and each line contains 4 points, see the discussion given on page 253. Hence the set of linear maps for this geometry is precisely $L_3(3)$. This Steiner association also shows that the group can be treated as a subgroup of S_{13} (or to be more precise a (maximal) subgroup of A_{13}), and, for example, $L_3(3)$ is isomorphic to the group generated by the permutations:

$$(1, 4, 6)(2, 3, 7, 10, 11, 8)(9, 13), \qquad (1, 2, 3)(4, 5, 6)(7, 8, 9)(10, 11, 12).$$
$$(12.8)$$

This pair of generators was constructed by the computer package GAP; in the presentation (12.7), if we map the first permutation in (12.8) to a and the second to b, then the relations in (12.7) hold. The maximal subgroups of this group are isomorphic to either $S_4 C_2 \rtimes (C_3 \times C_3)$ (26 copies in two conjugacy classes), $C_3 \rtimes C_{13}$ (144 copies in one conjugacy class), or S_4 (234 copies also in one conjugacy class). The first of these subgroups is called the *Hessian group*, it has connections with a particular cubic curve; see Coxeter and Moser (1984), page 98. Two further points. First, Brauer and Wong (see Huppert and Blackburn 1982b, page 343) have shown that if G is a finite non-Abelian simple group, a is an involution in G, and $C_G(a) \simeq GL_2(3)$, then G is isomorphic to $L_3(3)$ or M_{11} (Section 12.4); note the comments on page 104 and Problem 12.10. Second, Thompson (1968) has shown that, apart from some 2-dimensional linear groups and some Suzuki groups (see Web Section 12.6), $L_3(3)$ is the only non-Abelian simple group which is *minimal*, that is, all of its proper subgroups are soluble.

12.3 Unitary Groups

A number of simple groups, sometimes known as the *classical groups*, can be defined over the finite fields $\mathbb{F}_{q^2}$ using constructions from linear algebra similar to those given in the previous section. They involving various 'inner' or 'Hermitian' products, or more general bilinear forms. We work over $\mathbb{F}_{q^2}$ (where as usual q is a prime power) so that we can use an analogue of complex conjugation over $\mathbb{C}$. Note that if $x \in \mathbb{F}_{q^2}$, then $x^{q^2} = x$ (Problem 7.5), and so every element of $\mathbb{F}_{q^2}$ can be treated as a q^2th power. Hence it is easily checked that the map

$$\xi : a \to a^q \quad \text{for } a \in \mathbb{F}_{q^2} \backslash \{0\}$$

has the usual conjugation properties: It is an automorphism of $\mathbb{F}_{q^2}$, and ξ^2 is the identity map on $\mathbb{F}_{q^2}$. Because of this we usually write $\bar{a}$ for $a\xi (= a^q)$, also we let $\bar{0} = 0$.

Suppose V is a vector space of finite dimension n defined over $\mathbb{F}_{q^2}$. If $u = (a_1, \ldots, a_n) \in V$ where $a_i \in \mathbb{F}_{q^2}$, for $i = 1, \ldots, n$, the expression

$$f(u) = a_1 \bar{a_1} + \cdots + a_n \bar{a_n} \tag{12.9}$$

is called a non-singular *Hermitian form* over V. An $n \times n$ matrix A is called *unitary* if $A \in GL_n(q^2)$, for some $q (= p^m)$, and

$$f(uA) = f(u) \quad \text{for all } u \in V. \tag{12.10}$$

Note that forms are non-singular. A more general definition for f can be given, but a basis always exists for the underlying vector space V so that f is in the form (12.9).

The collection of all $n \times n$ unitary matrices over $\mathbb{F}_{q^2}$ is denoted by $GU_n(q)$, and is called the n-dimensional *general unitary group* over $\mathbb{F}_{q^2}$. It can be shown that the determinant of a general unitary matrix is a $(q + 1)$th root of unity, and the subset of $GU_n(q)$ of those matrices A with $\det A = 1$ forms a normal subgroup of order $o(GU_n(q))/(q + 1)$ which is called the n-dimensional *special unitary group* over $\mathbb{F}_{q^2}$; it is denoted by $SU_n(q)$. As in the linear case described in Section 12.2, we factor out the scalar matrices (that is, the centre) to obtain simple groups. $SU_n(q)$ has $d = (q+1, n)$ scalar matrices. Hence finally we define $U_n(q)$ to be the group formed from $SU_n(q)$ by factoring out its scalar matrices, it is called the n-dimensional *projective special unitary group*, or more usually the n-dimensional *unitary group*, over $\mathbb{F}_{q^2}$. These groups are simple except in the following cases: (a) $n = q = 2$, see example below; (b) $n = 2$ and $q = 3$, in this case $U_2(3) \simeq A_4$; and (c) $n = 3$ and $q = 2$, in this case $U_3(2)$ is isomorphic to the semidirect product $Q_2 \rtimes (C_3 \times C_3)$.

Example We consider the case $q = n = 2$. Let the field $\mathbb{F}_4$ have elements $\{0, 1, c, c + 1\}$ where $c^3 = 1$, $c^2 = c + 1$, and we work modulo 2. By direct calculation we see that $GU_2(2)$ is generated by the matrices

$$\begin{pmatrix} 1 & 0 \\ 0 & c \end{pmatrix}, \quad \begin{pmatrix} c & 0 \\ 0 & c^2 \end{pmatrix}, \quad \text{and} \quad \begin{pmatrix} 0 & c \\ c^2 & 0 \end{pmatrix},$$

and it has order 18, it is isomorphic to $C_3 \times S_3$. For instance, if $A = \begin{pmatrix} 1 & 0 \\ 0 & c \end{pmatrix}$ and $u = (a_1, a_2)$ (and so $uA = (a_1, ca_2)$), then $f(uA) = a_1\overline{a_1} + ca_2\overline{ca_2} = f(u)$ as $c\overline{c} = c^3 = 1$, therefore (12.10) applies. Further, the second and third matrices above generate $SU_2(2)$ which has order 6, and so is isomorphic to S_3 as it is not cyclic. Now we note that $d = (3, 2) = 1$ in this case, and so $U_2(2) \simeq SU_2(2) \simeq S_3$.

The 2-dimensional unitary groups do not give 'new' simple groups for we have

$$U_2(q) \simeq L_2(q),$$

although this does provide a further representation of the simple group $L_2(q)$. It is not difficult to show that

$$o\bigl(U_3(q)\bigr) = q^3\bigl(q^3 + 1\bigr)\bigl(q^2 - 1\bigr)/(q + 1, 3). \qquad (12.11)$$

The only 3-dimensional unitary groups in our chosen range are $U_3(2)$ and $U_3(3)$. The first of these was discussed above (it is not simple), and so we need to consider the second which is isomorphic to $SU_3(3)$, and has order $6048 = 2^5 \cdot 3^3 \cdot 7$, see (12.11). In this case, we work over the field $\mathbb{F}_9$ generated by 1 and c where $c^2 = c+1$ modulo 3, see Problem 12.1. First, it can be shown that the group $U_3(3)$ is generated by the matrices

$$\begin{pmatrix} c+2 & 1 & 1 \\ 1 & 2 & 0 \\ 1 & 0 & 0 \end{pmatrix} \quad \text{and} \quad \begin{pmatrix} 2c+1 & 2c+1 & 1 \\ c & 2 & 0 \\ 1 & 0 & 0 \end{pmatrix},$$

where the first matrix has order 8 and the second has order 7. Many other pairs of 3×3 matrices generate the group, see Problem 12.12. The group can also be represented as a permutation group and as the automorphism group of the Steiner System $S(2, 4, 28)$; no short presentations have been given but one has been constructed for $C_2 \rtimes U_3(3)$ as a Coxeter group, viz.:

$$\bigl\langle a_1, \ldots, a_5 : a_i^2 = (a_i a_j)^3 = (a_3 a_4)^8 = (a_2 a_3 a_4 a_5)^8 = e, a_1 = (a_3 a_4)^4 \bigr\rangle$$

for $1 \le i < j \le 5$ except when $i = 3$ and $j = 4$ in the second equation. After a fair amount of calculation, it has been shown that the subgroup of A_{28} generated by the two permutations

$(1, 5, 7, 3, 12, 24, 11)(2, 23, 4, 27, 13, 14, 26)(6, 20, 18, 8, 25, 21, 28)(9, 10, 17, 15, 22, 16, 19),$

$(3, 4)(5, 17, 7, 16, 8, 20, 6, 13)(9, 19, 11, 14, 12, 18, 10, 15)(21, 23, 26, 28, 24, 22, 27, 25)$

is, in fact, isomorphic to $U_3(3)$; see Hall and Wales (1968). (The group $U_3(3)$ is isomorphic to a maximal subgroup (of largest size) of the Hall–Janko simple group J_2 with order 604800; see the quoted paper for details.)

Using this second representation, it is not difficult to show that the group contains 63 involutions in a single conjugacy class $\mathcal{C}$. In this permutation representation, each element in $\mathcal{C}$ is a product of twelve disjoint 2-cycles, and so it fixes four of the 28 elements in the set $X = \{1, 2, \ldots, 28\}$. Now by Theorem 12.4, the Steiner system

S(2, 4, 28) also has 63 members, and a version of this system can be constructed using C as follows. For each element $\sigma \in C$, we form a block with four elements of X fixed by σ. After some computations, we obtained the following version of S(2, 4, 28).

Table 12.1 Steiner System S(2, 4, 28)

1,2,3,4							
1,5,18,28	1,6,14,23	1,7,15,22	1,8,19,25	1,9,16,27	1,10,17,24	1,11,20,21	1,12,13,26
2,5,14,24	2,6,19,26	2,7,18,27	2,8,15,21	2,9,20,23	2,10,16,25	2,11,13,28	2,12,17,22
3,5,16,22	3,6,17,28	3,7,20,25	3,8,13,23	3,9,19,24	3,10,15,26	3,11,14,27	3,12,18,21
4,5,20,26	4,6,16,21	4,7,13,24	4,8,17,27	4,9,15,28	4,10,18,23	4,11,19,22	4,12,14,25
5,6,7,8							
5,9,21,25	5,10,13,19	5,11,15,17	5,12,23,27	6,9,13,18	6,10,22,27	6,11,24,25	6,12,15,20
7,9,14,17	7,10,21,28	7,11,23,26	7,12,16,19	8,9,22,26	8,10,14,20	8,11,16,18	8,12,24,28
9,10,11,12							
13,14,21,22	13,15,25,27	13,16,17,20	14,15,18,19	14,16,26,28	15,16,23,24		
17,18,25,26	17,19,21,23	18,20,22,24	19,20,27,28	21,24,26,27	22,23,25,28		

The reader should check that each pair of elements in X occurs in exactly one member of the above Steiner system. Incidentally, this shows that the group $U_3(3)$ is doubly transitive, see Web Section 5.4. This fact can be used to show that the Steiner system given above is a valid one, and that the automorphism group of S(2, 4, 28) is isomorphic to $U_3(3)$. Lastly, we note that the maximal subgroups of $U_3(3)$ are isomorphic to either $C_8 \rtimes ES_2(3)$ (Problem 6.5, 28 copies), $L_2(7)$ (36 copies), an extension of S_4 by C_4 (63 copies), or $S_3 \rtimes C_4^2$ (also 63 copies). Details on this and the related groups can be found in Huppert (1967), pages 233 to 252, where a simplicity proof is given; see also Problem 12.12 and the ATLAS (1985).

12.4 Mathieu Groups

The smallest 'sporadic' group is the first Mathieu group M_{11}, it has order 7920. (The others with order less than 10^6 are M_{12} and M_{22}, the second and third Mathieu groups with orders 95040 and 443520, respectively, and J_1 and $J_2 = HJ$, the first and second Janko groups with orders 175560 and 604800, respectively; see Web Section 12.6.) As with all descriptions in this chapter, we give informally a number of definitions of M_{11}, but only sketch proofs; a good general account is given in Huppert and Blackburn (1982b). In fact, one of the best ways to work with M_{11} is as the stabiliser of a point in M_{12}, but as we are limiting our discussion to groups of order less than 10000 we cannot take this approach; see Web Section 12.6

In the 1860s and 1870s, É.L. Mathieu (1835–1890) and others were studying k-transitive groups for $k > 2$. A permutation group G defined on a set $X = \{1, \ldots, n\}$ is *k-transitive* if given two k-element subsets $Y = \{y_1, \ldots, y_k\}$,

$Z = \{z_1, \ldots, z_k\}$ of X there is an element $\alpha \in G$ with the property $y_i \alpha = z_i$ for $i = 1, \ldots, k$. For all n, S_n, S_{n+1} and A_{n+2} are n-transitive, so one might expect that other groups with high transitivity would exist, but this is not so. Apart from the symmetric and alternating groups, there are only two that are 4-transitive—M_{11} and M_{23}; two that are 5-transitive—M_{12} and M_{24}; and none that are n-transitive if $n > 5$. (The only known proof of these facts requires CFSG, but in the future it should be possible to give a proof without using the classification.) Mathieu discovered M_{11} and M_{12} in 1861, and the remaining three in 1873, their simplicity was established in 1896 by Cole and completed by Miller in 1900. This work was not widely accepted until the 1930s when Witt provided simpler and clearer characterisations, the 'M' notation is also due to him.

Our first definition of M_{11} is as a subgroup of A_{11} as follows. Scott (1964), page 286, Robinson (1982), page 202, and Curtis (2007), page 267, give slight variants on this definition; see also Coxeter and Moser (1984), page 99.

Definition 12.14 Let

$$\alpha_1 = (1, 2, 3)(4, 5, 6)(7, 8, 9), \qquad \beta_2 = (1, 2, 4, 6)(3, 9, 5, 8),$$

$$\alpha_2 = (1, 4, 7)(2, 5, 8)(3, 6, 9), \qquad \gamma_1 = (2, 8)(3, 5)(6, 9)(7, 11),$$

$$\beta_1 = (1, 8, 4, 9)(2, 5, 6, 3), \qquad \gamma_2 = (2, 3)(5, 6)(8, 9)(10, 11),$$

then M_{11} is the subgroup of A_{11} generated by the permutations $\alpha_1, \ldots, \gamma_2$ whose cyclic structures are listed above; it is called the *First Mathieu Group*. (The second Mathieu group M_{12} can be defined in a similar way, but as a subgroup of A_{12} and with the extra generator $\gamma_3 = (1, 4)(3, 11)(6, 10)(9, 12)$.)

The permutations α_1 and α_2 generate an elementary Abelian group A of order 9. Also β_1 and β_2 generate a copy B of the quaternion group Q_2 of order 8. (Note that $\beta_1^2 = \beta_2^2 = (\beta_1 \beta_2)^2$ all with order 2.) Further, we have

$$\beta_1^{-1} \alpha_1 \beta_1 = \alpha_2^{-1}, \qquad \beta_2^{-1} \alpha_1 \beta_2 = \alpha_2 \alpha_1,$$

$$\beta_1^{-1} \alpha_2 \beta_1 = \alpha_1, \qquad \beta_2^{-1} \alpha_2 \beta_2 = \alpha_2 \alpha_1^{-1}$$

by direct calculation. Hence $C = AB$ is a group of order 72 with A as a normal subgroup.

It is clear that B is the subgroup of C of those permutations which fix 7, and B acts transitively on the set $T = \{1, 2, 3, 4, 5, 6, 8, 9\}$. Hence C is transitive on $T \cup \{7\}$, and so C is 2-transitive on a 9-element set. Now applying methods given in Web Section 5.4 or in Robinson (1982), page 203, it can be shown that if we add the permutation γ_1 to C we obtain a 3-transitive group D of order 720, and then if we add the permutation γ_2 to D we obtain a 4-transitive group of order $7920 = 11 \cdot 10 \cdot 9 \cdot 8$ isomorphic to M_{11}. It is isomorphic because it has also been shown that all (sharply—see Web Section 5.4) 4-transitive groups defined on 9-element sets are isomorphic to M_{11}. In fact, only two sharply 4-transitive simple groups exist (up to isomorphism), they are A_6 and M_{11}.

The Mathieu group M_{11} has some notable maximal subgroups. The subgroup D (with order 720) defined above is maximal, it is isomorphic to an extension of the cyclic group C_2 by A_6. It can be shown that M_{11} contains eleven copies of this group in a single conjugacy class. Twelve copies (again in a single conjugacy class) of the linear group $L_2(11)$ also occur as maximal subgroups of M_{11}. The remaining maximal subgroups are isomorphic to $C_2 \rtimes C$ (55 copies, see paragraph below Definition 12.14), S_5 (66 copies), and an extension of S_4 by C_2 (165 copies). We can construct one of the copies of $L_2(11)$ as follows. Let

$$\phi = (1, 9, 4, 6, 2, 10, 7, 3, 11, 8, 5), \qquad \psi = (1, 8, 4, 9)(2, 5, 6, 3),$$
$$\theta = (1, 8, 9, 4, 7)(2, 6, 5, 10, 3), \qquad \nu = (1, 2)(3, 5)(4, 6)(7, 10),$$

then ϕ and ψ generate M_{11} as given above (Problem 12.14), ϕ and θ generate a subgroup of order 55, and ϕ, θ and ν generate a maximal subgroup H of order 660 isomorphic to $L_2(11)$. There is nothing special about the permutations ϕ, ψ, θ and ν, other choices give the same result but note that $\theta = \psi^2 \phi^5 \psi^2 \phi^4 \psi^2 \phi^5$ and $\nu = (\theta\phi)^{-1}\psi^2(\theta\phi)$. Further, H is doubly transitive on $\{1, 2, \ldots, 11\}$. (This can be checked by hand or by using a suitable computer program.) The index of H in M_{11} is 12, and this fact can be used to provide a 3-transitive representation of M_{11} on a 12-element set, see Huppert and Blackburn (1982b).

On the other hand, M_{11} is isomorphic to the automorphism group of the Steiner system $S(4, 5, 11)$, and we can use this fact to show that it is 4-transitive on an 11-element set. In Web Section 12.6, we give several methods for constructing the system $S(4, 5, 11)$. This is done by first constructing a version of $S(5, 6, 12)$ (the automorphism group for this system being isomorphic to M_{12}), and then several versions of $S(4, 5, 11)$ can be 'read-off'.

The ATLAS (1985) gives at least *ten* methods for constructing M_{12}, and most of these can be specialised to provide definitions of M_{11}; the reader should consult this work for further details. The group M_{11} can also be characterised as the unique (up to isomorphism) simple group of order 7920; for further details, see Huppert and Blackburn (1982b), pages 341 to 365. As a final pair of representations of M_{11} we give two of its (many) known presentations:

$$M_{11} \simeq \langle a, b, c \mid a^{11} = b^5 = c^4 = (ac)^3 = e, b^4 ab = a^4, c^3 bc = b^2 \rangle$$
$$\simeq \langle a, b, c, d \mid a^2 = b^2 = c^2 = d^2 = (ab)^5 = (bc)^3 = (bd)^4$$
$$= (cd)^3 = (abdbd)^3 = e \rangle;$$

The first of these is related to the second permutation representation given above with $a \mapsto \phi$, $b \mapsto \theta$, and $c \mapsto \psi$. In Problem 2.28, we showed that a simple group can be generated by its involutions, the second presentation is an example of this fact—in this case only four involutions are needed. As a computer challenge you could try to find four involutions in A_{11} which satisfy the relations given above and so generate another copy of M_{11} in A_{11}, see Problem 12.14. Further M_{11} properties can be found in this problem and at Web Section 12.6.

12.5 Problems

Problem 12.1 The field $\mathbb{F}_9$ of order 9 is used in a number of examples in this chapter. Here you are asked to determine its basic properties. It can be defined as a vector space of dimension 2 (with basis elements 1 and c) over the three element field $\mathbb{F}_3 = \{0, 1, 2\}$, with a vector multiplication which is defined using linearity by

$$c^2 = c + 1,$$

working modulo 3. For $1 \le n \le 8$ and $x \in \mathbb{F}_9$ determine

$$c^n, \quad \overline{x} = x^3, \quad x + \overline{x} \quad \text{and} \quad x^{-1},$$

that is, write each of these entities as linear combinations of 1 and c.

Problem 12.2 (i) Construct Steiner systems $S(r, s, t)$ in the cases where the parameters r, s, t equal (a) $2, 3, 9$, (b) $2, 4, 13$, (c) $3, 4, 8$, and (d)* $4, 5, 11$. For (d) try to construct the system without using Web Section 12.6.

(ii) Find two non-isomorphic Steiner systems for the parameters $2, 3, 13$, note this system has 26 members. One method is to find 13 members as discussed on page 251, then find two further 13-member sets each of which include consecutive pairs.

(iii) (The Dining Club) A dining club has the following rules:

(a) The Chairman, Secretary and Treasurer attend the first dinner.
(b) More people are invited to join.
(c) Only Club members may attend dinners.
(d) At least three members attend dinners.
(e) Each member meets all other members at the dinners.
(f) No member shall meet another member more than once.
(g) For every pair of members d_1 and d_2, there is at least one dinner that both d_1 and d_2 attend.

Using these rules answer the following questions.

1. What is the size of the club?
2. How many dinners are held?
3. How many members attend each dinner?
4. How many dinners does each member attend?

For further details, see O'Hara and Ward (1937), page 17—*per J.F. Bowers.*

Problem ◆ 12.3 You are given a Steiner system $S(r, s, t)$ for the parameters r, s, t. Extend Theorem 12.2 as follows. Suppose $a_1, \ldots, a_u \in X$ where $o(X) = t$ and $u < r$. Show that the number of blocks in $S(r, s, t)$ containing $a_1, \ldots, a_u$ is

$$v = \frac{(t - u) \cdots (t - (r - 1))}{(s - u) \cdots (s - (r - 1))}.$$

Hence show that v is independent of the particular elements a_i. Note that as v must be an integer for all u with $1 < u < r$, this result places constraints on the parameters r, s, t for which the system $S(r, s, t)$ can exist.

Problem 12.4 (i) Give a proof of Lemma 12.11.

(ii) Show that $A_6 \simeq L_2(9)$. This can be done using a method similar to that given in Problem 12.7*, or it can be done using a suitable computer algebra package, one method is as follows. First, show that A_6 can be given by the presentation

$$\langle a, b \mid a^5 = b^5 = (ab)^2 = (a^4b)^4 = e \rangle.$$

Begin by choosing two 5-cycles in A_6 which satisfy these conditions. Then, working in $SL_2(9)$ and its cosets modulo the centre, find two matrices in this group which also satisfy these conditions. See also Edge (1955).

Problem ♦ 12.5 Let p be a prime larger than 3, and let a be a primitive root modulo p (Appendix B). Show that the matrices

$$A = \begin{pmatrix} a & 0 \\ 0 & a^{p-2} \end{pmatrix} \quad \text{and} \quad B = \begin{pmatrix} p-1 & 1 \\ p-1 & 0 \end{pmatrix}$$

generate the group $SL_2(p)$. What are the orders of the matrices A and B in this group? See also Problem 3.19.

Problem 12.6 (Properties of $L_3(2)$) (i) Over the field $\mathbb{F}_2$ let

$$R = \begin{pmatrix} 0 & 1 & 0 \\ 1 & 0 & 1 \\ 0 & 1 & 1 \end{pmatrix} \quad \text{and} \quad S = \begin{pmatrix} 1 & 0 & 0 \\ 1 & 1 & 0 \\ 1 & 0 & 1 \end{pmatrix}.$$

By direct calculation, show that

$$R^7 = \left(R^4 S\right)^4 = I_3 \quad \text{and} \quad (RS)^3 = S^2,$$

and so prove that $\langle R, S \rangle \simeq L_3(2)$ using the presentation of this group given on page 261. Use the fact that $L_3(2) \simeq L_2(7)$, see Problem 12.7*.

(ii) Consider the 7-point 'geometry' described in the Fano Plane on page 253 related to the Steiner system $S(2, 3, 7)$. Using the coordinates given on page 254, show that if we apply the matrices R and S to this 'geometry' we 'transform the set of points to itself', and so using the correspondences given above they define permutations (in A_7) of the set $\{1, 2, \ldots, 7\}$. Show that the matrix R corresponds to a 7-cycle a and the matrix S corresponds to a 2-cycle $\times$ 2-cycle b, see (12.5) on page 261 with $p = 7$. Hence show that a and b generate an isomorphic copy of $L_3(2)$ in A_7.

(iii) Now using the permutation representation constructed in (ii) show that $L_3(2)$ contains subgroups isomorphic to (a) $F_{7,3}$, and (b) S_4.

Problem 12.7* (Simple Groups of Order 168) In this question, you are asked to prove that all simple groups of order 168 are isomorphic. These include the groups $L_3(2)$ and $L_2(7)$ discussed in this chapter and the problem above. One method is as follows. Suppose G is a simple group with $o(G) = 168$, and $P = \langle a \rangle$ is a Sylow 7-subgroup (so $o(a) = 7$).

(i)   Let $H = N_G(P)$. Show that $H \simeq \langle a, b \mid a^7 = b^3 = e, bab^2 = a^4 \rangle$ for some $b \in G$, that is, H is a non-Abelian group of order 21. Also show that $ba = a^4 b$.

(ii) Show that $J = N_G(\langle b \rangle) \simeq \langle b, c \mid b^3 = c^2 = e, cbc = b^2 \rangle$ for some $c \in G$, that is $J \simeq S_3$ and $cb = b^2 c$.

(iii) Prove that the set of left cosets of H in G can be written in the form

$$C = \{H, cH, acH, \ldots, a^6 cH\}.$$

Using this fact, deduce G is generated by a, b and c.

(iv) Define a map θ from G to S_8 as follows. Number the elements of C from 1 to 8, respectively. The left coset action of $g \in G$ permutes the elements of C and so defines an element of S_8, that is, $\{gH, gcH, \ldots, ga^6 cH\}$ is a permutation of C which we denote by $g\theta$. Show that (a) this defines an injective homomorphism from G to S_8, (b) $a\theta = (2,3,4,5,6,7,8)$, and (c) $b\theta = (3,6,4)(5,7,8)$.

(v)   Show that $c\theta$ can be expressed in three different ways as a product of four 2-cycles $\sigma_i = (1,2)(3, *)(*, *)(*, *)$ for $i = 1, 2, 3$, and if $L_i = \langle a\theta, b\theta, \sigma_i \rangle$ then $L_i \leq A_8$ again for $i = 1, 2, 3$.

(vi) Conclude that $L_1 = L_2 = L_3$, and so finally prove that all simple groups of order 168 are isomorphic.

We also have: If $o(G) = 168$ and G has no normal Sylow subgroup, then $G \simeq L_3(2)$; see Suzuki (1982), page 107.

Problem 12.8 Consider the group $L_2(11)$ (of order 660, Definition 12.5). Give a permutation representation of this group using the following method. Work on the projective line

$$\mathcal{P} = \{\infty, 0, 1, 2, \ldots, 10\} \quad \text{defined over } \mathbb{F}_{11},$$

and treat the symbol ∞ in the usual naive way. Given a matrix $A = \left(\begin{smallmatrix} a & b \\ c & d \end{smallmatrix}\right) \in SL_2(11)$, define a map $\theta_A : \mathcal{P} \to \mathcal{P}$ by

$$z\theta_A = \frac{az+b}{cz+d} \quad \text{for } z \in \mathcal{P}\backslash\{\infty\}, \quad \text{and} \quad \infty\theta_A = \frac{a}{c}.$$

Show that the map θ_A permutes the elements of $\mathcal{P}$ and so defines an element in S_{12}. Deduce that $L_2(11)$ is isomorphic to a subgroup J of S_{12}.

Secondly, let maps α and β on $\mathcal{P}$ be given by

$$z\alpha = z+1 \quad \text{and} \quad z\beta = -1/z \quad \text{for } z \in \mathcal{P}.$$

Prove that these two maps generate J, and so give a permutation representation of $L_2(11)$. Finally, show that the presentation given for this group on page 261, with

$p = 11$, is a valid one. As an extension to this problem, consider what is needed to establish the general case where 11 is replaced by a general prime p and $\mathcal{P}$ has $p + 1$ elements.

Problem 12.9 Using the permutation representation for $L_2(11)$ constructed in Problem 12.8, find examples of the maximal subgroups of $L_2(11)$. (Hint. Using some results from this and the previous chapters, show first that $L_2(11)$ has no proper subgroups of order greater than 66, then use a suitable computer algebra package, or try by hand.) Note that this group has an index 11 subgroup isomorphic to A_5, and this fact allows us to construct a degree 11 permutation representation for it.

Problem 12.10 The group $L_3(3)$ was discussed in Section 12.2 where we proved that it is simple. Show that $L_3(3)$ has 117 involutions, and using these involutions construct elements of the group of all possible orders, that is, of orders 1, 2, 3, 4, 6, 8, and 13. Show also that the group can be generated by two elements of order 3, and by three elements of order 2 (*cf.* Problems 2.28 and 3.20). One way to do this is to begin with the permutation representation (12.8) given on page 264.

Problem 12.11 (i) Continuing the previous problem show that

$$A = \begin{pmatrix} 0 & 0 & 1 \\ 0 & 2 & 0 \\ 1 & 0 & 0 \end{pmatrix}$$

is an involution in $L_3(3)$.

   (ii) Let H denote the centraliser of A in $L_3(3)$. Show that $o(H) = 48$ using the basic properties of the centraliser, and so by Problem 12.10 show that the collection of involutions in $L_3(3)$ forms a single conjugacy class in $L_3(3)$.

   (iii) Using Problem 12.6 deduce $H \simeq GL_2(3)$. Only two simple groups have this property; they are $L_3(3)$ and M_{11}, see Huppert and Blackburn (1982b).

Problem 12.12 (i) The group $U_3(3)$ contains elements of order 1, 2, 3, 4, 6, 7, 8, and 12. Using the matrices A and B given on page 266 find elements of this group having each of these orders.

   (ii) Prove that $U_3(3)$ contains 63 elements of order 2. You may assume that each matrix of order two in the matrix representation of $U_3(3)$ given on page 266 has the form

$$C = \begin{pmatrix} a_2 & a_1 & c_1 \\ a_3 & c_2 & \overline{a_1} \\ c_3 & \overline{a_3} & \overline{a_2} \end{pmatrix},$$

where the overline denotes conjugation ($\overline{x} = x^3$), $a_i \in \mathbb{F}_9$ and $c_i \in \{0, 1, 2\}$, for $i = 1, 2, 3$. (Hint. Use the standard formulas for $\det C$ and C^{-1}.) You could also try to prove the assumption.

(iii) If b is an involution in $U_3(3)$, show that $J = C_{U_3(3)}(b) \simeq C_4 S_4$, where $C_4 \triangleleft J$, using the following method. Apply the permutation representation of $U_3(3)$ given on page 267, choose an involution b which is represented by a product of twelve disjoint 2-cycles in S_{28}, see the comments on the construction of the Steiner system for this group on page 267. Secondly, find an element of order 4 in $U_3(3)$ defined on the same symbols as b which commutes with b (note that its square will equal b). Now using (ii) and the fact that the permutation group on the symbols not used in b is isomorphic to S_4, construct the remaining elements of J.

This problem is not hard but it will take time to complete and the use of a suitable computer algebra package will help.

Problem 12.13 (Simple Groups of Order 20160) By Theorems 2.15 and 11.7, the simple groups A_8, $L_4(2)$ and $L_3(4)$ all have order 20160, $A_8 \simeq L_4(2)$ (Conway and Sloane 1993), but $L_3(4)$ is not isomorphic to either of the other two groups, and in this problem you are asked to prove this fact. There exist infinitely many pairs of non-isomorphic simple groups with the same order, but no triples; this is a consequence of CFSG, see the ATLAS (1985) for details. There are a number of ways to establish the required non-isomorphism property. It can be done by showing that A_8 contains elements of order 15 whilst $L_3(4)$ does not, or by showing that $o(\text{Aut}(A_8)) = 8!$ whilst $o(\text{Aut}(L_3(4)) \geq 3 \cdot 8!/2$ (Scott 1964, page 314). We suggest two further methods (i) and (ii) that you can try.

(i) Show first that A_8 has non-conjugate involutions; then, using the following method, prove that the involutions of $L_3(4)$ form a single conjugacy class. We have

(a) a non-scalar matrix A in $SL_3(4)$ is an involution in $L_3(4)$ if and only if A^2 is scalar (that is, diagonal with identical diagonal entries);
(b) A^2 is scalar if and only if $(C^{-1}AC)^2$ is scalar for all non-singular matrices $C \in SL_3(4)$;
(c) by (b), A can be replaced by a similar matrix in one of the following three rational canonical forms:

$$B_1 = \begin{pmatrix} a_1 & 0 & 0 \\ 0 & a_2 & 0 \\ 0 & 0 & a_3 \end{pmatrix}, \quad B_2 = \begin{pmatrix} b_1 & 0 & 0 \\ 0 & 0 & 1 \\ 0 & b_2 & b_3 \end{pmatrix}, \quad B_3 = \begin{pmatrix} 0 & 1 & 0 \\ 0 & 0 & 1 \\ 1 & c_1 & c_2 \end{pmatrix},$$

where the entries belong to the field $\mathbb{F}_4$, and a_1, a_2, a_3, b_1, and b_2 are non-zero.

Show that the matrix A, see (a), can only be of type B_2 in (c), $b_2 = b_1^{-1}$ and $b_3 = 0$, hence the entries b_1, b_2, b_3 of B_2 equal either $1, 1, 0$; $c, c^2, 0$ or $c^2, c, 0$ where c is a multiplicative generator of the field $\mathbb{F}_4$. Now consider the squares of these matrices and use Problem 5.21(iii).

(ii) The second method is as follows. Firstly, show that both $L_4(2)$ and $L_3(4)$ have Sylow 2-subgroups consisting entirely of upper triangular matrices (Problem 3.15). (Note that $L_4(2) \simeq SL_4(2)$, and by Theorem 12.7 $L_3(4) \simeq SL_3(4)/C_3$, so in the second case work by identifying the three matrices in each coset.) Secondly, show that the centres of both groups also consist of upper triangular matrices, and all matrices in each centre have non-zero entries in their upper right-hand corners. Finally, by considering the orders of these centres, show that the Sylow 2-subgroups of the two groups in question are not isomorphic.

Problem 12.14 (i) Show that the two permutation definitions of M_{11} given on pages 268 and 269 define the same group. This can be done by a computer search, or by showing that $\alpha_1, \ldots, \gamma_2$ can be defined in terms of ϕ and ψ, and vice versa. To start with note that α_1 equals the conjugate of $\phi\psi$ by $\theta^{-1}\phi^{-1}$.

(ii) Find examples of the Sylow subgroups of M_{11} for each of the primes dividing its order. (Hint. Using the notation of Section 12.4, we have $\delta = \beta_1\beta_2\gamma_2$ has order 8.)

(iii) In M_{11}, choose an involution a and show that its centraliser in M_{11} is isomorphic to an extension of C_2 by S_4, see Problems 4.4 and 12.9.

(iv) Find four involutions in A_{11} which generate M_{11} and satisfy the relations given in the second presentation of M_{11} on page 269. (Hint. One method starts with γ_1 and γ_2 in Definition 12.14.)

Problem 12.15 (Simplicity of M_{11}) (i) Suppose G is a transitive subgroup of S_p, $o(G) = pnt$ where

$$n > 1, \quad n \equiv 1 \ (\mathrm{mod}\ p), \quad t < p \text{ and } t \text{ is prime.}$$

Apply Problem 6.12(ii) to show that G is simple. Method. Using the quoted problem we can take $n = n_p$ and $t = [N_G(P) : P]$ where P is a Sylow p-subgroup of G. Suppose $K \lhd G$ and $K > \langle e \rangle$. Show that there exists a Sylow p-subgroup Q of G which is a subgroup of K. Now use Theorem 6.9(ii) and the quoted problem again to show that $o(K) = pn_ps$ where $s \mid t$ and $s > 1$, and so deduce $K = G$; see Chapman (1995).

(ii) Use (i) to show that M_{11} is simple.

The same method works for the simplicity of M_{23} with $p = 23$ and $t = 11$. The Mathieu group M_{23} can be defined using the Steiner system $S(4, 7, 23)$ or as a subgroup of S_{24}, see either Scott (1964), pages 287–289, or the ATLAS (1985), page 71.

A number of the problems above are 'project-like', and so no specific final project will be given.

Appendices A to E

Appendix A: Set Theory

The elementary ideas and notation of set theory underlie all the work in this book. Here we give a outline of the basic theory and terminology which is sufficient to cover the material presented in the previous chapters. The reader requiring a more thorough treatment should consult Halmos (1974b).

A.1 Sets and Maps

Set theory is based on the following notions:

(a) *element* (or *member*)—we use lower case letters $x, y, \ldots$ for elements,
(b) *membership* (or *belonging*)—denoted by $\in$,
(c) *set* (or *collection*)—we use upper case letters $X, Y, \ldots$ for sets.

Using logical constructions, all statements can be built up from the basic proposition

$$x \in X, \tag{A.1}$$

that is, the element x is a member of the set X. In an axiomatic development of set theory, these notions are taken as primitives which obey certain axioms and rules. For our purposes, we take an informal approach, assuming that the reader has an intuitive idea of the notion of an element (or member) belonging to a set (or collection), and this is expressed formally in (A.1).

We also use the following notation, terminology and constructions.

(i) $X = Y$, X *equals* Y—every element in the set X also belongs to the set Y, and vice versa. The negation is written $X \neq Y$.
(ii) $\emptyset$, the *empty set*, is the (unique) set with no elements.
(iii) $Y \subseteq X$, Y is a *subset* of X—every element in the set Y also belongs to the set X (this is equivalent to $X \supseteq Y$, the set X contains the set Y). For all X, the

H.E. Rose, *A Course on Finite Groups,*
Universitext,
DOI 10.1007/978-1-84882-889-6_13, © Springer-Verlag London Limited 2009

empty set $\emptyset$ and X are subsets of X. We write $Z \subset X$ for $Z \subseteq X$ and $Z \neq X$, and say that Z is a *proper* subset of X.

(iv) $X \cap Y$, the *intersection* of X and Y, is the largest set that contains all those elements which belong both to X and to Y (note $X \cap (Y \cap Z) = (X \cap Y) \cap Z$ and $X \cap Y = Y \cap X$).

(v) $X \cup Y$, the *union* of X and Y, is the smallest set that contains every element of X and every element of Y (note $X \cup (Y \cup Z) = (X \cup Y) \cup Z$ and $X \cup Y = Y \cup X$), see Problem A.1. A union $X \cup Y$ is called *disjoint* if $X \cap Y = \emptyset$; this is written as $X \overset{.}{\cup} Y$.

(vi) $Y \backslash X$, the *complement* of X in Y, is the subset of Y of those elements that do not belong to X (note $Y \backslash Y = \emptyset$ and $X \backslash (X \backslash Y) = X \cap Y$).

(vii)   If $x_1, \ldots, x_n$ are elements in some set X, then the (unordered) set containing these elements is denoted by $\{x_1, \ldots, x_n\}$, and so $\{x\}$ denotes the set with the single element x. We use the notation $(x_1, \ldots, x_n)$ for the *ordered set* whose first element is x_1, second element is $x_2, \ldots$, and last element is x_n. Infinite versions of these constructions are also used.

(viii) If P is a property which is either true or false for elements of a set X, then we define the new set Y by

$$Y = \{x \in X : P(x)\}.$$

The set Y is the subset of X containing those elements x belonging to X for which $P(x)$ is true.

There is another construction which is important in group theory—*Cartesian product* named after the seventeenth century French mathematician and philosopher René Descartes. Suppose $X_1, \ldots, X_n$ is a finite (or infinite) collection of sets. We form the new set of *ordered n*-tuples

$$\{(x_1, \ldots, x_n) : x_1 \in X_1, \ldots, x_n \in X_n\}.$$

This is denoted by $X_1 \times \cdots \times X_n$ and called the *Cartesian product* of X_i, for $i = 1, \ldots, n$. Note that the *n*-tuples are ordered so, for instance, $X_1 \times X_2 \neq X_2 \times X_1$, unless $X_1 = X_2$. We write X^n for $X \times \cdots \times X$ with n copies of the set X.

Example Let X, Y and Z be subsets of the set of real numbers $\mathbb{R}$ given by

$$X = \{x \in \mathbb{R} : 0 \leq x \leq 1\}, \qquad Y = \{y \in \mathbb{R} : 1 \leq y \leq 2\},$$
$$Z = \{z \in \mathbb{R} : 0 \leq z \leq 2\}.$$

The following statements are valid; the reader should check them.

(i) $X \subseteq Z$, $X \cup Y = Z$, and $X \cap Y = \{1\}$,

(ii) $Z \backslash X = \{x \in \mathbb{R} : 1 < x \leq 2\}$,

(iii) $\{z \in Z : z$ is a positive rational number$\}$ denotes the set of rational numbers z satisfying $0 < z \leq 2$,

(iv) $X \times Z = \{(x, y) : 0 \leq x \leq 1, 0 \leq y \leq 2\}$ denotes a closed rectangle in $\mathbb{R}^2$.

Relations, Functions and Maps

An n-ary, or n variable, *relation* R on a set X is by definition a subset of the Cartesian product X^n; relations are called *binary* if $n = 2$, *ternary* if $n = 3, \ldots$. So for example, if $X = \mathbb{R}$, the binary relation $x \leq y$, where $x, y \in \mathbb{R}$, is defined as the subset of the set of all ordered pairs $(x, y) \in \mathbb{R} \times \mathbb{R}$ for which there exists a non-negative z with the property $y = x + z$.

There are a number of important classes of relations. One is the class of equivalence relations; they are defined as follows.

Definition A.1 Let X be a set and let $\sim$ be a binary relation on X.

(i) The relation $\sim$ is called an *equivalence relation* if the following three axioms
     hold for all $x, y, z \in X$:
     (a) it is reflexive—$x \sim x$,
     (b) it is symmetric—if $x \sim y$ then $y \sim x$, and
     (c) it is transitive—if $x \sim y$ and $y \sim z$, then $x \sim z$.
(ii) For $z \in X$, let

$$E_z = \{x \in X : x \sim z\}.$$

The subset $E_z \subseteq X$ is called the *equivalence class* of z for the relation $\sim$.
    We have

(i) $x \in E_z$ if and only if $x \sim z$, and
(ii) X is a *disjoint* union of its equivalence classes under $\sim$.

This union is disjoint because if $x \in E_{z_1}$ and $x \in E_{z_2}$, then $x \sim z_1$ and $x \sim z_2$, and so $z_1 \sim z_2$ and $E_{z_1} = E_{z_2}$ by (b) and (c) above. The simplest example of an equivalence relation is equality; in this case, the equivalence class of x is the singleton set $\{x\}$. In fact, equivalence relations are generalisations of equality. See also the congruence subsection on page 285.

The second important class of relations contains the *maps*. We use the words *map*, *mapping*, *transformation* and *function* interchangeably, see also (vi) below. As noted earlier, see page 68, we write maps 'on the right', that is, in algebraic contexts, we write $a\theta$, rather than $\theta(a)$, because we read from left to right. Formally, we define a map as follows:

Definition A.2 Given sets X and Y, a *map* $\phi : X \to Y$ is a relation on the set $X \times Y$ with the extra property.

For each $x \in X$, there exists a *unique* $y \in Y$ such that (x, y) belongs to the relation ϕ.

Note that a map or function must be *single-valued*. An equivalent, and more intuitive, definition is to say that:

There is some *rule*, or *procedure*, denoted by ϕ which, given $x \in X$, generates a *unique* element $x\phi \in Y$, and this holds for *all* $x \in X$.

For example, suppose $X = Y = \mathbb{R}$ and $x\phi = \sqrt{x}$ for $x \in X$. As this stands ϕ is *not* a function, but it is if we always take the positive square-root, that is, if we define $x\phi = \sqrt[+]{x}$.

If $X = X_1 \times \cdots \times X_n$, the map $\phi : X \to Y$ is called an *n-argument* or *n-variable function* from X to Y, and if $(x_1, \ldots, x_n) \in X$, the elements $x_1, \ldots, x_n$ are called its *arguments* or *variables*.

The basic definitions and facts about maps are as follows.

(i)   If X and Y are sets and $\phi : X \to Y$ is a map, then X is called the *domain* of ϕ, and Y is called the *codomain* of ϕ. The set $\{x\phi : x \in X\}$ is a subset of Y which is called the *image*, or *range*, of ϕ, and it is denoted by $\operatorname{im}\phi$. With a slight 'abuse of notation' we write $Z\phi$ for the image of Z under the map ϕ where $Z \subseteq X$; that is, $Z\phi = \{y \in Y : z\phi = y$ for some $z \in Z\}$. Also, $x\phi$ is called the *image* of x; and if $z\phi = y$, then z is called a *preimage* of y. If $Z \subseteq \operatorname{im}\phi$, then the set of *all* preimages of elements of Z under the map ϕ is denoted by $Z\phi^{-1}$.

(ii) If $\phi : X_1 \to Y_1$ and $\psi : X_2 \to Y_2$ are maps, then we say that they are *equal*, written $\phi = \psi$, if $X_1 = X_2$ and $x\phi = x\psi$ for all $x \in X_1$. Note that we do not require the codomains to be equal, only that the domains and images agree.

(iii) A map $\phi : X \to Y$ is called *injective* (or sometimes *one-to-one*) if $x\phi = y\phi$ implies $x = y$, for all $x, y \in X$; that is for each z in the image of ϕ there exists exactly one preimage w in the domain X with $w\phi = z$.

For example, if $X = Y = \mathbb{R}$ and $x\phi_1 = x^3$ for all $x \in X$, then ϕ_1 is injective because every real number has a unique real cube root, but if $x\phi_2 = x^2$ for all $x \in X$, then ϕ_2 is not injective because, for instance, both 2 and -2 are mapped to 4 by ϕ_2.

(iv) A map $\phi : X \to Y$ is called *surjective* (or sometimes *onto*) if for all $y \in Y$, there is at least one $x \in X$ satisfying $x\phi = y$, or equivalently, Y equals the image of ϕ.

In the examples given below (iii), ϕ_1 is surjective whilst ϕ_2 is not (for instance, -1 has no preimage).

(v)   A map is called *bijective* (or sometimes a *one-to-one correspondence*) if it is both injective and surjective. If a bijection exists between two sets (that is, there is a bijective map between them), then they have the same *cardinality*, or to put this another way, they have the 'same number' of elements. This applies in both the finite and infinite cases. We use $o(X)$ to denote the cardinality (order) of X throughout, that is both for sets *and* for groups. Some authors use the word *size* for the cardinality of a set and reserve the word 'order' for groups.

The map ϕ_1 given below (iii) is an example of a bijection. As most of the groups discussed in this book are finite, cardinality problems do not arise—the cardinality of a finite set is just the (finite) number of its elements. Problems can arise if the group in question is neither finite nor countably infinite, the interested reader should consult the text by Halmos referred to at the beginning of this Appendix.

(vi) Given a non-empty set X, the map $\iota : X \to X$ defined by $x\iota = x$ for all $x \in X$ is called the *identity map*—each $x \in X$ is 'identified by itself'. It is, of course, a bijection on X.

(vii) If Y is a set and $X = Y^2$, then the map $\mu : X \to Y$ is called a (binary) *operation* on Y (but note that it is a ternary relation). For every pair (y_1, y_2) in Y^2, there is always a unique y_3 in Y with $(y_1, y_2)\mu = y_3$. For instance, if $Y = \mathbb{R}$, we can take μ to be the usual addition operation $+$.

We come now to the important notion of *composition* of functions which is widely used throughout the book.

Definition A.3 Given two maps $\phi : X \to Y$ and $\psi : Y \to Z$, a third map, denoted by $\phi \circ \psi : X \to Z$ and called the *composition* of ϕ and ψ, is given by

$$x(\phi \circ \psi) = (x\phi)\psi.$$

Note that (a) as ϕ maps $X \to Y$ and ψ maps $Y \to Z$, the composite $\phi \circ \psi$ maps $X \to Z$, and (b) we read all algebraic formulas and propositions from left to right both here and throughout this book. In our opinion, this is especially helpful when discussing composition.

Theorem A.4 (i) *Composition is associative.*
(ii) *If $\phi : X \to Y$ and $\psi : Y \to Z$ are injective (surjective) maps, then $\phi \circ \psi : X \to Z$ is also injective (surjective, respectively).*

Proof (i) If $\theta : Z \to W$, then both $(\phi \circ \psi) \circ \theta$ and $\phi \circ (\psi \circ \theta)$ map X to W and we have using Definition A.3

$$x\big((\phi \circ \psi) \circ \theta\big) = \big(x(\phi \circ \psi)\big)\theta = \big((x\phi)\psi\big)\theta, \quad \text{and}$$
$$x\big(\phi \circ (\psi \circ \theta)\big) = (x\phi)(\psi \circ \theta) = \big((x\phi)\psi\big)\theta.$$

This holds for all $x \in X$, and so the result follows.
(ii) First part. We have, if $x(\phi \circ \psi) = y(\phi \circ \psi)$, then $(x\phi)\psi = (y\phi)\psi$ which implies, as ψ is injective, that $x\phi = y\phi$. This now gives $x = y$ because ϕ is also injective; the second part is similar. $\qquad\square$

We usually drop the symbol 'o' and write $\phi\psi$ for $\phi \circ \psi$. Some examples are given in Problem A.4.

If $\phi : X \to Y$ is a bijection, see above, then we can define the *inverse* map ϕ^{-1} as follows:

$$\text{if } y \in Y, \quad \text{then let} \quad y\phi^{-1} = x \quad \text{where } x \text{ is given by } x\phi = y.$$

Note that ϕ^{-1} is a map from Y to X because ϕ is both injective and surjective.
Also if $X = Y$, then

$$\phi \circ \phi^{-1} = \phi^{-1} \circ \phi = \text{the identity map } \iota \text{ on } X. \tag{A.2}$$

The first of these equations follows by definition; for the second, we note that if $x\phi^{-1} = z$ and $z \in X$, then $z\phi = x$, and so

$$x\left(\phi^{-1} \circ \phi\right) = \left(x\phi^{-1}\right)\phi = z\phi = x,$$

and this holds for all $x \in X$. Referring to the example below (iii) on page 280 where ϕ_1 is defined, we see that ϕ_1^{-1} is the (unique) real cube root function.

Ordered sets play some role in group theory, so we will give a brief outline here.

Definition A.5 (i) A *partial order* $\preceq$ on a set X is a binary relation on X which satisfies the following three axioms. For $x, y, z \in X$, the relation $\preceq$ is

(a) reflexive—$x \preceq x$,
(b) transitive—if $x \preceq y$ and $y \preceq z$, then $x \preceq z$, and
(c) antisymmetric—if $x \preceq y$ and $y \preceq x$, then $x = y$.

  (ii) If $\preceq$ satisfies (a), (b), (c), and it is

(d) symmetric—$x \preceq y$ or $y \preceq x$, for $x, y \in X$, then the order $\preceq$ is called *linear* or *total*.

  (iii) A total order on X is called a *well-order* on X if every non-empty subset of X has a least element in X. Induction can be applied to well-ordered sets.

Example 1 If X is a set and $\mathcal{P}(X)$ is the set of all subsets of X, and we define an order on $\mathcal{P}(X)$ by: for $Y_1, Y_2 \in \mathcal{P}(X)$, let $Y_1 \preceq Y_2$ if and only if $Y_1 \subseteq Y_2$, then $\preceq$ is a partial order on X.

Example 2 If X is the set of integers $\mathbb{Z}$, then the usual inequality relation $\leq$ is a linear order on $\mathbb{Z}$, and it is a well-order on the set of non-negative elements of $\mathbb{Z}$.

A.2 Problems

Problem A.1 Let X, Y and Z be sets. Prove

(i) $X \cup (Y \cap Z) = (X \cup Y) \cap (X \cup Z)$,
(ii) $X \cap (Y \cup Z) = (X \cap Y) \cup (X \cap Z)$,
(iii) $(X \backslash Y) \cup (Y \backslash X) = (X \cup Y) \backslash (X \cap Y)$.

Propositions (i) and (ii) are instances of De Morgan's Laws (named after the nineteenth century English mathematician Augustus De Morgan), they can both be extended finitely and infinitely, so for example, we have

$$X \cup (Y_1 \cap \cdots \cap Y_n) = (X \cup Y_1) \cap \cdots \cap (X \cup Y_n).$$

The expression on either side of the equation in (iii) is called the *symmetric difference* of X and Y, it is sometimes denoted by $X \triangle Y$.

Problem A.2 For each of the following sets X with relations $\sim$ determine which define systems with equivalence relations, and describe the equivalence classes.

(i) X denotes the collection of all human beings alive at noon today, and (a) $x \sim_1 y$ stands for 'x and y have the same ancestor' (assume that a person can be his, or her, own ancestor), and (b) $x \sim_2 y$ stands for 'x and y have the same mother';
(ii) $X = \mathbb{R}$, and (a) $a \sim_1 b$ stands for '$x < y$ and $y < x$', and (b) $a \sim_2 b$ stands for '$x < y$ or $y < x$';
(iii) X is the set of all triangles in the real plane, and $x \sim y$ stands for x is similar (that is, has the same angles) to y.

Problem A.3 Let X and Y be arbitrary sets.

(i) Is $X \times \emptyset$ non-empty if X is non-empty?
(ii) Show that $(X \times Y) \cap (Y \times X) \neq \emptyset$ if and only if $X \cap Y \neq \emptyset$. If X and Y have n elements in common, how many elements do $X \times Y$ and $Y \times X$ have in common?

Problem A.4 For the following maps $\phi : X \to Y$ determine whether they are injective, surjective or bijective.

(i) $X = \mathbb{Z} \times (\mathbb{Z} \setminus \{0\})$, $Y = \mathbb{Q}$, and $(a, b)\phi = a/b$;
(ii) $X = \mathbb{Z} \times \mathbb{Z}$, $Y = \mathbb{Q}$, and $(a, b)\phi = a + b$;
(iii) $X = W \times Z$, $Y = Z$ and $(x, y)\phi = y$—maps of this type are called *projections*;
(iv) $X = \mathbb{R}$, $Y = \mathbb{R}^+$, the set of positive real numbers, and $a\phi = 2^a$;
(v) $X = \mathbb{Z}$, $Y = \mathbb{Q}$ and $x\phi = x/2$;
(vi) $X = \mathbb{R}$, $Y = \mathbb{Z}$, and $a\phi = [a]$—where the square brackets denote integer part.

Problem A.5 Let Y be a finite set and let $\psi : Y \to Y$. Show that if ψ is injective, then it is also surjective, and vice versa. Is this still true if Y is infinite?

Problem A.6 Let M be a set with m elements and N be a set with n elements. How many functions are there that map M to N, and how many of these are (i) bijections, or (ii) injections, or (iii)* surjections.

Problem A.7 Show that the following sets are bijective, that is, they have the same cardinality.

(i) $X \times (Y \times Z)$ and $(X \times Y) \times Z$ where X, Y and Z are sets;
(ii) $\mathbb{Z}^+$ and $GL_2(\mathbb{Z})$; and (iii)* $\mathbb{Z}$ and $\mathbb{Q}$.

Problem A.8 Prove that $\phi : X \to Y$ is a bijection if and only if there is another map $\psi : Y \to X$ with the properties (a) $\phi \circ \psi$ is the identity map on X, and (b) $\psi \circ \phi$ is the identity map on Y.

Problem A.9 Give examples of operations on a set X which are associative but not commutative, and vice versa. (Hint. Try X as a 2-element set.)

Appendix B: Number Theory

Some aspects of elementary number theory play a vital role in group theory, especially in the finite case. In particular, the prime numbers, Euclid's algorithm, congruences and primitive roots are widely used. Here we give a brief introduction to those aspects of number theory that appear in this book including the four topics listed above; for further details the reader should consult, for example, Rose (1999) where full proofs can be found.

B.1 Divisibility and the Euclidean Algorithm

Number theory begins with the study of the integers $\mathbb{Z}$. This system is a group under addition $+$. It also has both a linear order $\leq$ and a multiplication operation, but the non-zero integers do not form a group under this second operation because it has no inverses. This leads to the first result

Theorem B.1 (Division Algorithm) *Suppose $a, b \in \mathbb{Z}$ and $b \neq 0$. There exist unique integers c and d satisfying*

$$a = bc + d \quad and \quad 0 \leq d < |b|,$$

where the vertical bars denote the absolute value.

This is a consequence of the fact that the positive integers are well-ordered. Some notation: we say b *divides* a if $d = 0$ in the division algorithm, and we write $b \mid a$ in this case, its negation is written $b \nmid a$. The basic properties are

(i)   for all $a \in \mathbb{Z}$, $a \mid a$, $a \mid 0$, $1 \mid a$ and $-1 \mid a$;
(ii) $0 \mid a$ if and only if $a = 0$;
(iii) if $c > 0$ and $a \mid c$, then $c \geq a$;
(iv) if $a \mid b$ and $a \mid c$, then $a \mid bx + cy$ for all x and y in $\mathbb{Z}$.

As immediate consequence of this result is

Theorem B.2 (Euclidean Algorithm) *Suppose $a, b \in \mathbb{Z}$, and at least one of a or b is non-zero. There exists a unique integer c which satisfies*

$$c > 0, c \mid a, c \mid b, \quad and \ if \quad d \mid a \ and \ d \mid b, \quad then \quad d \mid c.$$

This follows from the Division Algorithm (Theorem B.1) and (iv) above. The unique integer c given by this algorithm is called the *greatest common divisor* or GCD of a and b, and it is denoted by (a, b); also we set $(0, 0) = 0$. Some authors use HCF (highest common factor) for GCD. The basic properties are given in Problem B.1. We define the *least common multiple* or LCM of a and b to equal $ab/(a, b)$ provided both a and b are non-zero. This algorithm is named after the Greek geometer Euclid of Alexandria who was working in the third century BC.

There are two important corollaries of this algorithm, the first is

Theorem B.3 *The equation*

$$ax + by = n$$

has integer solutions x and y if and only if $(a, b) \mid n$.

In Theorem B.2, $c = (a, b)$ is the smallest positive integer of the form $ax + by$.
The second corollary is

Theorem B.4 *If $a \mid bc$ and $(a, b) = 1$, then $a \mid c$.*

This is proved by applying Theorem B.3 to the given equation.
Using Theorem B.4, we define the primes and unique factorisation in $\mathbb{Z}$.

Definition B.5 (i) An integer $p > 1$ is called *prime* or a *prime number* if its only positive divisors are 1 and p, otherwise it is called *composite*. Two integers are called *coprime* if they have no common positive divisor except 1.

(ii) We use the symbol π to denote a collection of primes, that is, a subset of the set of all prime numbers. So a π-*number* is a positive integer whose prime factors are included in π. Also we write π' for the complement of π in the set of all prime numbers.

There are infinitely many primes, this was first proved by Euclid. (If $p_1, \ldots, p_k$ is a list of all primes, consider the expression $p_1 \cdots p_k + 1$.) Unique factorisation in $\mathbb{Z}$ now follows directly from Theorem B.4, we have

Theorem B.6 (Unique Factorisation) *Every positive integer has a unique (except for the order) representation as a product of primes.*

Congruences

We begin with

Definition B.7 Given integers a, b and m where $m > 0$, the expression

$$a \equiv b \pmod{m} \tag{B.1}$$

is called a *congruence*, and stands for $m \mid a - b$. The number m is called the *modulus*, and b is called a *residue* of a modulo m.

This is equivalent to: a and b have the same remainder after division by m. Sometimes we write $a \equiv b(m)$ for (B.1). This was introduced by Gauss in 1801, and is one of the origins for the notions of coset and factor group.

Note that the congruence (B.1) is an equivalence relation on $\mathbb{Z}$, see Definition A.1; some elementary properties are given in the problem section. Also Theorem B.3 can be rewritten as

Theorem B.8 *Suppose* $a, b, m \in \mathbb{Z}$ *and* $m > 0$. *The congruence*

$$ax \equiv b \pmod{m}$$

has (a, m) *solutions* x *with* $0 \leq x < m$ *if* $(a, m) \mid b$, *and none otherwise.*

The inverse operation of the group $(\mathbb{Z}/p\mathbb{Z})^*$ is given by this result. If $m = p$ and $0 \leq a < p$, then $(a, p) = 1$, that is the condition in Theorem B.8 is always satisfied in this case, and the solution is unique.

Some congruence results used in this book are listed below. The first, which follows from Theorem B.8 (Problem B.6), is

Theorem B.9 (Chinese Remainder Theorem) *Given* $m_1, \ldots, m_k \in \mathbb{Z}$, *coprime in pairs, and* $c_1, \ldots, c_k \in \mathbb{Z}$, *the congruences*

$$x \equiv c_i \pmod{m_i}, \quad i = 1, \ldots, k,$$

have a common solution x *which is unique modulo the product* $m_1 \cdots m_k$.

The second result, which had a formative influence on the early development of group theory and was first discovered in the seventeenth century by Fermat (and probably first proved by Euler in the eighteenth century), is

Theorem B.10 (Fermat's Theorem) *If* p *is prime and* $a \in \mathbb{Z}$, *then* $a^p \equiv a \pmod{p}$.

The first proofs of this result used the multinomial theorem, but nowadays a more group-theoretic proof is used; see the next result and Problem B.5.

Euler extended this result to all positive moduli, to state it we need first to define a useful number-theoretic function which is named after him.

Definition B.11 If $n > 1$, $\phi(n)$ denotes the number of integers a satisfying the conditions: $1 \leq a < n$ and $(a, n) = 1$; and we set $\phi(1) = 1$.

The function $\phi(n)$ is called the *Euler function*, it is the order of the group $(\mathbb{Z}/n\mathbb{Z})^*$. Using Unique Factorisation in $\mathbb{Z}$ (Theorem B.6) it can be shown that if n has the prime factorisation $n = p_1^{s_1} \cdots p_k^{s_k}$, then

$$\phi(n) = \prod_{i=1}^{k} p_i^{s_i - 1} (p_i - 1).$$

For example, if p is prime, then $\phi(p) = p - 1$ as every positive integer less than p is coprime to p. Note also that ϕ is *multiplicative*, that is, $\phi(n)\phi(m) = \phi(mn)$ when $(m, n) = 1$. Using this function Euler extended Fermat's Theorem to

Theorem B.12 (Euler's Theorem) *If* $(a, m) = 1$, *then* $a^{\phi(m)} \equiv 1 \pmod{m}$.

Problem 2.8 when it is applied to the group $\mathbb{Z}/m\mathbb{Z}$ provides a proof of this result, another is given in Problem B.5. The last congruence result is

Theorem B.13 *Suppose f is a polynomial defined over the field $\mathbb{Z}/p\mathbb{Z}$, and has degree n. The polynomial congruence $f(x) \equiv 0 \pmod{p}$ has at most n roots in $\mathbb{Z}/p\mathbb{Z}$.*

This is proved by induction on n; use Theorem B.8 when $n = 1$, and consider possible linear factors of f in the inductive step. This result can also be proved purely algebraically, see Problem 7.5(iii).

Primitive Roots

The basic question here is: In which cases, if any, is the group $(\mathbb{Z}/m\mathbb{Z})^*$ cyclic? The answer is useful in the theories of cyclic groups and automorphisms (Section 4.4).

Definition B.14 (i) Given integers c and m with $(c, m) = 1$ and $m > 1$, the *order* of c modulo m, $\text{ord}_m(c)$, is the least positive integer t satisfying the congruence

$$c^t \equiv 1 \pmod{m}.$$

(ii) The integer c is called a *primitive root* modulo m if $\text{ord}_m(c) = \phi(m)$.

By Theorem B.12, $\text{ord}_m(c)$ always exists and is not greater than $\phi(m)$; in fact, it is a divisor of $\phi(m)$. Also $\text{ord}_m(c)$ is the order of the cyclic group generated by c in $(\mathbb{Z}/m\mathbb{Z})^*$. The following result was first proved by Gauss.

Theorem B.15 *Primitive roots exist for each prime modulus p.*

This is derived by establishing a stronger result: p has $\phi(p - 1)$ primitive roots modulo p, and one proof of this (which is very elegant) uses the Möbius function μ; see Chapter 5 in Rose (1999). There are a number of unsolved problems concerning primitive roots, for example, for large primes p only weak estimates are known for the least positive primitive root of p.

Not all integers have primitive roots, for example, 8 does not have one because $\phi(8) = 4$ and all odd integers have order 2 modulo 8. Using these results we have

Theorem B.16 *Suppose p is an odd prime and let $r > 0$. The only integers larger than 1 which have primitive roots are $2, 4, p^r$ and $2p^r$.*

This result follows from Theorem B.15 (Problem B.8), and it implies that the group $(\mathbb{Z}/m\mathbb{Z})^*$ is cyclic when $m = 2, 4, p^r$ or $2p^r$. In the remaining cases, it can be shown that $(\mathbb{Z}/m\mathbb{Z})^*$ is a direct product of cyclic groups. For $m = 2^r$ with $r > 2$, we have $(\mathbb{Z}/m\mathbb{Z})^* \simeq C_2 \times C_{2^{r-2}}$, and the remaining cases use the Fundamental Theorem of Abelian Groups (Theorem 7.12).

B.2 Problems

Problem B.1 Prove the following properties of the GCD.

(i) $(a, b) = (b, a)$,
(ii) $(a, 0) = |a|$, the absolute value of a,
(iii) if $a \mid b$ and $b \mid c$, then $ab \mid c(a, b)$,
(iv) $(a, a + b) \mid b$,
(v) if $(a, m) = (b, m) = 1$, then $(ab, m) = 1$,
(vi) if $(a, b) = d$, then $a/d, b/d \in \mathbb{Z}$ and $(a/d, b/d) = 1$.

Problem B.2 If $m = p_1^{r_1} \cdots p_k^{r_k}$ and $n = p_1^{s_1} \cdots p_k^{s_k}$ where some of the r_i or s_j may be zero, then

$$(m, n) = \prod_{i=1}^{k} p_i^{\min(r_i, s_i)} \quad \text{and} \quad LCM(m, n) = \prod_{i=1}^{k} p_i^{\max(r_i, s_i)}.$$

Problem B.3 If a and b are positive with $(a, b) = 1$, show that the equation $ax + by = n$ is soluble in non-negative integers x and y if $n \geq (a - 1)(b - 1)$.

Problem B.4 Show that for fixed r the following equation only has a finite number of positive integer solutions $\{n_1, \ldots, n_r\}$

$$1 = \frac{1}{n_1} + \cdots + \frac{1}{n_r}.$$

This is used in Problem 5.23.

Problem B.5 Prove Theorem B.12 using the same method as for Theorem 2.8.

Problem B.6 Prove the following congruence properties.

(i) If $a \equiv b \pmod{m}$ and $c \equiv d \pmod{m}$, then $a + c \equiv b + d \pmod{m}$, $ac \equiv bd \pmod{m}$, and $a \equiv b \pmod{m'}$ if $m' \mid m$.
(ii) If $a \equiv b \pmod{m}$, then $(a, m) = (b, m)$, $ar \equiv br \pmod{mr}$, and $a^t \equiv b^t \pmod{m}$ when $t \geq 0$.
(iii) Prove Theorem B.9 as follows: let $m = m_1 \cdots m_k$, $m_i' = m/m_i$ and m_i^* satisfy $m_i' m_i^* \equiv 1 \pmod{m_i}$ for $i = 1, \ldots, k$, and set $x = c_1 m_1' m_1^* + \cdots + c_k m_k' m_k^*$.

Problem B.7 Let the number-theoretic function χ be given by: $\chi(1) = \chi(2) = 1$, $\chi(4) = 2$, $\chi(2^{n+3}) = \phi(2^{n+3})/2$, $\chi(p^n) = \phi(p^n)$, p an odd prime, and if $m = p_1^{r_1} \cdots p_k^{r_k}$ then $\chi(m) = LCM(\chi(p_1^{r_1}), \ldots, \chi(p_k^{r_k}))$. Show that

$$a^{\chi(m)} \equiv 1 \pmod{m} \quad \text{if } (a, m) = 1.$$

Problem B.8 (i) Using the Binomial Theorem, show that if $r > 1$ and p is an odd prime, then

$$(1 + ap)^{p^{r-2}} \equiv 1 + ap^{r-1} \pmod{p^r}.$$

(ii) Prove that if $p \nmid a$, then $1 + ap$ has order p^{r-1} modulo p^r.

(iii) Use (ii) to show that if p is an odd prime and $r > 0$, then p^r has a primitive root. (Hint. Use Theorem B.15 and show that if c is a primitive root then $c^{p-1} \not\equiv 1$ (mod p^2), if this fails for c try $c + p$.)

(iv) Using Problem B.7 and (iii) prove Theorem B.16.

*   *   *   *   *   *

Appendix C: Data on Groups of Order 24

Data on groups of order 24 are presented in the four tables below, some notes concerning this data are given on pages 292 and 293.

Table C.1 Data on groups of order 24

GROUP	No. of elts of order 2	No. of conj. cls of elts	No. of subgps	No. of subgp conj. cls	No. of max. subgps	No. of normal subgps	Sylow 2-subgps
C_{24}	1	24	8	8	2	8	C_8
$C_{12} \times C_2$	3	24	16	16	4	16	$C_4 \times C_2$
$C_6 \times C_2^2$	7	24	32	32	8	32	C_2^3
$D_3 \times C_4$	7	12	26	16	6	11	$C_4 \times C_2$ by 3
$D_4 \times C_3$	5	15	20	16	4	12	D_4
$Q_2 \times C_3$	1	15	12	12	4	12	Q_2
$D_6 \times C_2$	15	12	54	32	10	21	C_2^3 by 3
$Q_3 \times C_2$	3	12	22	16	6	13	$C_4 \times C_2$ by 3
$A_4 \times C_2$	7	8	26	12	6	6	$(C_2^3)^*$
D_{12}	13	9	34	16	6	9	D_4 by 3
Q_6	1	9	18	12	6	9	Q_2 by 3
S_4	9	5	30	11	8	4	D_4 by 3
$F_{3,8}$	1	12	10	8	4	7	C_8 by 3
$SL_2(3)$	1	7	15	7	5	4	$(Q_2)^*$
E	9	9	30	16	6	9	D_4 by 3

Table C.2 Data on groups of order 24

GROUP	No. of Sylow 3-subgps	Subgroups of order 12	Maximal subgps of order 6	Centre	Class No.	Derived subgp (first)
C_{24}	1	C_{12}	–	C_{24}	24	$\langle e \rangle$
$C_{12} \times C_2$	1	C_{12} by 2, $C_6 \times C_2$	–	$C_{12} \times C_2$	24	$\langle e \rangle$
$C_6 \times C_2^2$	1	$C_3 \times C_2^2$ by 7	–	$C_6 \times C_2^2$	24	$\langle e \rangle$
$D_3 \times C_4$	1	C_{12}, D_6, Q_3	–	C_4	12	C_3
$D_4 \times C_3$	1	$C_{12}, C_6 \times C_2$ by 2	–	C_6	15	$(C_2)^*$
$Q_2 \times C_3$	1	C_{12} by 3	–	C_6	15	$(C_2)^*$
$D_6 \times C_2$	1	$C \times C_2, D_6$ by 6	–	C_2^2	12	C_3
$Q_3 \times C_2$	1	Q_3 by 2, $C_6 \times C_2$	–	C_2^2	12	C_3
$A_4 \times C_2$	4	A_4	C_6 by 4	C_2	8	C_2^2
D_{12}	1	C_{12}, D_6 by 2	–	$(C_2)^*$	9	C_6
Q_6	1	C_{12}, Q_3 by 2	–	$(C_2)^*$	9	C_6
S_4	4	A_4	D_3 by 4	$\langle e \rangle$	5	A_4
$F_{3,8}$	1	C_{12}	–	$(C_4)^*$	12	C_3
$SL_2(3)$	4	–	C_6 by 4	$(C_2)^*$	7	$(Q_2)^*$
E	1	$C_6 \times C_2, D_6, Q_3$	–	$(C_2)^*$	9	C_6

Table C.3 Data on groups of order 24

GROUP	Frattini subgp.	Fitting subgp.	Sockel See (17)	Largest factor group(s)	Smallest symmetric supergp.	Is nil-potent
C_{24}	C_4	'G'	C_6	C_{12}	S_{11}	Yes
$C_{12} \times C_2$	C_2	'G'	$C_6 \times C_2$	C_{12} by 2, $C_6 \times C_2$	S_9	Yes
$C_6 \times C_2^2$	$\langle e \rangle$	'G'	'G'	$C_6 \times C_2$ by 7	S_9	Yes
$D_3 \times C_4$	C_2	C_{12}	C_6	D_6	S_7	No
$D_4 \times C_3$	$(C_2)^*$	'G'	C_6	$C_6 \times C_2$	S_7	Yes
$Q_2 \times C_3$	$(C_2)^*$	'G'	C_6	$C_6 \times C_2$	S_{11}	Yes
$D_6 \times C_2$	$\langle e \rangle$	$C_6 \times C_2$	$C_6 \times C_2$	D_6 by 3	S_7	No
$Q_3 \times C_2$	C_2	$C_6 \times C_2$	$C_6 \times C_2$	Q_3 by 2, D_6	S_9	No
$A_4 \times C_2$	$\langle e \rangle$	$(C_2^3)^*$	$(C_2^3)^*$	A_4	S_6	No
D_{12}	$(C_2)^*$	C_{12}	C_6	D_6	S_7	No
Q_6	$(C_2)^*$	C_{12}	C_6	D_6	S_{24}	No
S_4	$\langle e \rangle$	$(C_2^2)^*$	$(C_2^2)^*$	D_3 (order 6)	S_4	No
$F_{3,8}$	$(C_4)^*$	C_{12}	C_6	Q_3	S_{11}	No
$SL_2(3)$	$(C_2)^*$	$(Q_2)^*$	C_2	A_4	S_8	No
E	$(C_2)^*$	$C_6 \times C_2$	C_6	D_6	S_7	No

Table C.4 Data on groups of order 24

GROUP	Degree pattern See (18)	Autom. group order	Outer autom. group	Expo-nent	Abelian invar-iants	2-Euler count See (19)
C_{24}	$1, \overset{24}{..}, 1$	8	C_2^3	24	3, 8	6144
$C_{12} \times C_2$	$1, \overset{24}{..}, 1$	16	$D_4 \times C_2$	12	2, 3, 4	768
$C_6 \times C_2^2$	$1, \overset{24}{..}, 1$	336	$GL_3(2) \times C_2$	6	2, 2, 2, 3	0
$D_3 \times C_4$	$1, \overset{8}{..}, 1, 2, 2, 2, 2$	24	$C_2 \times C_2$	12	2, 4	576
$D_4 \times C_3$	$1, \overset{12}{..}, 1, 2, 2, 2$	16	$C_2 \times C_2$	12	2, 2, 3	768
$Q_2 \times C_3$	$1, \overset{12}{..}, 1, 2, 2, 2$	48	A_4	12	2, 2, 3	768
$D_6 \times C_2$	$1, \overset{8}{..}, 1, 2, 2, 2, 2$	144	S_4	6	2, 2, 3	0
$Q_3 \times C_2$	$1, 1, 1, 1, 2, \overset{5}{..}, 2$	48	$C_4 \times C_2$	12	2, 4	576
$A_4 \times C_2$	$1, \overset{6}{..}, 1, 3, 3$	24	C_2	6	2, 3	288
D_{12}	$1, 1, 1, 1, 2, \overset{5}{..}, 2$	48	$C_2 \times C_2$	12	2, 2	576
Q_6	$1, 1, 1, 1, 2, \overset{5}{..}, 2$	48	$C_2 \times C_2$	12	2, 2	576
S_4	$1, 1, 2, 3, 3$	24	$\langle e \rangle$	12	2	216
$F_{3,8}$	$1, \overset{8}{..}, 1, 2, 2, 2, 2$	24	$C_2 \times C_2$	24	8	4608
$SL_2(3)$	$1, 1, 1, 2, 2, 2, 3$	24	C_2	12	3	384
E	$1, 1, 1, 1, 2, \overset{5}{..}, 2$	24	C_2	12	2, 2	576

Notes on Tables C.1 to C.4

(1) We write H^k as a short-hand for $H \times H \times \cdots \times H$ with k factors.

(2) 'H by k' stands for 'k isomorphic copies of the subgroup H'.

(3) If '$(H)^*$' occurs twice or more in a row(s), this means that the corresponding subgroups are identical. For example, the derived and Frattini subgroups of $D_4 \times C_3$ are identical.

(4) The group $Q_2 \times C_3$ is the only group in the tables that is Hamiltonian: It is not Abelian and all of its subgroups are normal; see Problem 7.13 and Note (20).

(5) The tabulated maximal subgroups have order 12 (except for $SL_2(3)$), or order 8 (all Sylow 2-subgroups are maximal), or in three cases ($A_4 \times C_2$, S_4 and $SL_2(3)$) they have order 6.

(6)   All largest factor groups have order 12 except for S_4 where the order is 6. This reflects the fact that all groups of order 24 except S_4 have at least one normal subgroup of order 2.

(7)   The centre is defined on page 32, and the derived subgroup on page 37, only S_4 and $SL_2(3)$ have non-neutral second derived subgroups, they are isomorphic to $C_2 \times C_2$ for S_4 and C_2 for $SL_2(3)$.

(8)   Nilpotency and supersolubility are defined on page 212 and page 227, respectively. The two non-Abelian nilpotent groups listed above ($D_4 \times C_3$ and $Q_2 \times C_3$) both have nilpotency class 2. All groups in the table are soluble, and all but $A_4 \times C_2$, S_4 and $SL_2(3)$ are supersoluble; see Problem 10.26.

(9)   The exponent (Table C.4) of a group G is the smallest positive integer n such that $g^n = e$ for all $g \in G$ (Definition 2.19(iv)).

(10) The Abelian invariants (Table 4) of G are defined as the orders of the elements of a minimal set of generators of 'G made Abelian', that is, of G factored by its derived subgroup G'.

(11) The Frattini subgroup $\Phi(G)$ of G is the intersection of the maximal subgroups of G, it is a normal subgroup of G, see page 217.

(12) The Fitting subgroup $F(G)$ of G is the (unique) maximal, nilpotent, normal subgroup of G; see page 221. If G is nilpotent then $F(G) = G$.

(13) The groups in the tables with the maximum number of elements of the specified order are as follows: $D_6 \times C_2$ has 15 elements of order 2, $A_4 \times C_2$, S_4 and $SL_2(3)$ have 8 elements of order 3, Q_6 has 14 elements of order 4, $C_6 \times C_2^2$ has 14 elements of order 6, $F_{3,8}$ has 12 elements of order 8, $Q_2 \times C_3$ has 12 elements of order 12, and C_{24} is the only group with elements (8 in all) of order 24.

(14) Cauchy's Theorem states that all the groups in the tables possess elements of order 2 and 3. For all other divisors n of 24, there is at least one group in the table which does not have an element of order n. The groups $C_6 \times C_2^2$, $D_6 \times C_2$ and $A_4 \times C_2$ do not have elements of order 4, S_4 does not have an element of order 6, all groups in the table except C_{24} and $F_{3,8}$ do not have elements of order 8, $C_6 \times C_2^2$, $D_6 \times C_2$, $Q_3 \times C_2$, $A_4 \times C_2$, S_4, $SL_2(3)$ and E do not have elements of order 12, and no group except C_{24} has elements of order 24.

(15) By Corollary 5.27, the order of the inner automorphism group times the order of the centre equals 24, so $o(\text{Inn } G)$ ranges from 1 (for Abelian groups) to 24 (when $G = S_4$).

(16) With one exception all groups in these tables are *reverse Lagrange*, that is, they possess subgroups of all orders dividing 24. The exception is $SL_2(3)$ which has no subgroup of order 12.

(17) The *Sockel* of a group G is the product of all minimal normal subgroups of the group G, see Theorem 11.6. It is normal in G. The *Abelian, (non-Abelian, Sockel)* is the product of all Abelian, (non-Abelian, respectively), minimal normal subgroups of G, see Theorem 11.11. The sockel is the product of the Abelian and non-Abelian sockels. A number of properties have been established, for example, if G is a finite group and $\Phi(G) = \langle e \rangle$, then $F(G)$ equals the Abelian Sockel. For further details, see Scott (1964), pages 168 to 170.

(18) The *degree pattern* of a group G is a set of positive integers $(r_1, \ldots, r_m)$ where r_i is the degree of the ith linear representation of G (that is, the dimension of its underlying vector space). Note that $\sum_{i=1}^{m} r_i^2 = o(G)$. This aspect of G is mainly concerned with the character theory of G; see Huppert (1998), and Web Section 14.3.

(19) The 2-Euler count is the number of pairs of (possibly not distinct) elements of a group G that generate G.

(20) All groups except C_{24} in the tables can be represented as direct or semi-direct products with two factors; see Problem 8.3. Apart from those listed in this chapter we have $D_{12} \simeq C_2 \rtimes C_{12}$, $Q_6 \simeq Q_2 \rtimes C_3$, and $F_{3,8} \simeq C_8 \rtimes C_3$. Note also that $Q_2 \times C_3$ can also be treated as a semi-dihedral group, see page 131.

(21) Most of the data in these tables was computed using the computer package GAP. The numbers (labels) given to the groups by the GAP program 'Small-Groups' are: 2, 9, 15, 5, 10, 11, 14, 7, 13, 6, 4, 12, 1, 3, and 8, using the same order of the groups as in the tables above—so for example, the GAP SmallGroup(24, 1) is the thirteenth group in our tables, that is, $F_{3,8}$.

$$*\quad *\quad *\quad *\quad *\quad *$$

Appendix D: Numbers of Groups with Order up to 520

The table apposite gives the number of (isomorphism classes of) groups of each order between 1 and 520 except 512 which is a staggering 10494213; see Besche *et al.* (2001). The entries 1* indicate that the number of groups of this order is 1 even though the order is not prime, there are 73 in Table D.1, see page 117. The data in this table were given by the programs GAP and Magma.

Table D.1 Number of Groups with Order up to 520

+	0	40	80	120	160	200	240	280	320	360	400	440	480
1	1	1	15	2	1*	2	1	1	1*	2	1	13	1*
2	1	6	2	2	55	2	5	4	4	2	6	4	2
3	1	1	1	1*	1	2	67	1	1	3	1*	1	2
4	2	4	15	4	5	12	5	4	176	11	5	18	12
5	1	2	1*	5	2	2	2	2	2	1*	16	1*	1*
6	2	2	2	16	2	2	4	4	2	6	6	2	261
7	1	1	1*	1	1	2	1*	1*	2	1	1*	1*	1
8	5	52	12	2328	57	51	12	1045	15	42	46	1396	14
9	2	2	1	2	2	1*	1*	2	1*	2	1	1	2
10	2	5	10	4	4	12	15	4	12	4	6	34	10
11	1	1*	1*	1	5	1	1	2	1	1*	1*	1*	1
12	5	5	4	10	4	5	46	5	4	15	4	5	12
13	1	1	2	1*	1	1*	2	1	5	1	1*	2	1*
14	2	15	2	2	4	2	2	23	2	4	10	2	4
15	1*	2	1*	5	2	1*	1*	1*	1*	7	1*	1*	4
16	14	13	231	15	42	177	56092	14	228	12	235	54	42
17	1	2	1	1	1*	1*	1	5	1	1*	2	1	2
18	5	2	5	4	2	2	6	2	5	60	4	2	4
19	1	1	2	1	1	2	1*	1*	1*	1	1	5	1
20	5	13	16	11	37	15	15	49	15	11	41	11	56
21	2	1	1	1*	1	1*	2	2	1*	2	1	1	1*
22	2	2	4	2	4	6	2	2	18	2	2	12	2
23	1	4	1	1*	2	1	1	1*	5	1	2	1	1
24	15	267	14	197	12	197	39	42	12	20169	14	51	202
25	2	1*	2	1*	1*	6	1*	2	1*	2	2	4	2
26	2	4	2	2	6	2	4	10	2	2	4	2	6
27	5	1	1	6	1*	1	1*	1	1	4	1*	1	6
28	4	5	45	5	4	15	4	9	12	5	4	55	4
29	1	1*	1	1	13	1	1	2	1	1	2	1*	1
30	4	4	6	13	4	4	30	6	10	12	4	4	8
31	1	1	2	1	1	2	1	1	14	1*	1	2	1*
32	51	50	43	12	1543	14	54	61	195	44	775	12	***
33	1*	1	1	2	1	1	5	1	1	1*	1	1*	15
34	2	2	6	4	2	16	2	2	4	2	4	6	2
35	1*	3	1*	2	2	1*	4	4	2	1*	1*	2	1*
36	14	4	5	18	12	4	10	4	5	30	5	11	15
37	1	1*	4	1	1	2	1	1	2	1	1*	2	1*
38	2	6	2	2	10	4	2	4	2	2	6	2	4
39	2	1	1*	1*	1	1	4	1*	1	5	1	1	1*
40	14	52	47	238	52	208	40	1640	162	221	51	1213	49

Appendix E: Representations of $L_2(q)$ with Order < 10000

In this appendix, we list some representations for the linear groups $L_2(q)$, see Section 12.2. All representations are illustrative examples and are in no way unique. A presentation for $L_2(p)$, p a prime, was given in Chapter 12. In the first list, we give some presentations for the remaining cases using small order generators. They were taken from the ATLAS (1985) where further details can be found.

$$L_2(8) \simeq \langle a, b \mid a^7 = (a^2b)^3 = (a^3b)^2 = (ab^5)^2 = e \rangle,$$

$$L_2(9) \simeq \langle a, b \mid a^5 = b^5 = (ab)^2 = (a^4b)^4 = e \rangle \simeq A_6,$$

$$L_2(16) \simeq \langle a, b \mid a^{15} = (a^2b)^3 = (a^3b)^2 = (ab^9)^2 = (a^8b^2)^2 = e \rangle,$$

$$L_2(25) \simeq \langle a, b, c \mid a^5 = b^{12} = c^2 = (bc)^2 = (ac)^3 = (b^4abac)^3 = e, b^{10}ab^2 = b^{11}a^2ba = ab^{11}a^2b \rangle,$$

$$L_2(27) \simeq \langle a, b, c \mid a^3 = b^{13} = c^2 = (bc)^2 = (ac)^3 = e, b^{10}ab^3 = ab^{12}ab = b^{12}aba \rangle.$$

Secondly, each linear group in our chosen range can be given a permutation representation, that is, $L_2(q)$ is isomorphic to a subgroup of A_t for some suitable chosen t depending on q. In the list below which was constructed using the computer program GAP, $t = q + 1$, but in some cases smaller values of t are possible (for $q = 5, 7, 9$). The case $q = 11$ is discussed in Problem 12.8. These facts were first noted by Galois in his famous letter to a colleague written the night before he was mortally wounded in a duel; see Huppert (1967), page 214.

$$L_2(4) \simeq \langle (3, 4, 5), (1, 2, 3) \rangle \simeq A_5,$$

$$L_2(5) \simeq \langle (1, 2, 3, 4, 5), (3, 4, 5) \rangle \simeq A_5,$$

$$L_2(7) \simeq \langle (1, 2)(3, 4), (1, 3, 5)(2, 7, 6) \rangle \preceq A_7,$$

$$L_2(8) \simeq \langle (3, 9, 7, 4, 8, 5, 6), (1, 2, 3)(4, 5, 6)(7, 8, 9) \rangle,$$

$$L_2(9) \simeq \langle (1, 2, 3, 4, 5), (4, 5, 6) \rangle \simeq A_6,$$

$$L_2(11) \simeq \langle (1, 3)(2, 5)(7, 10)(9, 11), (2, 6, 8)(3, 9, 5)(4, 11, 10) \rangle \preceq M_{11},$$

$$L_2(13) \simeq \langle (3, 13, 11, 9, 7, 5)(4, 14, 12, 10, 8, 6), (1, 2, 9)(3, 8, 10)(4, 5, 12)(6, 13, 14) \rangle,$$

$$L_2(16) \simeq \langle (3, 16, 14, 13, 11, 4, 8, 6, 17, 15, 5, 12, 10, 9, 7),$$
$$(1, 2, 3)(6, 14, 8)(7, 11, 10)(9, 17, 15)(12, 16, 13) \rangle,$$

$$L_2(17) \simeq \langle (3, 17, 15, 13, 11, 9, 7, 5)(4, 18, 16, 14, 12, 10, 8, 6),$$
$$(1, 2, 11)(3, 13, 9)(4, 15, 6)(5, 8, 12)(7, 18, 16)(10, 14, 17) \rangle,$$

$$L_2(19) \simeq \langle (3, 19, 17, 15, 13, 11, 9, 7, 5)(4, 20, 18, 16, 14, 12, 10, 8, 6),$$
$$(1, 2, 12)(3, 11, 13)(4, 17, 6)(5, 14, 18)(7, 20, 18)(10, 19, 16) \rangle,$$

$$L_2(23) \simeq \langle (3, 23, 21, 19, 17, 15, 13, 11, 9, 7, 5)(4, 24, 22, 20, 18, 16, 14, 12, 10, 8, 6),$$
$$(1, 2, 14)(3, 12, 16)(4, 18, 9)(5, 20, 6)(7, 13, 11)(8, 23, 22)(10, 24, 19)(15, 21, 17) \rangle,$$

$$L_2(25) \simeq \langle (3, 25, 23, 6, 20, 18, 5, 15, 13, 4, 10, 8)(7, 26, 24, 22, 21, 19, 17, 16, 14, 12, 11, 9),$$
$$(1, 2, 5)(3, 4, 6)(7, 18, 14)(8, 22, 9)(10, 17, 12)(11, 25, 24)(15, 26, 19)(16, 23, 21) \rangle,$$

$$L_2(27) \simeq \langle (3, 27, 25, 23, 21, 19, 17, 16, 14, 12, 10, 8, 6)(4, 15, 13, 11, 9, 7, 5, 28, 26, 24, 22, 20, 18),$$
$$(1, 2, 4)(5, 8, 24)(6, 21, 10)(7, 16, 15)(9, 25, 28)(11, 13, 14)(12, 27, 23)(17, 26, 28) \rangle.$$

Bibliography

This bibliography not only includes all works directly cited in this book and the web sections, but also all those books and papers that were consulted during its preparation and composition.

Alperin, J.L. (1969) A classification of n-Abelian groups. Can. J. Math. **21**, 1238–1244.

Alperin, J.L., Bell, R.B. (1995) *Groups and Representations*. Springer, New York.

Andrews, G.E. (1976) *The Theory of Partitions*. Addison-Wesley, Reading.

Armstrong, M.A. (1988) *Groups and Symmetry*. Springer, New York.

Artin, E. (1948) *Galois Theory*. Notre Dame University Press, Notre Dame.

Aschbacher, M. (1986) *Finite Group Theory*. Cambridge University Press, Cambridge.

Aschbacher, M. (1994) *Sporadic Groups*. Cambridge University Press, Cambridge.

ATLAS (1985) See Conway et al.

Baker, A. (2002) *Matrix Groups, an Introduction to Lie Group Theory*. Springer, London.

Benson, D.J. (1991) *Representations and Cohomology 2—Cohomology of Groups and Modules*. Cambridge University Press, Cambridge.

Besche, H.U., Eick, B., O'Brien, E.A. (2001) The groups of order at most 2000. Electron. Res. Announc. Am. Math. Soc. **7**, 1–4.

Björner, A., Brenti, F. (2005) *Combinatorics of Coxeter Groups*. Springer, New York.

Brauer, R. (1954) On the structure of groups of finite order. In *Proc. of the 1954 International Mathematical Congress*, Amsterdam.

Brauer, R., Fowler, K.A. (1955) On groups of even order. Ann. Math. **62**(2), 565–583.

Bruck, R.L. (1966) *A Survey of Binary Systems*. Springer, Berlin.

Burnside, W. (2004) *Theory of Groups of Finite Order*. Dover, New York. Reprint of the 2nd edition (1911).

Cameron, P.J. (1976) *Parallelisms of Complete Designs*. Cambridge University Press, Cambridge.

Cameron, P.J. (1999) *Permutation Groups*. LMS Student Texts **45**. Cambridge University Press, Cambridge.

H.E. Rose, *A Course on Finite Groups*,
Universitext,
DOI 10.1007/978-1-84882-889-6, © Springer-Verlag London Limited 2009

Cameron, P.J., van Lint, J.H. (1991) *Graphs, Codes, Designs and Their Links*. Cambridge University Press, Cambridge.

Carter, R.W. (1972) *Simple Groups of Lie Type*. Wiley, London/New York.

Chandler, B., Magnus, W. (1982) *The History of Combinatorial Group Theory: A Case Study in the History of Ideas*. Springer, New York.

Chapman, R.J. (1995) An elementary proof of the simplicity of the Mathieu groups M_{11} and M_{23}. Am. Math. Mon. **102**, 544–545.

Close, F. (2006) *The New Cosmic Onion, Quarks and the Nature of the Universe*. CRC Press/Taylor and Francis Group, New York.

Cohen, H. (1993) *A Course in Computational Algebraic Number Theory*. Springer, New York.

Cohn, P.M. (1965) *Universal Algebra*. Harper & Row, New York.

Conway, J.H., Curtis, R.T., Norton, S.P., Parker, R.A., Wilson, R.A. (1985) *The Atlas of Finite Groups—The* ATLAS. Clarendon Press (OUP), Oxford.

Conway, J.H., Sloane, N.J.A. (1993) *Sphere Packing, Lattices and Groups*, 2nd edition. Springer, New York.

Coxeter, H.S.M., Moser, W.O.J. (1984) *Generators and Relations for Discrete Groups*, 4th edition. Springer, Berlin.

Curtis, C., Reiner, I. (1962) *Representation Theory of Finite Groups and Associative Algebras*. Wiley, New York.

Curtis, R.T. (2007) *Symmetric Generation of Groups*. Cambridge University Press, Cambridge.

Curtis, R.T., Wilson, R.A. (eds.) (1998) *The Atlas of Finite Groups: Ten Years On*. Cambridge University Press, Cambridge.

Dickson, L.E. (2003) *Linear Groups with an Exposition of Galois Field Theory*. Dover, New York. Reprint of the 1st edition (1900).

Dixon, J.D. (1967) *Problems in Group Theory*. Blaisdell, Waltham.

Doerk, K., Hawkes, T. (1992) *Finite Soluble Groups*. Walter de Gruyter, Berlin.

Dornhoff, L. (1971) *Group Representation Theory, I and II*. Marcel Dekker, New York.

Edge, W.L. (1955) The isomorphism between $LF(2, 3^2)$ and A_6. J. Lond. Math. Soc. **30**, 172–185.

Feit, W., Thompson, J.G. (1963) Solvability of groups of odd order. Pac. J. Math. **13**, 775–1029.

Finkelstein, L., Rudvalis, A. (1973) Maximal subgroups of the Hall–Janko–Wales group. J. Algebra **24**, 486–493.

Fuchs, L. (1970) *Infinite Abelian Groups I*. Academic Press, New York.

Fuchs, L. (1973) *Infinite Abelian Groups II*. Academic Press, New York.

Gardiner, C. (1980) *A First Course in Group Theory*. Springer, New York.

Gorenstein, D. (1974) *Reviews on Finite Groups*. American Mathematical Society, Providence.

Gorenstein, D. (1982) *Finite Simple Groups*. Plenum Press, New York.

Gorenstein, D. (1983) *The Classification of Finite Simple Groups I. Groups of Non-characteristic 2 Type*. Plenum Press, New York.

Gorenstein, D., Lyons, R., Solomon, R. (1994) *The Classification of the Finite Simple Groups, Number 1, 2, ...* American Mathematical Society, Providence.

Gross, F. (1987) Conjugacy of odd order Hall subgroups. Bull. Lond. Math. Soc. **19**, 311–319.

Hall, M., Wales, D.B. (1968) A simple group of order 604,800. J. Algebra **9**, 417–450.

Hall, P. (1928) A note on soluble groups. J. Lond. Math. Soc. **3**, 98–105.

Hall, P. (1937) A characteristic property of soluble groups. J. Lond. Math. Soc. **12**, 198–200.

Halmos, P. (1974a) *Finite-Dimensional Vector Spaces*. Springer, New York. Reprint of the 1958 edition.

Halmos, P. (1974b). *Naive Set Theory*. Springer, New York.

Herstein, I.N. (1964) *Topics in Algebra*. Blaisdell, Waltham.

Hughes, D.R., Piper, F.C. (1985) *Design Theory*. Cambridge University Press, Cambridge.

Huppert, B. (1967) *Endliche Gruppen I*. Springer, Berlin.

Huppert, B. (1998) *Character Theory of Finite Groups*. Walter de Gruyter, Berlin.

Huppert, B., Blackburn, N. (1982a) *Finite Groups II*. Springer, Berlin.

Huppert, B., Blackburn, N. (1982b) *Finite Groups III*. Springer, Berlin.

Humphreys, J.F. (1996) *A Course in Group Theory*. Clarendon Press (OUP), Oxford.

Isaacs, I.M. (2006) *Character Theory of Finite Groups*. AMS/Chelsea, Providence. Reprint of the 1976 edition.

Isaacs, I.M. (2008) *Finite Group Theory*. American Mathematical Society, Providence.

James, G., Liebeck, M. (1993) *Representations and Characters of Groups*. Cambridge University Press, Cambridge.

Kaplansky, I. (1969) *Infinite Abelian Groups*. University of Michigan Press, Ann Arbor.

Kelley, J.L. (1955) *General Topology*. Van Nostrand, Princeton.

Khukhro, E.I. (1993) *Nilpotent Groups and Their Automorphisms*. Walter de Gruyter, Berlin.

Kleidman, P., Liebeck, M. (1990) *The Subgroup Structure of the Finite Classical Groups*. Cambridge University Press, Cambridge.

Kochendörffer, R. (1966) *Group Theory*. McGraw-Hill, London.

Kurosh, A.G. (1955) *The Theory of Groups, Volumes 1 and 2*. Chelsea, New York. Translated and edited from the Russian 2nd edition by K.A. Hirsch.

Kurzweil, H., Stellmacher, B. (2004) *The Theory of Finite Groups*. Springer, New York.

Lang, S. (1993) *Algebra*, 3rd edition. Addison-Wesley, Reading.

Ledermann, W., Weir, A.J. (1996) *Introduction to Group Theory*, 2nd edition. Addison-Wesley/Longman, Harlow/Essex.

Lyndon, R.C., Schupp, P.E. (1976) *Conbinatorial Group Theory*. Springer, Heidelberg.

MacDonald, I.D. (1968) *The Theory of Groups*. Clarendon Press (OUP), Oxford.

MacLane, S., Birkhoff, G. (1967) *Algebra*. MacMillan, New York.

Michler, G. (2006) *Theory of Finite Simple Groups*. Cambridge University Press, Cambridge.

Moody, J. (1994) *Groups for Undergraduates*. World Scientific, Singapore.

O'Hara, C.W., Ward, P.G. (1937) *An Introduction to Projective Geometry*. Clarendon Press (OUP), Oxford.

Ol'shanskii, A.Yu. (1983) Groups of bounded period with subgroups of prime order. Algebra Log. **21**, 369–418.

Ore, O. (1938) Structures and group theory II. Duke Math. J. **4**, 247–269.

Passman, D. (1968) *Permutation Groups*. Benjamin, New York.

Robinson, D.J.S. (1982) *A Course in the Theory of Groups*. Springer, New York.

Rose, H.E. (1999) *A Course in Number Theory*, 2nd edition. Clarendon Press (OUP), Oxford.

Rose, H.E. (2002) *Linear Algebra, a Pure Mathematical Approach*. Birkhäuser, Basel.

Rose, J.S. (1978) *A Course in Group Theory*. Cambridge University Press, Cambridge.

Rotman, J.J. (1965) *The Theory of Groups, an Introduction*, 1st edition. Allen and Bacon, Boston.

Rotman, J.J. (1994) *An Introduction to the Theory of Groups*, 4th edition. Springer, New York.

Schenkman, E. (1965) *Group Theory*. Van Nostrand, New York.

Schmidt, R. (1994) *Subgroup Lattices of Groups*. Walter de Gruyter, Berlin.

Scott, W.R. (1964) *Group Theory*. Prentice-Hall, Englewood Cliffs.

Short, M.W. (1992) *The Primitive Soluble Permutation Groups of Degree less than 256*. Lecture Notes in Mathematics **1519**. Springer, Berlin.

Smith, G., Tabachnikova, O. (2000) *Topics in Group Theory*. Springer, London.

Solomon, R. (2001) A brief history of the classification of the finite simple groups. Bull. Am. Math. Soc. **38**, 315–352.

Stewart, I. (1989) *Galois Theory*. Chapman and Hall, London.

Suzuki, M. (1982) *Group Theory I*. Springer, New York.

Suzuki, M. (1986) *Group Theory II*. Springer, New York.

Thompson, J.G. (1968) Nonsolvable finite groups all of whose local subgroups are solvable. Bull. Am. Math. Soc. **74**, 383–427.

Vaughan-Lee, M. (1993) *The Restricted Burnside Problem*. Clarendon Press (OUP), Oxford.

van der Waerden, B.L. (1948) *Modern Algebra*. F. Ungar Publishing, New York. Translation of *Moderne Algebra*, 2nd edition (1937).

van der Waerden, B.L. (1985) *A History of Algebra, from al-Khwārizmī to Emmy Noether*. Springer, Berlin.

Wehrfritz, B.A.F. (1999) *Finite Groups, a Second Course on Group Theory*. World Scientific, Singapore.

Weinstein, M. (1977) *Examples of Groups*. Polygonal, Passaic.

Weyl, H. (1952) *Symmetry*. Princeton University Press, Princeton/New Jersey.

Wielandt, H. (1964) *Finite Permutation Groups*. Academic Press, New York.

Willard, S. (1970) *General Topology*. Addison-Wesley, Reading.

Williams, W.S.C. (1994) *Nuclear and Particle Physics*. Clarendon Press (OUP), Oxford.

Wussing, H. (1984) *Genesis of the Abstract Group Concept*. MIT Press, Cambridge. Translation of the German text.

Zassenhaus, H.J. (1958) *The Theory of Groups*, 2nd edition. Chelsea, New York.

Notation Index

Three notation indexes are given below: 1—Symbol index; 2—Index of names of classes of groups; and 3—Index of names of individual groups.

1—Symbol Index

$\odot$, 11

T_1, 12

$(G, \odot)$, G, H, J, K, 13

$\langle X \rangle$, 13, 25

$\langle e \rangle$, 13, 24

a, b, c, d, g, h, j, k, 13

ab, e, g^{-1}, 15

$o(G)$, $o(X)$, 16

$\mathbb{Z}$, 18, 21

$\simeq$, 17, 69

$\mathbb{Q}, \mathbb{R}, \mathbb{C}, \mathbb{Q}^*, \mathbb{R}^*, \mathbb{C}^*$, 18

$\mathbb{Q}^+, \mathbb{R}^+$, 18

F^* (non-zero elements of F), 18

$a \equiv b \pmod{m}$, $a \equiv b(m)$, 18, 285

$\mathbb{Z}/m\mathbb{Z}$, 18, 74

$(\mathbb{Z}/p\mathbb{Z})^*$, 18

T_2, T_n, 18

$GL_n(F)$, 19, 53

$SL_n(F)$, 19, 53

$UT_n(F)$, 19, 54

D_n, 20, 58

S_X, S_n, 20, 49

A_n, 20, 50

C_n, 22, 80

$H \leq G$, $H < G$, 24

$\langle g \rangle$, 26

$o(g)$, 26

XY, gH, Hg, 27

$[G : H]$, 29

$\triangleleft$, 30

K, 30

$H \vee J$, 31

$\langle H, J \rangle$, 31

$Z(G)$, 32, 102

$[g, h]$, $[G, H]$, 37, 210

G', 37

core(H), 39, 101

H^*, 39

$\begin{pmatrix} 1 & 2 & \dots & n \\ a_1 & a_2 & \dots & a_n \end{pmatrix}$, 42

$(a_1, \dots, a_n)$ (cycle), 44

k-cycle, 44

sgn(σ), 47

$GL_n(q)$, $SL_n(q)$, 53

$L_n(q)$, 54, 255

$\langle A \mid R \rangle$, $\langle g_1, \dots \mid x_1 = e, \dots \rangle$, 58

$\mathbb{Z}$, 58

Q_n, 59

$IT_n(F)$, $IZT_{n,r}(F)$, 63

H.E. Rose, *A Course on Finite Groups*,
Universitext,
DOI 10.1007/978-1-84882-889-6, © Springer-Verlag London Limited 2009

$\phi|_H$, 70

$\mathrm{im}\,\phi$, 70, 280

$\ker\phi$, 71

G/K, 73

$\preceq$ (isomorphic to subgroup), 75

$\mathrm{Aut}\,G$, $\mathrm{Aut}(G)$, 81

$\mathrm{Inn}\,G$, $\mathrm{Inn}(G)$, 81

$\mathrm{Out}\,G$, $\mathrm{Out}(G)$, 82

char, 89

$\backslash g$, 91$f\!f$

$\backslash$ (action), 92

$\mathcal{O}\{x\}$, $\mathcal{O}_G\{x\}$, 94

$\mathrm{stab}_G(x)$, 95

$\mathrm{fix}(G, X)$, 98

$\mathcal{C}\ell\{x\}$, $\mathcal{C}\ell_G\{x\}$, 102

$C_G(x)$, 102

$h(G)$, 103

$C_G(X)$, 104

$N_G(H)$, 105

$v(n)$, 117

$F_{m,n}$, $F_{n,m,r}$, 130

$ES_1(p)$, $ES_2(p)$, 132

$\times$, 139

$G_1 \times \cdots \times G_n$, 140

$\mathrm{part}(n)$, 151

$\rtimes$, 151

$K : A$, 152

wr, $\wr$, 156

$\mathrm{Hol}(G)$, 162

$PGL_n(q)$, 170

$G^{r,s,t}$, 170

E, 177

$\lhd\lhd$, 206, 224

$\mathcal{D}_i(G)$, $\mathcal{D}_i$, 210

$\mathcal{Z}_i(G)$, $\mathcal{Z}_i$, 210

$\Phi(G)$, 217

$F(G)$, 221

$O_p(G)$, $O_\pi(G)$, 221, 237

$F^*(G)$, 222

$\mathrm{Sz}(2^n)$, 222

$^2F_4(2)'$, 222

$\mathrm{Gal}(F)$, 229, 230

$G^{(n)}$, 234

π, π', 236, 285

$\mathrm{S}(r, s, t)$, 251

$\mathrm{Aut}(\mathrm{S}(r, s, t))$, 253

$\mathbb{F}_q$, 254

$PSL_n(q)$, 255

$E_{ij}(r)$, 255

$GU_n(q)$, $SU_n(q)$, $U_n(Q)$, 265

M_{11}, M_{12}, ..., 267, 268

J_1, J_2, 267

$\in$, 277

$=$, 277

$\emptyset$, 277

$\subseteq$, $\subset$, 277

$\cap$, 278

$\cup$, 278

$\backslash$ (complement), 278

$\{x \in X : P(x)\}$, 278

$\{x_1, \ldots, x_n\}$, 278

$(x_1, \ldots, x_n)$, 278

$\times$, 278

X^n, 278

$\phi : X \to Y$, 279

$Z\phi$, 280, 281

$Z\phi^{-1}$, 280, 281

ι, 281

$\circ$, 281

$\preceq$ (partial order), 282

Δ, 282

$a \mid b$, 284

$a \nmid b$, 284

(a, b), 284

GCD, LCM, 284

$\equiv$, mod, 285

$\phi(n)$, 286

$\mathrm{ord}_m(c)$, 287

2—Notation for Classes of Groups

In the first table below, we list the primary definitions and notations for the infinite classes of groups discussed in this book. *Note that in all cases the subscript n is used to indicate the nth group in the class*, and so is only indirectly related to the order of the group. The column called 'Primary definition' gives the group's 'main' representation. The symbol F denotes an arbitrary field, if F is finite then the symbol F is replaced by q the order of the field.

Group symbol	English name	Primary definition	Page
A_n	Alternating	Even permutations on n symbols	50
$C_n \simeq \mathbb{Z}/n\mathbb{Z}$	Cyclic	Integers with addition modulo n	58, 79
D_n	Dihedral	Symmetries of regular n-gon	58
$F_{r,s}$ or $F_{r,s,t}$	Frobenius or metacyclic	$\langle a, b \mid a^r = b^s = e, b^{-1}ab = a^t \rangle$ where $t^s \equiv 1 \pmod r$	131
$GL_n(F)$	General linear	Non-singular $n \times n$ matrices over F	53
$IT_n(F)$	Uni-upper triangular	Subgroup of $UT_n(F)$ with each diagonal element 1	63
$IZT_{n,r}(F)$	–	Subgroup of $IT_n(F)$ with r superdiagonals zero	63
$L_n(F)$ or $PSL_n(F)$	Linear or Projective special linear	$SL_n(F)$ factored by its centre	255
Q_n	Dicyclic or generalised quaternion	$\langle a, b \mid a^n = b^2 = (ab)^2 \rangle$	59
S_n	Symmetric	All permutations on n symbols	49
$SL_n(F)$	Special linear	Subgroup of $GL_n(F)$ of matrices A with $\det A = 1$	53
$Sz(2^n)$	Suzuki	See Web Section 14.3	–
$T_n \simeq C_2^n$	n copies of '2'	–	18 and 151
$U_n(F)$	Unitary	See page 265	265
$UT_n(F)$	Upper triangular	Subgroup of $GL_n(F)$ of matrices with zeros below main diagonal	54
$(\mathbb{Z}/n\mathbb{Z})^*$	–	$\{a \in \mathbb{Z} : 0 < a < n \text{ and } (a,n) = 1\}$ with multiplication mod n	18

3—Notation for Individual Groups

In the second table below, we list some notation for individual groups, that is, groups that do not form a part of an infinite family.

Group	English name	Operation	Page
$\mathbb{C}$	Complex numbers	Addition	18
$\mathbb{C}^*$	Non-zero complex numbers	Multiplication	18
C_{p^∞}	Infinite group with only finite subgroups	[Abelian]	132
E	Exceptional	See Section 7.3	177
ES_i for $i = 1, 2$	Extra special	See Problems 3.18 and 5.5	132
J_1, J_2	Janko	–	267
M_{11}, M_{12}	Mathieu	See Section 11.4	268
Q_2	Quaternion	$\langle a, b \mid a^4 = e, a^2 = b^2, b^{-1}ab = a^3 \rangle$	118
$\mathbb{Q}$	Rational numbers	Addition	18
$\mathbb{Q}^*$	Non-zero rational numbers	Multiplication	18
$\mathbb{Q}^+$	Positive rational numbers	Multiplication	18
$\mathbb{R}$	Real numbers	Addition	18
$\mathbb{R}^*$	Non-zero real numbers	Multiplication	18
$\mathbb{R}^+$	Positive real	Multiplication	18
$\mathbb{Z}$	Integers, or infinite cyclic group	Addition	18, 58

Index

2-Euler count, 293
4-group, 19

A

Abel, N, 12
Abelian, 2, 12
  group, finite, 146*ff*
    elementary, 88
    invariant, 292
    maximal, 109
Action, 91*ff*
  conjugate element, 101
    subgroup, 105
  coset, 100
  intransitive, 94
  natural, 93
  permutation, 94
  restricted, 98
  right multiplication, 93
  transitive, 94
Algorithm, division, 284
  Euclidean, 284
Alphabet, 21, 56
Alternating (group), 20, 50, 61 (A_4), 158
Antisymmetric, 282
Argument, 280
  Frattini, 128
Arithmetic, modular, 18
Associative, 2, 11
Automiser, 106
Automorphism, 69
  group, 81*ff*
  inner, 82
  outer, 82
  of a Steiner system, 252

B

Basis Theorem, 147
Belong ($\in$), 277
Bertrand's Postulate, 52
Bijection, 280
Binary (relation), 279
  tetrahedral (group), 176
Block (of a Steiner system), 251
Bruhat Decomposition Theorem, 55
Burnside, W. 27, 74, 130, 220
  Normal Complement Theorem, 130
  $p^r q^s$-theorem, 231, 239

C

Cancellation, 15
Canonical (rational form), 262
Carbon60, 23
Cardano, G. 243
Cardinality, 16, 280
Cartesian (product), 278
Cauchy, A. 42, 114
  Theorem, 114
Cayley, A. 1, 42, 48, 72
  number, 119
  Theorem, 42, 72
Central, 109
  series, 210
Centraliser, 101*ff*, 104, 108
Centre, 32, 87, 102
  higher, 210
Centreless, 32
CFSG, 1
Characteristic (of a field), 254
  subgroup, 30, 89
Characteristically simple, 161
Chief series, 207
Chord–tangent process, 22

H.E. Rose, *A Course on Finite Groups*,
Universitext,
DOI 10.1007/978-1-84882-889-6, © Springer-Verlag London Limited 2009

Class Equations, 103
  equivalence, 279
  number 103
Closure, 2, 11
Cocycle Identity, 198
Codomain, 280
Cohomology theory, 187
  second group, 197
Collection, 277
Combinatorial group theory, 55
Commutative, 12
Commutator, 37
  identities, 37
  subgroup, 37
Companion matrix, 262
Complement (set), 278
  Frobenius, i
  subgroup, 151
  p-, 239
Composite, 285
Composition factor, 188
  (of maps), 281
  series, 188, 190
Concatenation, 21, 56
  reduced, 56
Congruence, 18, 285
Conjugacy class, 30, 102
  action, 102, 105
Conjugate element, 30
  subgroup, 38, 105
Coprime, 285
Core, 39, 100
Correspondence Theorem, 78
Coset (left and right), 27
  action, 100
  double, 40
  enlargement, 87
  enumeration, 58
  multiplication, product, 73
Coxeter group, 64, 269
Cycle (permutation), 44
  of length k, 44
Cyclic, 26
  extension, 196, 203ff
  group, 58, 80

D
Degree pattern, 293
Dekekind, R. 161
  Modular Law, 38
De Morgan's Law, 282
Derived length, 234
  series, 234
  subgroup, 37, 85

Descartes, R. 243, 278
Dicyclic (group), 59
  generalised, 205
Dihedral (group), 4, 58, 64
Dining Club, 270
Direct product, 139, 140
  external, 140
  internal, 140, 141
Disjoint (sets), 278
Divide, 284
Division algebra, 119
  algorithm, 284
Domain, 280
Double cosets, 40
Dyck, W. von, 55

E
Element, 277
  neutral, 2, 4, 12
Elementary Abelian group, 27, 88, 162
Empty set, 277
  word, 21, 56
Endomorphism, 69
Equation, Class, 103
Equality (function), 280
  (group), 16
  (set), 277
Equivalence class, 279
  relation, 279
Equivalent (series), 191
Euclid, 284, 285
Euclidean Algorithm, 284
Euler's function, 288
  Theorem, 288
Even (permutation), 47ff
Exceptional group, 177ff
Exponent (of a group), 26, 291
  (rules), 16
Extension, 151, 196,
  cyclic, 196, 203ff
  problem, 196ff
  split, 151
External (direct product), 140
  (semi-direct product), 151
Extra-special (group), 132

F
Factor (of a direct product), 140
  group, 72ff
  pair, 197, 198
  (of a series), 188
Factorisable (group), 152
Factorisation, unique, 285
Fano plane, 253

Feit, W. 231
Fermat, P. 286
    theorem, 286
Ferrari, L. 243
del Ferro, S. 243
Field, finite, 254
    Galois, 53
    splitting, 230
Finite field, 254
    geometry, 253
Fitting, H. 221
    subgroup, 222
Fixed set, 98
Fontana (Tartaglia), N. 243
Frattini, G. 217
    Argument, 128
    subgroup, 217ff
Free (group), 21, 57
Friendly giant, 20
Frobenius complement, i
    group, 62, 131, 184, 241
Fully invariant subgroup, 30
Function, 279
    n-variable, 280
Fundamental theorem of Abelian groups, 146ff

G
Galois, È. 1, 29, 30, 229, 242, 246, 261, 295
    field, 53, 254
Gauss, C.F. 18, 28, 83, 285, 287
General linear group, 13, 19, 53
Generalised associativity law, 34
    dicyclic group, 205
    Fitting subgroup, 223
Generate (a group), 13, 25
Generating set, 13, 25
Generator, 21, 26, 58
Greatest common divisor (GCD), 284
Group, 12
Group attributes
    automorphism, 81ff
    cyclic Sylow subgroups, 131
    factor, 72ff
    free, 57
    fundamental (of a topological space), 23
    Hamiltonian, 161
    maximal normal, 189
    nilpotent, 129, 144, 210, 212
    of odd order, 231
    order 8, 118
    order 12, 156
    outer automorphism, 82, 250
    quotient, 72
    perfect, 86

presentation, 41, 55ff
product, 18, 140ff
simple, 33
soluble, 5, 230ff
supersoluble, 227
Group examples
    alternating, 20, 50, 158
    Chevalley, 249
    classical, 249
    Coxeter, 64
    cyclic, 58, 80
    dicyclic, 59
    dihedral, 4, 20, 58, 159
    elementary Abelian, 27, 88
    Exceptional, 177ff
    extra-special, 132
    Frobenius, 62, 130, 184, 241
    Galois, 213
    general linear, 13, 19, 52
        unitary, 265
    generalised dicyclic, 205
    Hessian, 265
    infinite cyclic, 21, 58
    Janko, 267
    Klein, 19
    linear, 52, 249, 254ff
    Mathieu, 250, 264ff
    metacyclic, 62, 130, 184, 241
    neutral, 13
    octahedral, 171
    projective general linear, 170
        special linear, 255
    quaternion, 59, 119, 159
    second cohomology, 197
    semi-dihedral, 131
    special linear, 19, 52, 172
        unitary, 265
    sporadic, 249, 267
    Suzuki, i, 223
    symmetric, 13, 49, 165
    tetrahedral (A_4), 61
        binary, 176
    Tits, 223
    unitary, 265
G-set, 92

H
Hall, P. 1, 236
    π-subgroups, 76, 236, 246, 247
    –Witt Identity, 37
Hamilton, W. 119
Hamiltonion (group), 119, 161, 292
Higher centre, 210
    commutator subgroup, 210

Hölder, O. 131, 190, 196
Holomorph, 162
Homomorphism, 67*ff*
    equation, 67
    natural, 76
Hypercentre, 211

I
Idempotent, 14
Identity (of a group), 12
    Cocycle, 198
    Hall–Witt, 37
    map, 281
Image, 280
Indecomposable, 143
Index (of subgroup), 29
Infinite (cyclic group), 58
Injective, 280
Internal (direct product), 141
    (semi-direct product), 151
Intersection, 278
Intransitive (action), 94
Invariant series, 188
    Abelian, 292
Inverse, 2
    of a bijection, 281
Involution, 4, 27, 39, 113
Isometry, 34
Isomorphism, isomorphic, 17, 67*ff*
    class, 17
    of a Steiner system, 252
    Theorems, 75*ff*
Iwasawa's Lemma, 7, 92, 257

J
Janko, Z. 267
Join (of subgroups), 31
Jordan, C. 29, 69, 190
    –Hölder Theorem, 190*ff*

K
k-cycle, 43
Kernel, 30, 71*ff*
Kirkman, T.P. 251
Klein (group), 19
k-transitive, 94, 268

L
Lagrange, J.-L. 1, 29
    reverse, 101
    Theorem, 30
Least common multiple (LCM), 284

Left coset, 27
    inverse, 13
    neutral element, 13
Length, of a cycle, 43, 44
    of a series, 188
    derived, 234
Letter, 56
Linear group, 52, 254
    order, 282
List, 56
Lower central series, 211

M
Map, mapping, 279
    identity, 281
    natural, 67
    structure-preserving, 67
Mathieu, É.L. 268
    group, 20, 250, 268
Matrix, non-singular, 52
    permutation, 62
    scalar, 256
    upper triangular, 54
Maximal, Abelian, 109
    normal subgroup, 189
    subgroup, 24
McKay, J. 114
Member, membership, 277
Metacyclic (group), 62, 131, 184
Minimal, normal subgroup, 235
    simple group, 264
Modulus, modulo, 18, 285
    arithmetic, 18
Monster, 20
Morphism, 67
Multiplication, 13
    coset, 73
    scalar, 93
    table, 36
Multiplicative, 286

N
n-argument, 280
Natural action, 93
    homomorphism, 76
    map, 67
N/C-theorem, 106
Neutral element, 2, 4, 12
    group, 13
    subgroup, 24
Nilpotency class, 210, 212
Nilpotent, 129, 144, 210*ff*
    with class r, 212
Noether, E. 74

Non-generator, 217
Normal (subgroup), 30
    closure, 39
    conditions, 30, 71
    minimal, 235
    properties, 36
    series, 188
Normaliser, 105, 109
Number, Cayley, 119
    class, 103
n-variable, 280

O

Octahedral (group), 170
Octonion, 119
Odd (permutation), 47ff
One-to-one, 280
    correspondence, 280
Onto, 280
Operation, 11, 13, 281
Orbit, 94ff
    compared with a cycle, 94
    of an element, 94
Order of an element, 26
    of a group, 16
    function, 35
    modulo an integer, 287
    partial, 282
Ordered set, 278
Outer, automorphism, 82
    group, 82, 250

P

Partial order, 282
    system, 12
Partition, 151
p-complement, 239
Perfect (group), 86
Permutation, 12, 42ff
    action, 94
    even and odd, 44, 47ff, 62
    group, 48ff
    matrix, 62
    product, 42
    representation, 97
p-group, 114ff
Poincaré, H. 23
    Theorem, 38
Preimage, 280
Presentation, 58
    of S_n, 64
p-radical, π-radical, 86, 133, 222
Primitive root, 287
Prime, prime number, 285

Product, 13
    cartesian, 278
    coset, 73
    direct, 139ff
    group, 19
    of permutations, 42
    semi-direct, 151ff
    wreath, 156
Projection, 283
Projective, General Linear Group, 170
    Special Linear Group, 255
    Unitary Group, 265
Proper, refinement, 188
    series, 188
    subgroup, 24
    subset, 277

Q

Quaternion, 119
    group, 59, 119, 159
Quotient (group), 73

R

Radical, p-, 86, 222, 237
    (type of root), 229
Range, 280
Reduced word, 21, 56
    concatenation, 56
Refinement (of a series), 188
Reflexive (order), 282
    (relation), 279
Regular, 60
Relation, 21, 57, 58, 279
    equivalence, 279
Relator, 58
Representation, 4, 9, 41ff
    permutation, 97
Residue, 285
Restricted, action, 98
Restriction (of a function), 70
Reverse Lagrange, 101
Right coset, 27
    multiplication (action), 93
Root, primitive, 287
Russell, B. 41

S

Scalar matrix, 256
    multiplication, 93
Schmidt, O. 246
Schreier, O. 82, 193
    Conjecture, 82
    Refinement Theorem, 195

Schur, I. 187
   multiplication, 250
Section, 197
Self-normalising, 106
Semi-dihedral (group), 131
Semidirect (product), 151*ff*
   internal, 151
   external, 153
Semigroup, 11
Semi-regular, 60
Semi-simple, 196
Series, 188*ff*
   central, 210
   chief, 207
   composition, 188*ff*
   derived, 234
   equivalent, 191
   invariant, 188
   lower central, 211
   normal, 188
   soluble, 230
   subnormal, 188
   upper central, 211
Set, 277
   fixed, 98
   generating 13, 25
   ordered, 278
   underlying, 13
Sign (of a permutation), 49
Simple (group), 33
   characteristically, 161
Single-valued, 279
Size, 16, 280
Sockel, 177, 293
Solubility conditions, 241*ff*
Soluble (group), 5, 116, 230
   series, 230
Solvable, 230
Special Linear Group, 53, 172
Sphere packing, 20
Split (extension), 152, 229
Splitting field, 230
Sporadic (group) 249
Stabiliser, 95*ff*
Steiner, J. 251
   system 250*ff*
      automorphism, 252
      isomorphism, 252
Structure-preserving map, 67
Subgroup, 24
   characteristic, 31, 89
   condition, 24, 25
   conjugate, 38, 105

derived (commutator), 37
   higher, 210
Frattini, 217
Fitting, 222
   generalised, 223
fully invariant, 31
identity, 24
Hall, 236, 247
lattice, 32, 169, 175, 180
maximal, 24
minimal normal, 235
n-derived, 234
neutral, 24
normal, 30
proper, 24
subnormal, 206
Sylow, 122
trivial, 24
unit, 24
Subnormal (series), 188
   subgroup 206, 224
Subset, 277
Supersoluble (group), 227
Surjective, 280
Suzuki, M, i, 52
   group i, 223
Sylow, L. 120
   p-subgroup, 81, 122
   subgroup, 1, 120*ff*
      properties, 133, 134
   system, 241
   theorems, 120*ff*
Sylow 1, ..., Sylow 5, 125
Symbol, 56
Symmetric, 279
   (group), 13, 48, 165
   difference, 282
Symmetry, 3
   of a tetrahedron, 61
   science of, 42

T
Tarski Monster, 132
Term (of a series), 188
Ternary (relation), 279
Tetrahedron, 61
Theorem, Basis (of Abelian Groups), 147
   Brahut Decomposition, 54
   Burnside's Normal Complement, i, 130
      $p^r q^s$-, i, 231, 242, 245
   Cauchy, 114
   Cayley, 72
   Correspondence, 77*ff*
   Euler, 286

Fermat, 35, 286
Fitting, 221
Freshman's, 79
Fundamental, of Abelian Groups, 146*ff*, 162
Isomorphism, First, 75
    Second, 77
    Third, 79
Jordan–Hölder, i, 190*ff*
Lagrange, 27, 29, 30
N/C, 106
Orbit-stabiliser, 95
Poincaré, 38
Schreier Refinement, 195
Schur–Zassenhaus, i, 187
Sylow, i, 120*ff*
Zassenhaus, 193
Thompson, J. 231, 242, 246
Three Subgroup Lemma, 225
Tits group, 223
Total (order), 282
Transfer, 68
Transformation, 279
Transitive, 279
    action, 94
    order, 282
Transposition (2-cycle), 44
Transvection, 55, 255
Trivial (homomorphism), 69

U
Underlying set, 13
Unimodular (matrix), 147
Union (of sets), 278
Unipotent (matrix), 63
Unique factorisation, 285
Unitary group, 265
    matrix, 265
Unordered (set), 278
Upper central series, 211
    triangular matrix, 19, 54

V
Variable, 280
Viergruppe (4-group), 19

W
Wedderburn, J.H.M. 143
Well-defined, 2, 11, (coset) 73
Well-order, 282
Wielandt, H. 119
Word, 21, 56
Wreath (product), 156

Z
Zassenhaus, H. 130, 187, 193
    Lemma, 193
    Schur (theorem), 187
Zero, 12